9th International Conference "Refrigeration and Food Technologies in the 21st Century" (RAFT2019)

IOP Conference Series: Materials Science and Engineering
Volume 826

St. Petersburg, Russia
13 - 15 November 2019

ISBN: 978-1-7138-3172-3
ISSN: 1757-8981

Printed from e-media with permission by:

Curran Associates, Inc.
57 Morehouse Lane
Red Hook, NY 12571

Some format issues inherent in the e-media version may also appear in this print version.

Institute of Physics
Dirac House, Temple Back
Bristol BS1 6BE UK

Phone: 44 1 17 929 7481
Fax: 44 1 17 920 0979

techtracking@iop.org

Additional copies of this publication are available from:

Curran Associates, Inc.
57 Morehouse Lane
Red Hook, NY 12571 USA
Phone: 845-758-0400
Fax: 845-758-2633
Email: curran@proceedings.com
Web: www.proceedings.com

TABLE OF CONTENTS

IOP Conf. Series: Materials Science and Engineering **826** (2020) 011001 doi:10.1088/1757-899X/826/1/011001

PREFACE

The 9th International Conference "Refrigeration and Food Technologies in the 21st Century" 13 – 15 November, 2019 in the School of Biotechnology and Cryogenic Systems of ITMO University. The conference has been held biennially since 2001 and has become a significant event for the Russian and European professional community in the last decade. The traditional format of the event is presentations and discussions of scientists and engineers working in the field of low-temperature and food technologies, as well as reports by specialists in the field of innovations in the food industry.

The purpose of the conference is to present modern approaches and methods for solving current scientific problems, and to exchange the latest research results obtained by Russian scientists working in higher education institutions, research institutes (including the Russian Academy of Sciences) and leading industry organizations, as well as scientists from universities in Ukraine, Kazakhstan, the Czech Republic and Finland.

In 2020, the conference was held under the auspices of the Saint Petersburg Committee of science and higher education and the Ministry of science and higher education of the Russian Federation. The organizing Committee of the conference includes representatives of leading academic organizations and industrial enterprises in Russia, as well as representatives of specialized foreign organizations. The conference sessions were prepared by the staff of the School of Biotechnology and Cryogenic Systems. The information and organizational sponsor of the conference was the International Academy of Refrigeration.

Within the framework of eleven sections, reports were presented on a wide range of issues in the following areas: low - temperature equipment and low-grade energy systems; reliability of materials for biotechnological equipment and low-temperature systems; automation of processes and production; cryogenic equipment and technologies of liquefied natural gas; air conditioning and life support systems; heat and cold engineering for the food industry, industrial ecology and technosphere safety, economic and administrative aspects of food quality assurance.

The conference was attended by 365 participants, including 125 representatives from 27 institutions of higher education, 16 participants from 7 research institutes (including the Russian Academy of Sciences) and leading sectoral organizations. During the conference, 193 reports were heard: 4 at the plenary session; 189 at the sessions of sections.

The conference helped Russian specialists to get acquainted with research works carried out in many regions of the country and abroad. In this collection, the best of the reported works were selected based on the results of the single-blind review. The main criteria for selecting articles were the relevance of the problem, the originality of the results obtained and their reliability.

With the kind regards,
Chair of the Committee
Igor V. Baranov

IOP Conf. Series: Materials Science and Engineering **826** (2020) 011002 doi:10.1088/1757-899X/826/1/011002

Peer review statement

All papers published in this volume of *IOP Conference Series: Materials Science and Engineering* have been peer reviewed through processes administered by the proceedings Editors. Reviews were conducted by expert referees to the professional and scientific standards expected of a proceedings journal published by IOP Publishing.

On the issue of calculating the force factors acting on the working bodies of a single rotor screw compressor

V Pronin, D Zhignovskaia, A Minikaev and D Yerezhep

Faculty of Cryogenic engineering, ITMO University, Lomonosova st., 9, St. Petersburg, 191002, Russia

E-mail: diana.zhignovskaya@gmail.com

Abstract. The subject of research is a single rotor screw compressor. For the proper compressor design, the forces and moments must be calculated. The cutter's teeth have a circular shape, therefore, the calculation is reduced to the determination of a torque. This power calculation is based on the law of pressure change in closed cavities of a single rotor screw compressor depending on the angle of a screw rotation and the law of change in the horizontal projection area of a cutter's tooth meshed with a screw, depending on the angle of a cutter rotation. The term "flat closed figure" in our task means the horizontal projection of the disk cutter's tooth part meshed with a screw, which takes the form of a triangle or trapezoid. The work shows a horizontal projection of the engagement rotor screw with a disk cutter. To determine the torque acting on the central screw, it is necessary to calculate the circumferential forces acting on the screw and the shoulders of the application forces. The calculation method described above allows one with a sufficient degree of accuracy to determine the forces and moments acting on the working bodies of a screw single-rotor compressor, and therefore, to calculate the deformation of the parts of a screw single-rotor compressor.

1. Introduction

For the correct design of a screw single-rotor compressor (SSC), it is necessary to be able to calculate the forces and moments acting on its working bodies. The main efforts arising in the working part of the machine are due to the pressure of the compressing substance. The closed working cavity (Figure 1) is limited by the cavity of a screw 1, the cylindrical part of a housing 2 and the tooth of a cutter 3. The total force $\boldsymbol{F}_{n_1}$, acting on a cavity of screw can be decomposed into the radial $\boldsymbol{F}_{r_1}$ and circumferential $\boldsymbol{F}_{t_1}$ force. Due to the symmetry of the working parts of the screw rotor compressor, the radial forces are compensated, and the circumferential forces create torque $\boldsymbol{M}_1$. A radial force $\boldsymbol{F}_{r2}$ will act on the cylindrical section 2, causing tensile stresses in it [1,2]. A force $\boldsymbol{F}_{a_1}$ directed parallel to the axis of rotation cutter and causing a corresponding bending moment will act on the cutter tooth. The listed forces are defined as the product of the excess pressure in the cavity $\boldsymbol{\Delta P_a}$ by the area of the corresponding projection. In addition to the indicated forces, an axial force $\boldsymbol{F}_{a_2}$ due to the difference in discharge and suction pressures will act on the end surface of the screw. However, in most cases, the screw rotor compressor is designed so that the cavity formed by a housing and the end surface on the discharge side is connected to the suction cavity to unload a screw from action of an axial force $\boldsymbol{F}_{a_2}$.

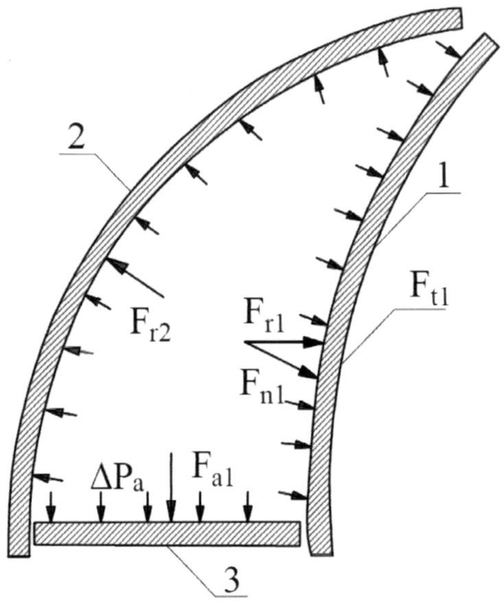

Figure 1. Closed working cavity of a single rotor screw compressor

2. Formulation of the problem

Thus, the task of power calculation of a screw single-rotor compressor is reduced, first of all, to the determination of forces $\boldsymbol{F}_{a_1}$ and $\boldsymbol{F}_{t_1}$, that are equal in magnitude. To determine them, it is necessary to know the law of pressure change in the closed cavities of the compressor, depending on the angle of rotation of the screw, to know the law of changing the horizontal projection area of the tooth of the cutter meshed with the screw, depending on the angle of rotation of the cutter [3].When changing the horizontal projection area of tooth cutter meshed with a screw, depending on the angle of cutter rotation [3].

In order to graphically represent the law of change in closed cavities of aerospace defense, we use:

$$^{P_s}/_{P(\alpha)} = \left[\left(V - V(\alpha)\right)/V\right]^m \quad (1)$$

where m - an average polytropic indicator; P_s – a suction pressure, Pa; $P(\alpha)$ – a pressure of the working medium in screw cavities depending on the rotation angle, Pa; V – a volume of the helical groove, m^3; $V(\alpha)$ – a part of the volume of a screw groove of the screw described by tooth cutter meshed with screw, depending on the angle of screw rotation, m^3.

It follows from (1) that $(\alpha) = {^{P_s V^m}}\big/_{[V - V(\alpha)]^m}$,

where α – an angle of screw rotation: $\alpha = \beta/i$; β – an angle of rotation of a disk cutter, deg; i – gear ratio determined by the ratio of the screw number z_1 to the number of the disk cutter z_2.

The pressure of the working medium in the cavity of a screw, depending on an angle of the cutter's rotation is found from:

$$p(\beta) = \left[P_s V^m / V - V_{(\beta)}\right]^m. \quad (2)$$

The volume of the screw groove of the screw is determined by the method based on the Gulden theorem, which assumes that the volume of the body obtained by rotating some flat closed figure about an axis lying in the plane of this figure and not intersecting it is equal to the product of the area of this figure by the length of the arc described by it center of gravity [4,5].

3. Materials and methods

The term "flat closed figure" in our task means the horizontal projection of a tooth disk cutter meshed with a screw, which takes the form of a triangle or trapezoid. The volume cavity of the screw is defined as the sum of the volumes $V = V_1 + V_2$, where V_1 – a volume of the part of the screw groove described by a flat closed figure in the form of a triangle, m^3; V_2 –the part volume of screw groove described by a flat closed figure in the form of a trapezoid, m^3.

Volume V_2 is determined:

$$V_2 = A_1 + (\beta_{1k} - \beta_{1n}) - A_2 \ln \frac{tg\left(\frac{\beta_{1k}}{2}+\frac{\pi}{4}\right)}{tg\left(\frac{\pi}{4}+\frac{\beta_{1n}}{2}\right)} - A_3(\sin\beta_{1k} - \sin\beta_{1n}). \quad (3)$$

There $\beta_{1k} = arcos\frac{D_{2n}}{D_{e2}} - \gamma$ – an angular gearing parameter, deg; $\gamma = \arcsin\frac{b}{D_{e2}}$ – a half of the angular width of the tooth disk cutter, deg; D_{2n} – a diameter of the initial circumference of the disk cutter, m (Fig.2); D_{e2} – an outer diameter of the disk cutter, m; b – a width of the tooth of the disk cutter, m; $A_1 = \frac{D_{OB}}{4i}(D_{e1} + D_{2n}), A_2 = \frac{b}{4i}\left(D_{2n}D_{e1} + \frac{D_{2n}^2}{2} + \frac{b^2}{6}\right), A_3 = \frac{b}{8i}(D_o^2 - \frac{b^3}{3})$- geometric constants, m^3. $D_o = D_{e2} \cdot cos\gamma$ - a diameter of the inscribed circle, m; β_{1n} – an angle of the beginning compression, that is, the angle at which the screw groove of the screw is cut off from the compressor suction cavity by the tooth of the disk cutter; $|\beta_{1n}| = |\beta'| - |\beta_1|$, where $\beta' = arcos\frac{D_{2n}}{D}, \beta_1 = arcsin\frac{b}{D}$ - angular parameters of engagement, deg; $D = (D_{2n} + 4l)^{1/2}$ – a diameter of inscribed circle, m; l – a half the length of a chopped part of a screw [6].

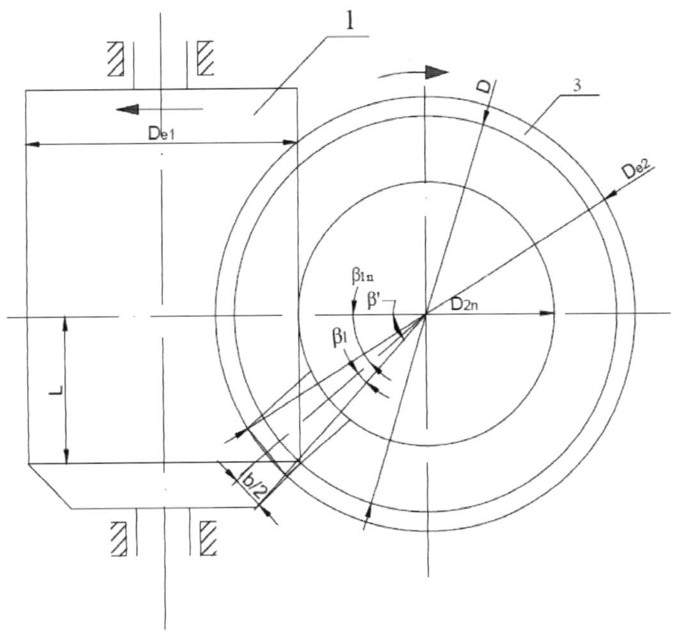

Figure 2. The horizontal projection of the engagement of the rotor screw with a disk cutter

In this reference frame, the values of the angles β', β_1, β_{1n} are negative therefore, the angle β_{1n} in (3) should be substituted with a minus sign.

Volume V_1 is calculated using:

$$V_1 = \beta_1(\beta_{2k} - \beta_{1k}) - \beta_2 ln\frac{tg\beta_{2k}/2}{tg\beta_{1k}/2} + \beta_3 ln\frac{sin\beta_{2k}}{sin\beta_{1k}} + \beta_4 ln\frac{cos\beta_{1k}}{cos\beta_{2k}} + \beta_5 ln\frac{tg\beta_{2k}}{tg\beta_{1k}} - \beta_6 ln\frac{tg(\beta_{2k}/2+\pi/4)}{tg(\beta_{1k}/2+\pi/4)} - \beta_7(sin\beta_{2k} - sin\beta_{1k}) - \beta_8(cos\beta_{2k} - cos\beta_{1k}),$$

where $\beta_{2k} = \beta_{1k} + 2\gamma$ – an angular gearing parameter, deg;

$$\left.\begin{array}{c} \beta_1 = (D_o b/8i)(D_{e1} + D_{2n}) \\ \beta_2 = (D_o b/8i)(D_{2n}D_{e1} + D_{2n}{}^2/2 + D_o{}^2/6 \\ \beta_3 = (D_o/16i)(D_{e1} + D_{2n}) \\ \beta_4 = (b^2/16i)(D_{e1} + D_{2n}) \\ \beta_5 = (D_{2n}{}^2/16i)(D_{e1} + D_{2n}/3) \\ \beta_6 = (b/8i)(D_{2n}D_{e1} + D_{2n}{}^2/2 + b^2/6 \\ \beta_7 = (b/16i)(D_o{}^2 + b^2/3) \\ \beta_8 = (D_o/16i)(D_o{}^2/3 + b^2) \end{array}\right\} \text{geometric constants of screw single compressor, m}^3$$

4. Discussion and results

As a result of calculations carried out according to (2), the law of pressure change in the closed cavities of the screw single-rotor compressor in Figure 3.

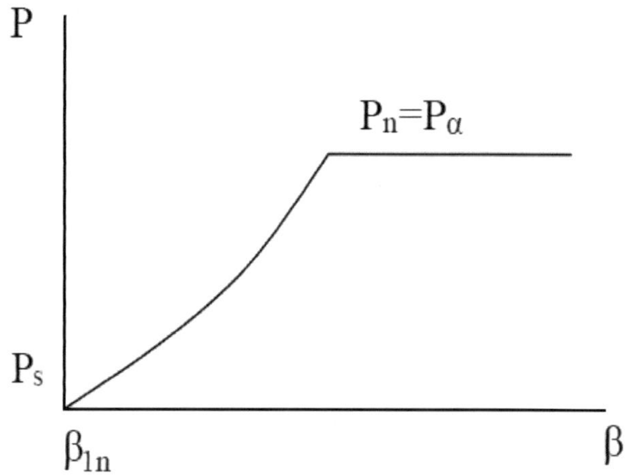

Figure 3. The law of pressure change in the closed cavities of a screw single-rotor compressor

The internal compression pressure $P(\alpha)$ by the condition coincides with the discharge pressure P_H. The axial force acting on the disk cutter is found by:

$$F_{a1} = \sum_{i=1}^{n} F_{a1i}\ , \tag{4}$$

where F_{a1} is a force acting on one of a teeth disk cutter meshed with a screw, N; n – the number of teeth disk cutter engaged with a screw [7,8].

The force acting on one tooth of disk cutter is determined by $F_{a1i} = S(\beta)[P(\beta) - P_s]$, where $S(\beta)$ – horizontal projection area of the tooth disk cutter meshed with a screw, depending on the angle rotation of the cutter, m^2.

The value $S(\beta)$ in the range interval of angles $\beta_1 \ll \beta \ll \beta_{1k}$ is determined by:

$$S(\beta) = \frac{D_o b}{2}\left(1 - \frac{D_{2n}}{D_o cos\beta}\right), \quad (5)$$

and in interval $\beta_{1k} \ll \beta \ll \beta_{2k}$ by $S(\beta) = \frac{D_o^2 ctg\beta}{8} - \frac{D_o D_{2n}}{4sin\beta} + \frac{D_o b}{4} + \frac{b^2}{8}tg\beta + \frac{D_{2n}^2}{8sin\beta cos\beta} - \frac{D_{2n} b}{4cos\beta}$.

Graphically, the results of calculations carried out by (5) and (6) are presented in Figure 4.

5. Conclusion

To determine the torque M_1 acting on the screw rotor, it is necessary to calculate the circumferential forces F_{t_1} acting on the screw and the moments of the application of forces L [9].

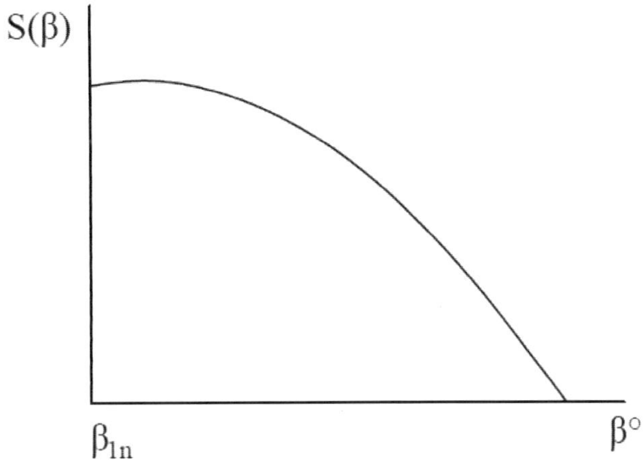

Figure 4. The schedule for changing the horizontal projection area of the tooth part of the disk cutter, which is in connection with the rotor screw

Torque M_1 is determined as follows:

$$M_1 = n\left[F_{t_1} L_{1(\beta)} + F_{t_2} L_{2(\beta)} + \ldots + F_{t_{1k}} L_{1k(\beta)}\right],$$

where n – number of disk cutters; $L_i(\beta)$ – moment of the force $F_{t_{1i}}$, m. The moment of the force $F_{t_{1i}}$ is calculated by $L_{i(\beta)} = \frac{D_{e1}+D_{2n}}{2} - r_{i(\beta)}$, where $r_{i(\beta)}$ – moment of force's application $F_{a_{1i}}$ relative to the axis of a cutter rotation, m.

$$r_{i(\beta)} = \left(y_c^2 + x_c^2 + \frac{D_{e2}^2}{4} - D_{e2}\sqrt{y_c^2 + x_c^2}\, cos\left[arcos\frac{y_c}{\sqrt{y_c^2+x_c^2}} - arcsin\frac{b}{D_{e2}}\right]\right)^{1/2},$$

by $\beta_{1k} \ll \beta \ll \beta_{2k}, \beta_{2k} = \beta_k + 2\gamma$.

$$r_{i(\beta)} = \left(\left[\frac{b^2 sin\beta}{6D_o(cos\beta - D_{2n}/D_o}\right]^2 + \left[\frac{D_o}{4} + \frac{D_{2n}}{4cos\beta}\right]^2 - \frac{b^2 cos^2\beta}{12D_o cos\beta(cos\beta - D_{2n}/D_o)}\right)^{1/2},$$

by $\beta_{1н} \ll \beta \ll \beta_{1k}$

IOP Conf. Series: Materials Science and Engineering **826** (2020) 012001 doi:10.1088/1757-899X/826/1/012001

There:

$$y_c = \frac{a(f+c)}{2(f+C+a)}; a = \frac{D_o}{2} - \frac{D_{2n}}{2cos\beta} + \frac{b}{2}tg\beta,$$

$$x_c = \frac{f(a+c)}{2(f+C+a)}; c = (a^2+f^2)^{1/2}; f = actg\beta,$$

where y_c, x_c – the coordinates in center of gravity of the tooth disk cutter horizontal projection meshed with the screw; *a, f, c* – are the lengths sides of the tooth disk cutter horizontal projection meshed with a triangle and meshed with the screw [10,11].

The calculation method described above allows one to determine with sufficient accuracy the forces and moments acting on the working parts of a single rotor screw compressor, and therefore, to calculate the deformation of the screw single rotor compressor parts, determine the safe gaps in the compressor working part and provide the necessary strength and durability of the most critical screw single rotor compressor components.

References

[1] Sakun I A 1970 *Screw compressors* (Leningrad: Mechanical Engineering)
[2] Pronin V A 1978 "*Investigation of a screw compressor with self-unloading support nodes*", PhD thesis, Leningrad.
[3] Pronin VA, Kuznetsov YL, Zhignovskaia DV 2018 Features of designing screw compressors for the oil and gas industry*AIP Conference Proceedings* **2007** 030017
[4] Stosic N, Smith I K, Kovacevic A and Zhang W M 2005 An Investigation of Liquid Injection in Refrigeration Screw Compressors, *ICCR the 5th International Conference on Compressors and Refrigeration*, Dalian.
[5] Heinz P. Bloch A 2006 Practical Guide to Compressor Technology, *2nd Edition «WILEY»* pp 4-30 ISBN: 978-0-471-92952-9.
[6] Fleming J, Tang Y and Cook G 1998 The helical twin screw compressor. Part 1: development, applications and competitive position. Proc. Instn. Mech. Engrs., Part C, *Journal of Mechanical Engineering Science* **212** pp 355–367.
[7] Tian F, Tao K and Shao J 2010 The Research on Meshing Pair Profile of Single-Screw Compressor, *IEEE Computer Society*, Wuhan, China.
[8] Tang, Y 1995 Computer aided design of twin screw compressors. PhD thesis, University of Strathclyde, Scotland.
[9] Xing Z 2000 Screw Compressors: Theory, Design and Application, (in Chinese), *China Machine Press*, Beijing, China.
[10] Excell, J 2013 The rise of additive manufacturing *The Engineer. Retrieved* 31–56
[11] Yerezhep D, Tukmakova A, Fomin V, Masalimov A, Asach A, Novotelnova A and Baranov A 2018 *J. Phys.: Conf. Ser.* 1015 032151. DOI: 10.1088/1742-6596/1015/3/032151.

IOP Conf. Series: Materials Science and Engineering **826** (2020) 012002 doi:10.1088/1757-899X/826/1/012002

Geo-ecological assessment of job-related diseases in mono-cities of the Russian Federation

E G Zinovyeva, S V Koptyakova, N R Balynskaya ,V V Chuprin and A I Luntsova

Nosov Magnitogorsk State Technical University, 38 Lenin Ave., Magnitogorsk, 455000, Russia

E-mail: Ekaterina_7707@mail.ru

Abstract. The relevance of the study is due to the importance of the health state and the quality of life among the population in connection with the adverse influence of a number of factors (environment, emission from large enterprises, food quality, and natural resources).
In the list of these factors affecting the health of employees working in industrial enterprises, harmful factors of the working environment are very important, the impact of which under certain conditions leads to high morbidity with temporary disability, to the development of work-related diseases. The results of a study investigating the level of work-related morbidity suggest that the number of employees in hazardous and harmful working conditions at industrial enterprises in Magnitogorsk increased to 42.95%. In 2017, the ratio of workplaces of industrial enterprises that do not meet sanitary and hygienic requirements for noise increased; the same variable (ratio) according to the standards of vibration, microclimate, lighting decreased; no deviations were detected in electromagnetic fields (EMF). Chronic work-related pathology in 2017 more often arose due to imperfection of technological processes (36.65% of cases), imperfection of workplaces (30.0%), structural shortcomings of labor facilities (16.65%) and sanitary facilities (6.65%), labour (working) contact with an infectious agent (10.0%). The results of the study can be used to improve programs at the federal and regional levels to improve the environmental situation.

1. Introduction

Preserving the health of employees is a priority of state policy in the field of labor relations, labor protection that means that an employer has to provide healthy and safe working conditions targeted at prevention of work-related diseases since the economic recovery of the country in general is associated with the able-bodied (employable) population. Of the list of industrial factors, working conditions of employees are the main risk factor in development of job-determined pathology. Assessing the level of harmful effects impacted on employees in the course of their labor activity, identifying certain factors of a labor process and the development of mechanisms to control them with the aim to reduce the levels of acceptable risks allow preserving the health of employees and lead to saving labor resources as well.

The aim of the paper is to give a geo-ecological assessment of the level of job-related morbidity in mono-industry towns of the Russian Federation based on the example of Magnitogorsk; to propose some measures ensuring healthy working conditions.

2. Materials and methods

Occupational morbidity is the frequency of cases of all job-related diseases (poisoning) recorded during the year among the working population, or groups that underwent a medical examination [1].

Issues of occupational morbidity among the population are widely covered both in Russian literature and in foreign one.

Theoretical and methodological issues of the sanitary-epidemiological well-being of the population are considered in the works of Ivanov S I, Belyaev E N [2], Butaev T M, Reshetnikov A V, Kucherenko V Z [3], Kuzmin V B [4], Zaitseva N V [5], Steblyanko V L [6], Zinovieva E [7; 8; 9] and others.

Theoretical and methodological issues regarding the quality of atmospheric air, soil and water, the level of occupational morbidity, emissions of pollutants are considered in the works of Butorina N N [10], Onishchenko G G [11], Mayorova T V [12; 13] etc.

Possibilities of using foreign experience in the sanitary-epidemiological well-being of the population in the Russian Federation are being studied by such authors as Smolensky V Yu [14], Toporkov I G, Leaves G D [15], Herbert R D, Jaakkola J J K [16], Alvarez-Herranz A [17], Aznar-Sánchez J A [18], Jiang K [19], Fu Y [20], Gani A [21], Pacca L [22] etc.

The theoretical base for the study in the view of the effective indicators is provided by the documents of legislative authorities, regulatory legal acts of the Federal Service for Supervision of Consumer Rights Protection and Human Well-being of the Russian Federation [23; 24], as well as the data from the Office of the Federal Service for Supervision of Consumer Rights Protection and Human Well-Being in the Chelyabinsk Region, Magnitogorsk city for 2014 - 2016 [25].

3. Assessment of occupational morbidity level in Magnitogorsk city

It is customary to study the geo-ecological basis for monitoring public health in the following areas: the state of the human environment and its impact on public health; the state of the medical and demographic situation and the incidence of mass non-communicable diseases (poisoning) in connection with the harmful effects of environmental factors on humans; level of occupational morbidity. In this article, we analyze the level of the occupational morbidity.

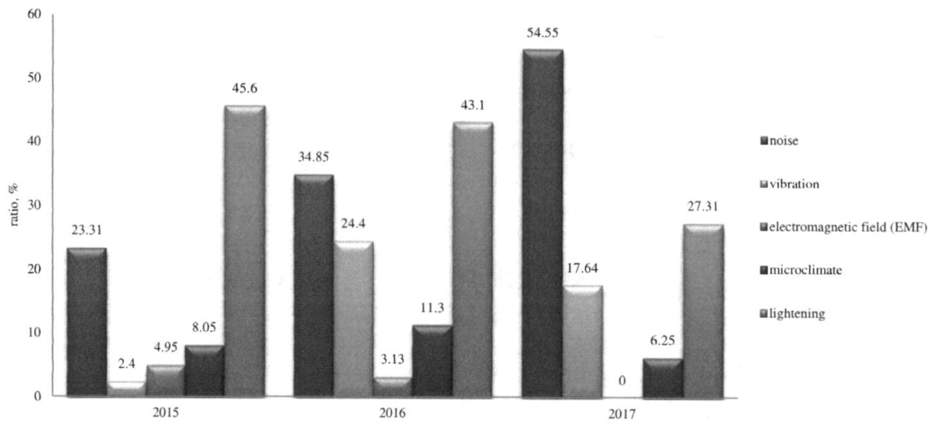

Figure 1. Ratio of workplaces at the industrial enterprises within the physical factors, %

The level of the occupational morbidity in Magnitogorsk is significantly impacted by some working conditions, as one of the main risk factors for development of a professionally determined pathology. At the industrial enterprises of Magnitogorsk (controlled by the territorial department) the number of workers employed in jobs with harmful and dangerous working conditions has increased and amounts to 42.95%. Over the past 3 years, there has been no tendency to a decrease in the ratio of employees who work in harmful and dangerous conditions, and there has even been a slight increase in this indicator, for example, in 2015 - 42.6%, in 2016 - 42.8%, in 2017 - 42.95%. The condition of the workplaces at the industrial enterprises in terms of certain physical factors in 2017 changed compared to 2016. So, in 2017, the ratio of jobs at the industrial enterprises, which do not meet the sanitary and hygienic requirements, increased concerning such a physical factor as noise, the ratio of jobs, which do not meet the sanitary and hygienic standards for vibration and microclimate, and lighting, decreased. Graphically it is represented in Figure 1.

The highest occupational morbidity rate per 10 thousand employees was recorded, as in previous years, at manufacturing enterprises - 3.93. The data are shown in table 1.

Table 1. Indicators of the occupational morbidity by the main types of economic activity

Types of economic activity	Indicator per 10.000 employees		
	2015	2016	2017
For all types of economic activity	1.72	1.81	3.19
Manufacturing enterprises:	2.54	2.12	3.93
Production of other non metallic mineral products	4.31	-	-
Metallurgical production	2.71	3.11	6.60
Manufacture of fabricated metal products	-	4.54	1.13
Machinery and equipment production	1.64	0.55	2.91
Recycling of secondary raw materials	11.16	11.16	-
Civil building	1.99	1.33	2.24
Health care and social services	-	1.37	4.11

Chronic job-related pathology in 2017 more often arose due to imperfection of technological processes (36.65% of cases), imperfection of workplaces (30.0%), structural shortcomings of labor facilities (16.65%) and sanitary facilities (6.65%), labour (working) contact with an infectious agent (10.0%). All this information is presented in Figure 2.

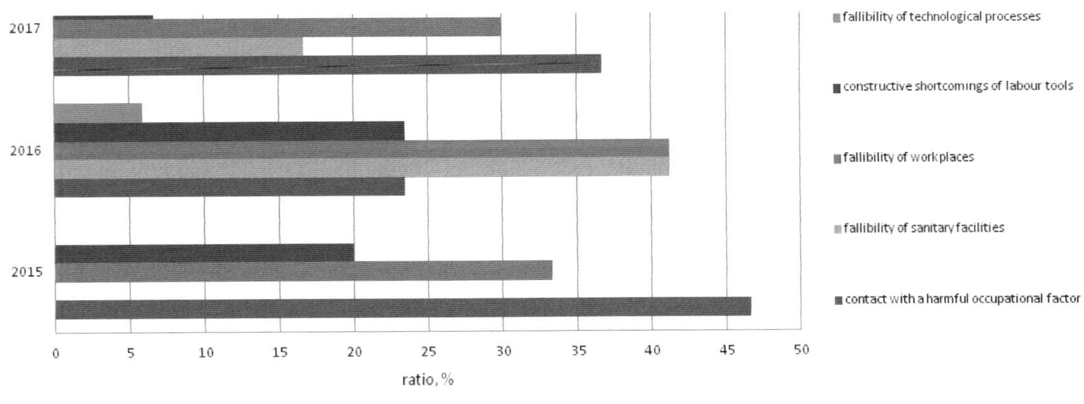

Figure 2. Circumstances and conditions caused chronic occupational diseases, %

The level of chronic occupational pathology detection, depending on the type of medical institutions and their specialization, practically has not changed. In 2017, Magnitogorsk Center for Occupational Pathology detected 96.67% of cases of occupational diseases (in 2016 - 95.24% of the total number of cases, in 2015 - 86.67%). The analysis showing the dependence between the level of occupational morbidity on a job, duration of contact with a harmful production factor and the age of an employee allows us to identify individual occupational groups that are most at risk of any occupational pathology.

The analysis of occupational pathology, first established in 2017, allows us to conclude that the length of job experience, exposure to a harmful occupational factor and the level of occupational morbidity are interconnected, and the maximum risk of an occupational disease at industrial enterprises is manifested among male workers exposed to harmful occupational factors over 30 years, whereas, among female workers this job experience is over 20 years. In the 20-30-year experience, the share of registered occupational diseases among male workers is 31.82%, with the experience of 30-40 years - 63.64%, whereas the figure among female workers in the group with 20-25-year experience is 66.7% of all occupational diseases distributed by a gender principle.

Workers aged at 50-60 are most at risk of occupational disease: occupational diseases among men at this age category are 90.91%, among women - 66.67% of all occupational diseases distributed by gender. Depending on the professions of workers, men and women are at greatest risk of acquiring occupational pathology - electric gas and gas welders, refractory workers, coke oven drivers, choppers. The share of occupational diseases of workers in these professions first documented in 2015–2017 is 36.21%.

Most cases of occupational pathology were detected during periodic medical examinations and all were of chronic forms, which, as a rule, were gained by workers with large job experience (more than 30 years) in contact with a harmful production factor and at the age over 50. Despite a decrease in the indicator of the validation of victims, the severity of the course of an occupational disease and the degree of losing professional suitability, it remains higher than the all-Russian one and is an indicator of low active detectability of occupational pathology under the current system of medical examination of mandatory groups of employees.

As of the end of 2017, in the Office of the Federal Service for Supervision of Consumer Rights Protection and Human Well-Being in the Chelyabinsk Region in Magnitogorsk, there were under control 175 industrial enterprises with a total number of 87,181 employees, including 37,452 people working in harmful conditions, which is shown in table 2.

Table 2. Number of industrial enterprises and number of employees working in Magnitogorsk within 2015-2017

Year	The number of employees in total, people		Including those who work in harmful and adverse working conditions, in person	Ratio of workers working in harmful and adverse working conditions, %
	total	deviation, %		
2015	87080	-5.4	37133	42.6
2016	86905	-0.2	37153	42.75
2017	87181	+0.32	37452	42.96

The main number of employees - 76788 people or 88.0% is concentrated at the enterprises of six types of economic activity: metallurgical production - 25759 people; mechanical engineering - 20637 people; civil building - 13426 people; manufacture of fabricated metal products - 8871 people; land transport activities - 5687 people; production of building materials - 2408 people. At the enterprises where the main number of employees is occupied - MMK JSC, MRK JSC, Ogneupor LLC, OSK LLC, MMK-METIZ JSC, as before, working conditions do not comply with the hygienic standards. Accordingly, the technological processes that cause non-compliance of working conditions with the sanitary norms and rules are as follows: coke production, iron and steel smelting, foundry, metal cutting - according to the presence of harmful substances in the air of the working area; metal processing, production of hardware and wire, chipping operations, - on production noise; vibration pressing in the manufacture of building materials, polishing, forging, hand molding - according to the level of general and local vibration.

The harmful effects of production factors are exacerbated by the lack of regulated breaks and working hours for employees at enterprises when they work in cooling or heating microclimate, as well as they work with vibration tools in conditions of high levels of noise connected with industrial and production processes.

4. Conclusion

Thus, in the field of ensuring healthy working conditions, it is advisable to propose actions to be fulfilled:

- to consider the development of a targeted program "Improving working conditions and medical support for the working population of Magnitogorsk city";
- to organize production monitoring of compliance with the sanitary rules, hygiene standards and some preventive measures;
- organize periodic medical examinations of those who are working in contact with harmful production factors;
- to carry out all planned actions to improve working conditions at the workplaces of industrial enterprises, as well as to develop some additional measures helping to bring working conditions at workplaces to the requirements of the hygienic standards (increase the efficiency of ventilation systems, replace technically obsolete equipment, upgrade technological equipment, repair technology for metallurgical furnaces and aggregates). For this reason, the primary task of any municipality is to monitor the environmental situation, provide labor protection, and medical examinations (periodic).

References

[1] Fomenko G A 2014 *Scientific and methodological foundations of creating an innovative system in the environmental sphere* (Moscow: Favorite)

[2] Belyaev E N, Fokin M V, Novikov S M, Prusakov V M, Shashina T A and Shayakhmetov S F 2013 Actual problems of improving the health risk assessment of the population to ensure sanitary and epidemiological well-being *Hygiene and Sanitation* [in Russian – Gigiena i Sanitariya] **5** 53–55

[3] Kucherenko V Z 2014 *Organizational and legal framework of the Federal Service for Supervision of Consumer Rights Protection and Human Welfare* (Moscow: GEOTAR-Media)

[4] Kuzmin S V, Gurvich V B, Dikonskaya O V, Malykh O L and Yarushin S V 2016 Methodology of assessing and managing risk to public health in the system of legislative regulation of the sanitary and epidemiological welfare of the population *Occupational medicine and industrial ecology* **1** 4–8

[5] Zaitseva N V, Popova A Yu, May I V and Shur P Z 2015 Methods and technologies of health risk analysis in the public administration system while ensuring the sanitary and epidemiological well-being of the population *Hygiene and Sanitation* [in Russian – Gigiena i Sanitariya] **94(2)** 93–98

[6] Steblyanko V L, Zinovyeva E G and Koptyakova S V 2019 Efficiency of monitoring the sanitary-epidemiological state of Magnitogorsk *Steel* [in Russian – Stal] **4** 80–83

[7] Zinovyeva E G and Koptyakova S V 2019 Integrated assessment of sanitary and epidemiological safety management efficiency of the population in Industry–based cities in the Russian Federation *Advances in economics, business and management research: proc. of the Int. Sci. Conf. Far East Con (ISCFEC 2018)* **47** 512–516

[8] Zinovyeva E G, Koptyakova S V, Balynskaya N R and Stepashkov P Y 2019 Assessment of monitoring efficiency for the state of sanitary and epidemiological well-being of the population of metallurgical monocities *IOP Conference Series: Earth and Environmental Science* 012050

[9] Limarev P V, Limareva Yu A and Zinovieva E G 2019 Comprehensive approach to research of

the consumer market in montygorods of the Russian Federation *Advances in economics, business and management research: proc. of the Int. Sci. Conf. Far East Con (ISCFEC 2018)* **47** 1056–1059
[10] Butorina N N, Bychkov A A, Samutin N M, Orlova I G and Pozdnyakov S A 2017 Modern approaches to the regulation of waste management of non-canalized facilities at the regional level *Ecology and Industry of Russia* [in Russian – Ekologiya i promyshlennost Rossii] **10** 68–71
[11] Onishchenko G G 2012 Results and prospects of ensuring the sanitary and epidemiological welfare of the population of the Russian Federation *Hygiene and Sanitation* [in Russian – Gigiena i Sanitariya] **4**
[12] Kolokoltsev V M, Vdovin K N, Maiorova T V and Ponomareva O S 2017 Ecological indicators in the system of non-financial reporting at industrial enterprises *CIS Iron and Steel Review* **13** 4–10
[13] Mayorova T V and Ponomareva O S 2015 Methodology for assessing the economic efficiency of environmental management of metallurgical enterprises *Vestnik of Nosov Magnitogorsk State Technical University* **4(52)** 112–116
[14] Smolensky V Yu, Toporkov V P, Karnaukhov I G, Toporkov A V and Kutyrev V V 2012 Organizational and methodological aspects of international cooperation in ensuring the sanitary and epidemiological well-being of the population *Problems of especially dangerous infections* [in Russian – Problemy osobo opasnyh infektsij] **2(113)** 5–13
[15] Leaves G D and Herbert R D 2002 Economic and environmental impacts of pollution control in a system of environment and economic interdependence *Chaos, Solitons & Fractals* **13(4)** 693–700
[16] Jaakkola J J K et al 2000 Use of health information systems in the Russian Federation in the assessment of environmental health effects *Environmental Health Perspectives* **108(7)** 589–594
[17] Alvarez-Herranz A, Balsalobre-Lorente D, Shahbaz M and Cantos J M 2017 Energy innovation and renewable energy consumption in the correction of air pollution *Energy Policy* **105** 386-397
[18] Aznar-Sánchez J A, Belmonte-Ureña L J, Velasco-Muñoz J F and Manzano-Agugliaro F 2018 Economic analysis of sustainable water use: A review of worldwide research *Journal of Cleaner Production* **198** 1120-1132
[19] Jiang K, You D, Li Z and Shi S 2019 A differential game approach to dynamic optimal control strategies for watershed pollution across regional boundaries under eco-compensation criterion *Ecological Indicators* **105** 229-241
[20] Fu Y, Cui X, Liu L, Zhao J and Zhang S 2019 Equilibrium cost of water environmental protection based on watershed sustainability *Journal of Hydrology* **579** 124216
[21] Gani A and Scrimgeour F 2014 Modeling governance and water pollution using the institutional ecological economic framework *Economic Modelling* **42** 363-372
[22] Pacca L, Antonarakis A, Schröder P and Antoniade A 2020 The effect of financial crises on air pollutant emissions: An assessment of the short vs. medium-term effects *Science of The Total Environment* **698** 133614
[23] Federal Law of 30.03.1999 no 52-FZ On the sanitary-epidemiological welfare of the population 1999 *Collected legislation of the Russian Federation* **14** 1650
[24] Federal Law of 21.11.2011 no 323-FZ On the basis of the protection of public health in the Russian Federation 2011 *Russian newspaper* **263**
[25] The impact of environmental factors on public health. Medico-demographic situation in the Chelyabinsk region, Available at: http://www.mineco174.ru/files/media/doklad/2014/6-1.htm

Study of convective heat transfer in small-diameter channels

V V Gerasyutenko[1], A V Sharkov[1], V A Korablev[1], A V Kayalajnen[2] and D A Minkin[3]

[1] ITMO University, 49, Kronverksky Pr., St. Petersburg, 197101, Russia
[2] The Russian State Scientific Center for Robotics and Technical Cybernetics (RTC), 21, Tikhoretskiy Pr., St. Petersburg, 194064, Russia
[3] St. Petersburg University of State Fire Service of EMERCOM of Russia, 149, Moskovskiy Pr., St. Petersburg, 196105, Russia

E-mail: viktoriya.gerasyutenko@mail.ru

Abstract. This paper investigates the convective heat transfer in a tube made of stainless steel 12X18H9T with outer diameter of 2 mm and inner diameter of 1.5 mm. For the research an experimental setup has been designed and assembled. The experimental setup consists of a fluid pumping system and flow-measuring device, working area, electrical power system and temperature measuring system. The experiment studies of heat transfer intensity in small diameter channels are performed. The study compares experimental data obtained with the theoretical ones. The paper presents the graphs of the Nusselt criterion dependence on the Reynolds criterion and the dependence of the convective heat transfer coefficient on the Reynolds criterion.

1. Introduction

The study of convective heat transfer in the channels of cooling systems in electronic equipment is an actual problem to solve. Miniaturization of electronic equipment elements and increase of power dissipation leads to a reduction of the size of electronic equipment cooling systems. In modern power modules for industrial applications, the level of power dissipation per square centimeter of a casing area reaches several hundred watts. Liquid cooling systems can become a solution to heat dissipation problems.

The main conditions that determine the use of a cooling system for devices are the maximum allowable operating temperature of a device and the maximum amount of heat generated per unit surface area of a device. For example, laser diodes generate heat fluxes with density which is more than 5 kW / cm^2 at the temperature difference of 30-40°C [1]. To calculate the parameters of cooling systems, it is necessary to know the intensity of convective heat transfer [2-5]. Therefore, this paper aims to study the convective heat transfer in tubes with an internal diameter of 1.5 mm. Small diameter channels require the use of highly purified liquids.

2. Experimental setup for study of convective heat transfer in tubes

The experimental setup for studing convective heat transfer in small diameter tubes was designed and assembled. The scheme of this experimental setup is shown in Figure 1.

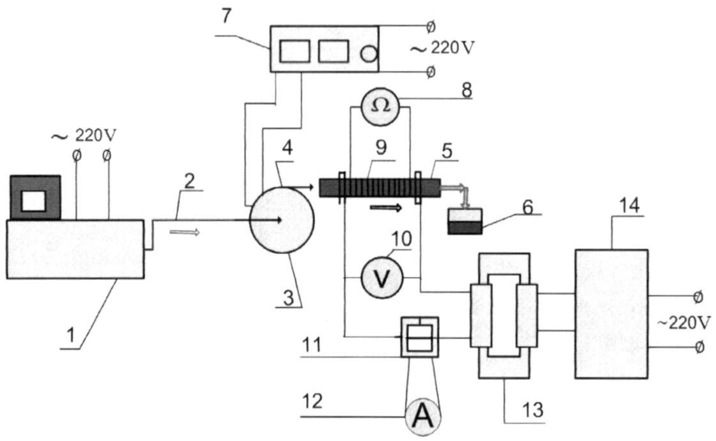

Figure 1. Schematic image of experimental setup

The experimental setup consists of the elements, which are indicated in Figure 1 by numbers: 1 - thermostat; 2 - connecting tube; 3 - pump; 4 - connecting tube; 5 - working area; 6 - measuring cylinder; 7 - power supply; 8 - universal voltmeter for measuring electrical resistance; 9 - thermistor; 10 – voltmeter measures the voltage going to tube heating; 11 - measuring transformer; 12 - ampermeter; 13 - step - down transformer; 14 - laboratory autotransformer.

In this paper, distilled water was used as a highly purified liquid. Distilled water is poured into the thermostat 1. The thermostat has an embedded system for external circulation. Water is pumped through working area 5 by the use of pump 4 and embedded system in thermostat for fluid pumping. The flow control occurs by changing current-voltage characteristics of the pump with the help of power supply 7 (Figure 1). The volume flow liquid is measured by measuring cylinder 6 and a stopwatch. It is determined how long it takes a fixed volume of liquid to fill the measuring cylinder. Then, the volume flow is calculated by the equation:

$$G_v = \frac{V}{\tau}, \quad (1)$$

where V - volume, τ- the time of liquid outflow.

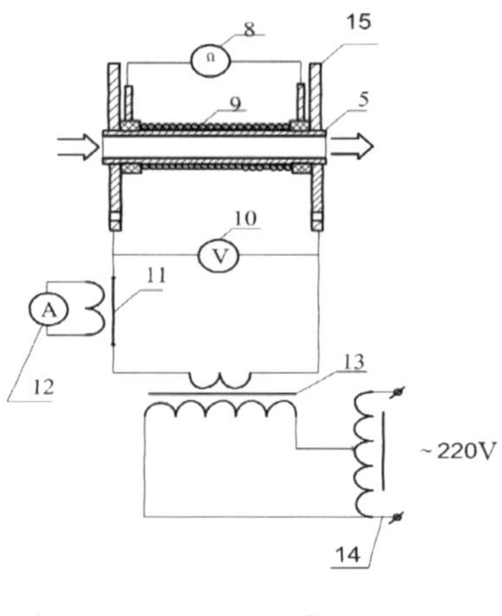

Figure 2. Schematic image of the main area power supply system

The volumetric flow rate and the corresponding voltage at the power source supplied to the pump are calibrated. Figure 2 shows the electrical power system of working area 5. Heating of working area 5 is conducted by electric current. For this purpose electric current is supplied to terminals 15. The terminals are soldered to the tube through which the liquid is pumped. Alternating current 220V goes to transformer 13 from laboratory autotransformer 14. Then, the electric current is measured with measuring transformer 11 and amperemeter 12. The voltage supplied to terminals 15 is measured by a voltmeter.

The temperature is measured indirectly. To do this, the electrical resistance of thermoresistor 9 is calibrated and thermoresistor 9 reeled on the tube. Increasing the temperature of the liquid pumped heats the tube from 30 to 95°C. Simultaneously, the corresponding value of the electrical resistance is read using a universal voltmeter. After graduating the thermoresistor, the working area is heated by electric current. The values of the electrical resistance before and after heating and temperature corresponding to these resistances are taken.

3. Preparation of the working area

The paper has investigated a tube made of stainless steel 12X18H9T with an outer diameter of 2 mm and an inner diameter of 1.5 mm. The terminals for electric current are soldered to the tube tested. Wrapping of the thermoresistor is performed manually. The tube is fixed in a machining station which movable part reels the thermoresistor. Then the area with the thermoresistor is covered with a layer of glue.

The thermoresistor is a thin wire covered with an electrical-insulating varnish. After reeling the thermoresistor on the tube, the ends of the thermoresistor are cleaned. Thermoresistor copper wire is soldered to the wires located on Teflon tubes. The wires, on which the electrical resistance is read by a universal voltmeter, are soldered to a wire fixed to the Teflon tubes. Plastic foam is glued to the thermoresistor and terminals to reduce the temperature fluctuations on the thermoresistor and decrease the loss of thermal energy. The tube is connected to the pump through an adapter and a silicone tube. The outer diameter of the tube is measured by a micrometer. The inner diameter of the tube is measured by a microscope. For this purpose, the number of visible divisions under microscope for the outer and inner diameter of the tube is measured.

The proportion and the inner diameter of the tube are calculated by the equation:

$$d_1 = \frac{d_2 n_1}{n_2}, \tag{2}$$

where d_1 is the inner diameter of the tube; d_2 is the outer diameter of the tube; n_1 – the number of visible divisions under a microscope for the inner diameter of the tube; n_2 - the number of visible divisions under a microscope for the outer diameter of the tube.

4. Calculation of convective heat transfer coefficient

Electrical power used to heating the tube:

$$P = UI \text{ [W]} \tag{3}$$

Convective heat transfer coefficient:

$$\alpha = \frac{P}{S\Delta t} \text{ [W / m}^2\cdot\text{K]} \tag{4}$$

The temperature difference between the mean temperature of the inner surface of wall tube and the mean fluid temperature:

$$\Delta T = \Delta T_{\mathrm{t}} - \Delta T_{\mathrm{sur}} - \Delta T_{\mathrm{heat}} \text{ [K]}. \tag{5}$$

The temperature difference between the mean temperature of the tube and the mean temperature of the liquid at the inlet of the tube:

$$\Delta T_{\mathrm{t}} = \frac{\Delta \mathrm{R}}{\mathrm{R}_0 \alpha}\ [\mathrm{K}] \tag{6}$$

Differential temperature on the wall:

$$\Delta t_{\mathrm{w}} = \frac{q_v}{4\lambda_w}\left[2ln\left(\frac{r_2}{r_1}\right) + \left(\frac{r_1}{r_2}\right)^2 - 1\right] = \bar{t}_{w2} - \bar{t}_{w1}\ [\mathrm{K}]\ [6] \tag{7}$$

Volumetric power density:

$$q_v = \frac{P}{V} = \frac{4P}{\pi(d_2^2 - d_1^2)L}\ [\mathrm{W/m^3}] \tag{8}$$

The temperature on the inner surface of the wall:

$$\bar{t}_{w1} = \bar{t}_{w2} - \Delta t_{\mathrm{w}}\ [\mathrm{K}] \tag{9}$$

The mean liquid temperature:

$$\bar{t}_f = t_{in} + \frac{F}{2c_p G_v \rho}\ [\mathrm{K}] \tag{10}$$

The convective heat transfer coefficient:

$$\alpha = \frac{F}{S(\bar{t}_{w2} - \bar{t}_f)}\ [\mathrm{W/\ m^2 \cdot K}] \tag{11}$$

Nusselt number:

$$Nu = \frac{\alpha \cdot d}{\lambda} \tag{12}$$

The calculation of the temperature difference between mean fluid temperature and the mean temperature at the inlet:

$$\Delta T_{\mathrm{heat}} = \frac{P}{2c\rho G_v}\ [\mathrm{K}] \tag{13}$$

The program Scilab was used for data processing of the results and compared them with the theoretical ones. The equations of Mikheev, Hausen, Ramm, Kutateladze, for heat transfer at movement of liquid in pipes, are used for calculation and given below.
The equation for heat transfer in the transient behavior of fluid motion ($2 \cdot 10^3 \leq Re_f \leq 10^4$): [7]

$$\overline{Nu_f} = K \cdot M;\ M = Pr_f^{0,43}\left(\frac{Pr_f}{Pr_\omega}\right)^{0,25} \tag{14}$$

The equation for heat transfer in the transient regime of fluid motion [8]:

$$\overline{Nu_f} = 0{,}0225 Re_f^{0,8} Pr_f^{0,4} \cdot f;\ f = 1 - \frac{6 \cdot 10^5}{Re_f^{1,8}} \tag{15}$$

The equation for heat transfer (for Re from 2300 and for the entire region of turbulent fluid flow) [9]:

$$\overline{Nu_f} = 0{,}12\left(Re^{\frac{2}{3}} - 125\right) Pr^{\frac{1}{3}}\left({}^{\mu}/_{\mu_w}\right)^{0,14} \tag{16}$$

The equation for turbulent behavior:

$$\overline{Nu_f} = 0{,}021\ Re_{df}^{0,8} Pr_f^{0,43}\left(\frac{Pr_f}{Pr_w}\right)^{0,25} \cdot \varepsilon_l \tag{17}$$

where ε_l is the change of mean heat transfer coefficient along the length of the tube [10].
The equation for the turbulent behavior:

$$Nu_f = \frac{0{,}023 \cdot Pr \cdot Re^{0,8}}{1 + 2{,}14 Re^{-0,1} \cdot (Pr^{2/3} - 1)}\ [11] \tag{18}$$

After data processing , the graphs were built which are shown in Figure 3 and Figure 4.

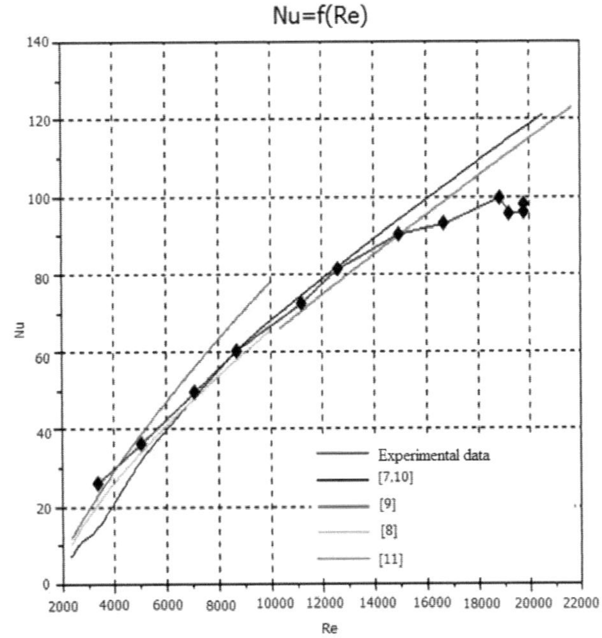

Figure 3. Dependency graphs of Nusselt number on Reynolds number Nu (Re) according to experimental and theoretical data.

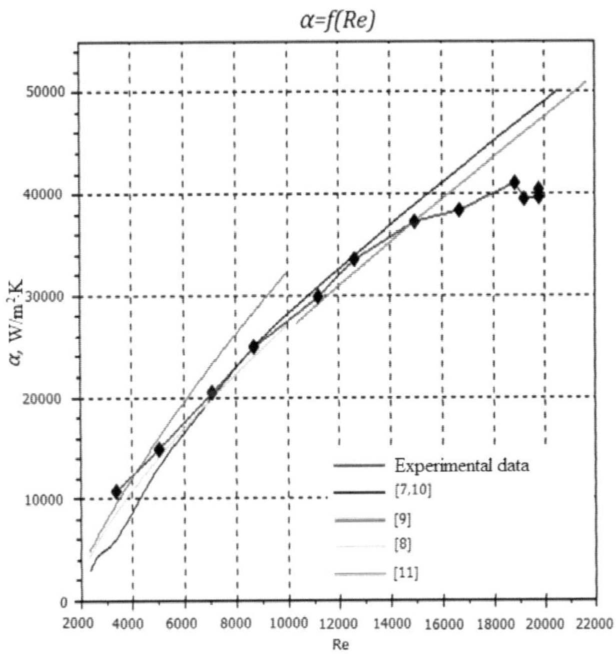

Figure 4. Dependency graphs of the convective heat transfer coefficient on the Reynolds criterion α (Re) according to experimental and theoretical data.

5. Conclusion

The experimental setup has been designed and assembled to investigate heat transfer in small diameter tubes. The working area was fabricated for the study of convective heat transfer in tubes with an internal diameter of 1.5 mm. The intensity of convective heat transfer in the liquid flow inside a smooth tube of circular cross-section is investigated. The dependency graphs of the convective heat transfer coefficient on Reynolds number α(Re) are compared according to experimental and theoretical data. Moreover, the dependency graphs of Nusselt number on Reynolds number Nu(Re) are compared according to experimental and theoretical data. The experimental data has revealed no contradictions with the dependencies previously known.

References

[1] Bezotosnyi V V, Krokhin O N, Oleshchenko V A , Pevtsov V F, Popov YU M and Cheshev E A 2015 Thermal condition of high-power laser diodes *Bull. Of the Lebedev Phys. Inst.* **42(3)** 88-92 DOI: 10.3103/S1068335615030057

[2] Kadirova D K 2017 Microchannel heat exchange systems applications to ensure thermal conditions *Bull. of the Science Week – 2017 of Dagestan St. Tech. University* 35 – 37

[3] Konovalov D A, Kozhukhov N N and Drozdov I G 2016 Modeling of processes of heat and mass transfer in microchannel heat exchangers of space-system engineering control systems *Bull. of the Reshetnev University* **17(1)** 137-146

[4] Belyaev A V, Varavva A N, Dedov A V and Komov A T 2012 *Proceedings of the VIII school-seminar of Kazan National Research Technical University named after A N Tupolev* 24 – 27

[5] Malakhovskiy S A, Varavva A N, Dedov A V, Zakharov E M and Komov A T 2007 Experimental research of hydrodynamics and heat transfer in the channels of small diameter *Bull. of the Moscow Power Engineering Institute* **(1)** 51-56

[6] Isachenko V P, Osipova V A and Sukomel A S 1981 *Heat Transfer* (Moscow: Energy Publishing)

[7] Dul'nev G N and Semyashkin E M 1968 *Heat Transfer in Radioelectronic Devices* (Leningrad: Energy Publishing)

[8] Andreev V A 1971 *Heat Exchangers for Viscous Liquids. Fundamentals of Calculation and Design* (Leningrad: Energy Publishing)

[9] Tarasov F M 1964 *Thin Layer Heat Exchangers* (Moscow – Leningrad: Engineering)

[10] Miheev M A and Miheeva I M 2010 *Fundamentals of Heat Transfer* (Moscow: Publishing House Bastet)

[11] Kutateladze S S 1979 *Fundamentals of Heat Transfer Theory (*Moscow: Atom Publishing)

IOP Conf. Series: Materials Science and Engineering **826** (2020) 012004 doi:10.1088/1757-899X/826/1/012004

Choice and justification of the optimum cooling temperature for HTS cable

A V Glushaev[2], E V Sokolova[1], D V Kravchenko[1], D N Kirilov[1], D A Khrustovskaia[1] and L V Ivanov[1]

[1]Information Technologies, Mechanics and Optics University, 49, Kronversky Ave., St. Petersburg, 197101, Russia
[2]JSC "D.V. Efremov Institute of Electrophysical Apparatus", 3, Doroga na Metallostroy, St. Petersburg, 196641, Russia

E-mail: lorencia48155@gmail.com

Abstract. At the first stage of designing a high-temperature superconducting (HTS) cable and its cryogenic refrigeration system (CRS), it is necessary to choose the optimum cooling temperature. The operating temperature is selected not only according to the critical parameters of the HTS conductor, but also taking into account other design and operational characteristics - geometry, length, operating current and the need to have subcooled nitrogen to prevent the occurrence of steam inclusions. Normally the range of cooling temperatures is limited by the critical temperature of the HTS conductor ($T_{HTS} \leq$ 104K) and the triple point temperature of nitrogen ($T_{LN} \geq$ 63.15K). The choice of a certain temperature value will enable us to approximately evaluate the parameters of the CRS and overall characteristics of the cable. The article presents an analysis of the dependence of the electric power magnitude consumed by the CRS on the choice of the cooling temperature value and cable geometry. The analysis may be useful for determining the characteristics of the HTS cable operating at different temperature levels.

1. Introduction

To use superconducting equipment, the effective and reliable CRS is required which can maintain the superconductor temperature at a given level. The complexity and high price of effective CRS prevent the widespread use of high-temperature superconducting (HTS) cable line (CL) in the power grid. As a rule, the CRS use liquid nitrogen as a coolant due to its properties: high dielectric strength, wide range of operating temperatures (especially at elevated pressure), high specific heat capacity, low cost and fire safety.

Replacement of liquid nitrogen in a saturation condition (77 K) by subcooled liquid nitrogen with a temperature below 77 K increases the permissible density of critical current [1]. An increased current density makes it possible to reduce the cost and weight of the cable. However, as the operating temperature level decreases, the thermal load on the CRS and the energy consumption for cooling of the HTS CL increase. Excessive power consumption in the CRS due to relatively low temperature level of heat removal from the subcooled nitrogen can seriously impair the competitive ability of the HTS CL in comparison with alternative technologies of energy transmission. The choice and justification of the optimum cooling temperature of the HTS CL, taking into account the compactness and effectiveness of the CRS, are one of the most important scientific and technical problems of HTS cable commercialization.

Today, subcooled liquid nitrogen is used for cooling in most HTS cables. The nitrogen temperature at the inlet to the cryostat is lowered to T_{LN}= 65 K [2]. The choice of this temperature range is determined by the fact that it provides high current density of HTS tapes and reduces the cost of cable manufacture. At the same time, capital and operating costs associated with the CRS are increased.

The wider use of the HTS tapes forms a tendency of gradual reduction in their price. In the future it will be possible to increase the specific consumption of HTS material and, thus, to rise the cooling temperature and simplify the CRS.

2. HTS cable model description

In this investigation, an attempt was made to analyze the dependence of the cryostat size and the power used by CRS on the choice of the cooling temperature. It is necessary to determine the cooling temperature at which the minimum energy consumption in the CRS will be achieved. The size of a cryostat affects the energy efficiency of the CRS (see Figure 1).

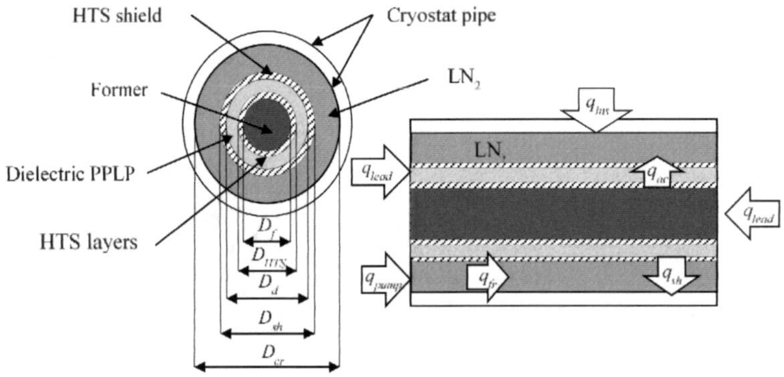

Figure 1. HTS cable model.

To select the CRS parameters, the design of a power cable with a cold dielectric described in the literature [3] was used. Sumitomo's 1st generation DI-BSCCO Type HT tape with a width of 4.5 mm and a thickness of 0.34 mm is used as a current-carrying element. The flow part of the cable is defined by an annular channel (See Figure 1). It is assumed that the temperature of the HTS cable varies along its length. The temperature gradient (ΔT) along the cable length is taken equal to 10 K. The critical current density of the HTS tape decreases with increasing temperature. The number of tapes in layers of the cable increases, so the diameter and weight of the cable also depend on the choice of the operating temperature of the CRS. In this case, a HTS cable with two layers is considered. Two layers of the HTS tapes ensure an even current distribution [4]. The HTS cable diameter is made up of the former diameter (D_f), the thickness of the power HTS layer, the insulation and the thickness of the HTS shield, which are proportional to the cryostat temperature. The required number of tapes is determined by the current density in the HTS cable and the critical current of the HTS conductor. The former diameter is selected for laying two HTS layers. The tapes are placed evenly across the layers. The diameter is selected based on the possibility of placing the tapes of the first layer along the perimeter of the cable without gaps.

The cryostat diameter (D_{cr}) is selected based on the condition of minimizing hydraulic losses and ensuring a given temperature gradient (ΔT) along the cable length.

3. Thermal load on the cryogenic refrigeration system

The thermal load on the cryogenic system of the HTS cable consists of the following components:

- Thermal leak through current leads.

- Thermal leak through thermal insulation.
- AC losses.
- Dielectric losses.
- Hydraulic losses.
- Pumping losses.
- Losses in contacts.

3.1. Thermal leak through current leads

Current leads - devices to transfer electric energy to the low temperature zone. One end of the current lead is in contact with the electric power source at ambient temperature, the other is connected to the HTS cable. The design of Current leads should provide a minimum thermal load (q_{lead}) on the CRS. To calculate the intensity of thermal leak through current leads, you can use the expression [5]:

$$q_{lead}^{in} = I \cdot (2 \cdot \lambda \cdot \Delta T_{in} \cdot \rho)^{1/2},\ q_{lead}^{out} = I \cdot (2 \cdot \lambda \cdot \Delta T_{out} \cdot \rho)^{1/2}$$

$$q_{lead} = q_{lead}^{in} + q_{lead}^{out}$$

where ΔT_{in}, ΔT_{out} – temperature difference at the initial and final sections of the current lead, respectively, K;
$\rho = f(T)$– average integral resistivity of copper in the temperature range, Om·m;
$\lambda = f(T)$ – average integral thermal conductivity of copper in the temperature range, W/(m·K);
I – operating current, A.

3.2. Thermal leak through thermal insulation

Thermal leak through insulation (q_{ins}) mainly provides thermal radiation. For modern flexible cryostats, this type of thermal load is from 1 to 3 W/m [6]. The amount of thermal leak through the insulation varies in length since there is a gradient ΔT between the input and output of the cable. The temperature gradient shall be taken into account when calculating the value ofq_{ins}. The thermal leak from heat transfer by radiation is proportional to the diameter and length of the cable. It is shown that the cable diameter depends on the operating temperature of the cable. The thermal leak by radiation can be described by the following expression:

$$q_{in} = \varepsilon \cdot \sigma \cdot (T_R^4 - T_{in}^4) \cdot S_{cr},\ q_{out} = \varepsilon \cdot \sigma \cdot (T_R^4 - T_{out}^4) \cdot S_{cr}$$

q_{in} and q_{out} are thermal leaks to the cryostat at the initial temperature T_{in} and T_{out} respectively, where ε – Emissivity of the surfaces; σ – Stefan-Boltzmann Constant, $5.67 \cdot 10^{-8}$ $W \cdot m^{-2} \cdot K^{-4}$; T_R – Room temperature, K; $S_{cr} = f(T)$ – the area of the inner pipe of the cryostat, m^2. The traditional way to reduce thermal leak from radiation is to use multilayer insulation (MLI) with high reflectivity [7]. In a first approximation, the average thermal leak to the cryostat can be calculated as the arithmetic mean between the thermal leak determined for the initial and final temperature:

$$q_{ins} = \left(\frac{q_{in} + q_{out}}{2} \right) \cdot (n+1)^{-1}$$

where n is the number of layers MLI.

3.3. AC and dielectric losses

AC losses are converted into heat and distributed along the entire length of the cable. This heat is to be diverted by the CRS. The sum of the specific energy losses during AC transmission in one phase can vary from 0.1 to 5 W/m, depending on the operating current. The AC losses in a superconductor are usually much lower than resistive losses in a normal conductor with the same operating current. However, minimization of thermal load associated with the movement of alternating current is technically important, since thermal energy is dissipated in the low-temperature zone of the cable. Complete loss modeling in the HTS cable requires consideration of the method of winding HTS tapes onto the former. In a simplified model, this conductor can be described as a superconducting tube. This is the so-called monoblock model [8]. It takes into account neither the method of tape winding nor changes in the properties of HTS tapes during cable manufacturing. In this case, the energy loss in the current-carrying layer can be calculated by the formula:

$$q_{ac} = \frac{f \cdot \mu_0 \cdot I_c^2}{2\pi \cdot b^2}\left((2 - R \cdot b) \cdot R \cdot b + 2 \cdot (1 - R \cdot b) \cdot \ln(1 - R \cdot b)\right) \text{ [W/m]}$$

$$b = \frac{D_{HTS}^2 - D_f^2}{D_{HTS}^2}, \quad R = \frac{I_p}{I_c}$$

where $R=I_p/I_c$ is the ratio of the transport to the critical current;
μ_o is the magnetic constant;
f is the frequency of the electric current.
The AC losses for HTS shield (q_{sh}) are calculated similarly. Dielectric losses depend on the operating voltage of the cable and are usually insignificant; therefore they are not taken into account in this analysis.

3.4. Hydraulic and pumping losses

The heat dissipation (q_{fr}) in the cable path due to irreversible energy losses (friction of the coolant along the channel walls) is determined by: flow rate, coolant temperature, design features of the cooling channel (roughness of the channel wall, turbulence of the flow due to swirling), the presence of local hydraulic resistances. For a given temperature gradient along the cable length, the corresponding coolant flow rate (G_v) is determined and the optimal external diameter (D_{cr}) is selected. With this diameter, q_{fr} will be minimal. The pressure head (Δp) created by the cryogenic pump is converted into thermal energy. Additional heating of the liquid in the pump path (q_{pump}) occurs because the pump coefficient of performance (COP_{pump}) usually does not exceed 70%. The thermal load from hydraulic and pumping losses can be calculated using the following formulas:

$$q_{fr} + q_{pump} = \frac{\Delta p \cdot G_v}{COP_{pump}}$$

$$q_{fr} = \Delta p \cdot G_v$$

3.5. Actual Performance of cryogenic refrigeration system

The thermal load of the HTS cable shall be removed using a cryogenic refrigerator. The ratio between the thermal load and the power consumed determines the coefficient of performance, which shows the relationship between the power consumed and the temperature cooling level:

$$\sum q = q_{lead} + q_{ins} + q_{ac} + q_{sh} + q_{pump} + q_{fr}$$

$$COP = \frac{\sum q}{W}.$$

The minimum value of the power consumed by the refrigerator can be determined using the Carnot cycle [9]:

$$COP_{car} = \frac{T_x}{T_0 - T_x}, \quad W_{\min} = \frac{\sum q}{COP_{car}}.$$

To compare the performance of real cryogenic refrigerator with the Carnot refrigerator, it is possible to use a figure of merit (*FOM*):

$$FOM = \frac{COP}{COP_{car}}.$$

For the modern cryogenic refrigerator, this indicator is within 25% [10].

4. Thermal analysis of HTS cable

A model for finding the optimal operating temperature with minimized power consumption for the HTS cable is presented in the previous section. The analysis was based on the cable with an operating current of 2500 A and a length of 2.5 km. The results of thermal analysis of the HTS cable are presented in Figures 2 and 3. As you can see in Figure 2, minimum power consumption in the CRS is at an average cable temperature of 88 K. Figure 3 illustrates the individual contribution of each type of thermal load. At low temperatures, the influence of AC losses is great. As the cooling temperature of the HTS cable increases, the thermal load on the CRS decreases. But at higher temperatures, the proportion of thermal leak by radiation increases, since the calculated dimensions of the cryostat are enlarged at the increased cable cooling temperature.

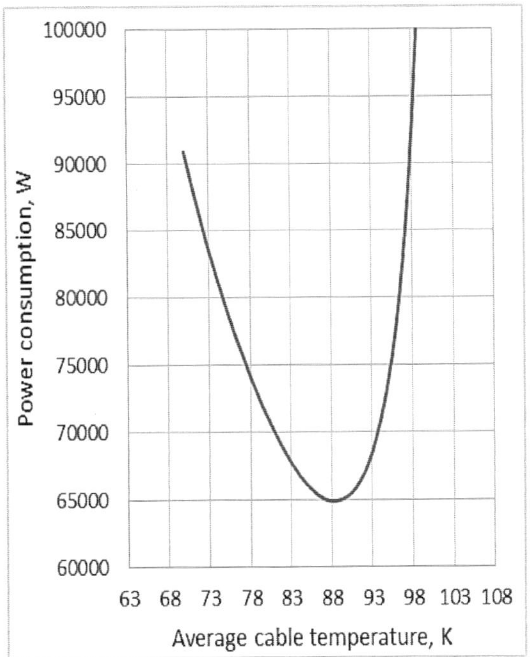

Figure 2. Power consumption of CRS to compensate for all thermal load.

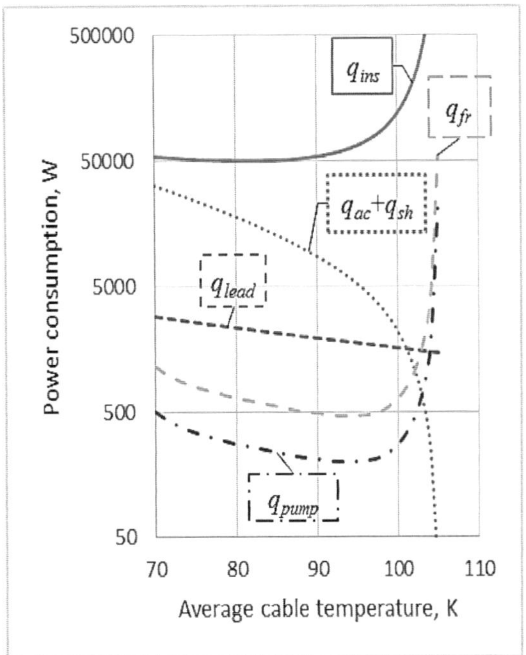

Figure 3. Power consumption of CRS depending on the type of thermal load.

It is useful to show the negative impact of low temperatures on the energy efficiency of the HTS CL. Compare the operational characteristics of two cables with cooling temperatures T_{in}=65 K, T_{out}=75 K and T_{in}=83, T_{out}=93 K, respectively (see table 1).

Table 1. Parameters of HTS cable.

Cable temperature, K	Number of tapes	Cable diameter, mm	Hydraulic losses, kPa	COP_{car}	Σq, W	W, kW
65-75	19	64	321	0.276	6283	91
83-93	51	86	211	0.38	6205	65

Due to an increase in the cooling temperature of the cable, the critical current of the HTS tapes has decreased, so the number of the tapes has almost doubled, and the outer diameter of the cable has become 25% larger. But, despite the increase in the cable size, the intensity of thermal load has scarcely changed. At the same time, due to an increase in the temperature level of cooling, the power consumption of the CRS has decreased by 28%. Taking into consideration the duration of the HTS cable operational campaign, such energy savings significantly increase the investment prospects of such projects.

5. Conclusions

The decrease in power consumed by the CRS is related to the fact that the COP of cryogenic refrigerators increases with an increase in temperature. The HTS tapes have not been mass-produced so far. Generally the tapes are produced on preliminary request. As the introduction of HTS technologies in urban power grids grows and the production of the HTS tapes increases, a significant reduction in prices can be expected. When performing a full economic assessment of HTS projects, it is necessary to take into account the purchase costs for the CRS and HTS conductor. The increased temperature of the coolant increases the dimensions and cost of the cable, but reduces the dimensions and cost of the CRS, as well as power consumption for cooling. If these two factors are taken into account, the optimum cooling temperature of the cable is in the range of 65-75 K at present, since the cost of the HTS conductor is very high.

The choice of the temperature range of 65-75 K is based on the assumption that the coolant flows out of the cryostat in the CRS at atmospheric pressure. Under such conditions, a maximum temperature of 75 K guarantees a single-phase state of the coolant. However, the single-phase state of the coolant along the length of the cable can be ensured at higher temperatures by increasing the pressure in the cable path. When choosing a temperature of 83–93K, the overpressure in a closed-loop system of the CRS should be at least 0.6 MPa, which will require an increase in the wall thickness of the cryostat.

References

[1] Fleck U, Vogel D and Ziegler B 2002 Cooling of HTS applications in the temperature range of 66 K to 80 K *Advances in Cryogenic Engineering* **47** 188-195

[2] Iwakuma M, et al. 2015 Development of cooling system for 66/6.9kV-20MVA REBCO superconducting transformers with Ne turbo-Brayton refrigerator and subcooled liquid nitrogen *IOP Conference Series: Materials Science and Engineering* **101** 597-602

[3] Malozemoff A P, Yuan J and Rey C M 2015 Superconductors in the Power Grid *High-temperature superconducting (HTS) AC cables for power grid applications* (Cambridge UK: Woodhead Publishing) pp 133–185

[4] Zheng Y B, Wang Y S, Pi W and Ju P 2014 Current distribution among layers of single phase HTS cable conductor *Phys. C* **507** 59-64

[5] Iwasa Y 2009 *Case Studies in Superconducting Magnets, Design and Operational Issues Second Edition* (New York USA: Springer Science+Business Media)

[6] Tonnesen O and Ostergaard J 2004 High Temperature Superconductivity 2 *High temperature superconducting cable* ed A V Narlikar (Berlin, Heidelberg: Springer) pp 537-567

[7] Fesmire J E, Augustynowicz S D and Demko J A 2002 Thermal Insulation Performance of Flexible Piping for Use in HTS Power Cables *Advances in Cryogenic Engineering* **47** 1525-32

[8] Vellego G and Metra P 1995 An analysis of the transport current losses measured on HTSC single-phase conductor prototype *Supercond. Sci. Technol.* **8** 476-483

[9] Demko J A 2015 Superconductors in the Power Grid *High-temperature superconducting cable cooling systems for power grid applications* ed C Rey (Cambridge UK: Woodhead Publishing) pp 263–28

[10] Ter Brake H J, Wiegerinck G F 2002 Low-power cryocooler survey *Cryogenics* **42** 705–718

IOP Conf. Series: Materials Science and Engineering **826** (2020) 012005 doi:10.1088/1757-899X/826/1/012005

Mathematical model of compressor-condenser unit

I V Tatarenko, V V Mitropov and A A Nikitin

ITMO University, 49, Kronverksky av., St. Petersburg, 197101, Russia

E-mail: lavrtat@mail.ru

Abstract. The article aims to consider a mathematical model of a compressor-condenser unit, based on the piston compressor with frequency control. The accepted initial data are temperatures and relative humidity of environment and a cooled room, geometrics of the main unit elements. CoolProp library was used for calculation of thermodynamic and thermophysical parameters. The flowchart of calculation of the compressor-condenser unit is provided in the article. The evaluation of the impact of change in environmental relative humidity on the heat transfer coefficient of the condenser has been made.

1. Introduction

Since the late 20th century the issue of working substances in refrigerating engineering and industry continues to be relevant. To control the distribution of substances that deplete the ozone layer, the Montreal Protocol was developed. It is an international treaty designed to protect the ozone layer. It was done to prohibit spread and exploitation of numerous chemical ozone-depleting working substances according to the plan of Montreal Protocol's developers as represented in the document [1].

The Montreal Protocol was agreed on the 16th of September, 1987, and entered into force on January 1, 1989. It was a suggestion on the ozone layer recovery by 2050, subject to the conditions of the agreement, signed by all countries. Nowadays European Union countries start using ozone-safe equipment, the most of objects put into operation are protected against carbon dioxide that is an alternative refrigerant [2].

In January 2019 the Kigali amendment to the Montreal Protocol was accepted. There was information about future rejection of many groups of cooling agents that are applicable in refrigerating and low-potential engineering [3].

For this reason calculation and modelling of compressor refrigerating machine's static characteristics are highly relevant. In the modelling there is a possibility to determine many properties of the refrigerating machine in different operating modes, as well as in changing the operating refrigerating machine to an alternative working substance [4].

The precise geometrical parameteres of components are needed in creating the mathematical model of the compressor-condenser unit. Authors [5] came to this conclusion; they found characteristics of the refrigerating machine working on the mixture of the cooling agent using mathematical modelling. Authors [5] concluded that the main affecting things for system's characteristic are initial conditions, geometrical parametres of machines, the type of refrigerating machine's components, such as a condenser and an evaporator.

Many researchers in the field of refrigeration and air conditioning seek to reduce energy consumption and focus on researching the capacity regulation methods for one of the main

refrigerating machine's components, namely the compressor. Capacity regulation reduces cyclic losses of machines
and increases reliability of the compressor by reducing pressure difference when working with partial capacity [3].

There are a lot of capacity regulation methods, namely: a bypass method, implementation of additional dead space, steam throttling on absorbing and smooth variation of shaft speed. The characteristics in using the above methods of regulation comressor's capacity are described by authors Tassou S.A. and Qureshi T.Q. [8] and authors of [9-10].

2. Research methods

This paper considers the mathematical model of the compressor-condenser unit, based on the piston compressor with frequency control.

Data of calculating the characteristics of the one-step compressor-condenser unit based on a piston compressor obtained using implemented subprograms, namely characteristics of the compressor - subprogram XAR_CM and characteristics of the condenser - subprogram XAR_CD, which are consolidated into a single program – XAR_CMCD. The flowchart of the program XAR_CMCD is depicted in figure 1.

The known input signals are the temperature of the air in the room and the ambient temperature. Using these entry conditions and having the geometrical parameters of the compressor and the air condenser, there is a possibility to calculate all the parameters of the compressor-condenser unit at target temperatures both by the air and by the working substance.

The main thermodynamic and thermophysical properties of normal working substance are calculated using the CoolProp property library [12]. This library is supported by the majority of programs and programming languages used in the mathematical modelling. It has free distribution.

The main equations and algorithms of components of the compressor-condenser unit with the piston compressor implemented on ECM are based on material contained in paper [6].

There are some assumptions during modelling:

- overheating cooling agent on absorption is constant for all concerned operating modes of the compressor-condenser unit.

- all components of the compressor-condenser unit are connected in series, and mass flow through the components is unchanged.

- overcooling of the refrigerant lower than the condensing temperature for all operating modes is equal to zero.

It should be noted that for calculation of the heat-transfer coefficient for condensation inside horizontal tubes the model equation of Ackers and Rosson (1960) is used in the form [7]:

$$\mathrm{Nu} = 0{,}1 \cdot \mathrm{Pr}^{1/3} \left(\frac{r}{c_p \, \Delta t} \right)^{1/6} \mathrm{Re}^{2/3} \left(\frac{\rho_l}{\rho_v} \right)^{1/3}, \tag{1}$$

where ρ_l, ρ_v are the density of the liquid and vapor in kg/m^3; c_p is heat capacity in kJ/(kg·K), r – specific heat of vaporization in kJ/kg.

Inconsidering the convective heat transfer in the flow around the tube stack on the side of the air, a universal formula is used. It takes into account the type of the tube stack, the initial data for the calculation are the same. Accordingly, by comparing the contacted computations with the calculations performed by the formula of D. M. Ioffe for staggered tube bank [11], the error of the calculations did not exceed 1 %:

$$\mathrm{Nu} = c\, \mathrm{Re}^{n} \left(\frac{L}{d_{eqv}} \right)^{m}, \tag{2}$$

where L is depth of heat exchanger in m, d_{eqv} is equivalent diameter in m.

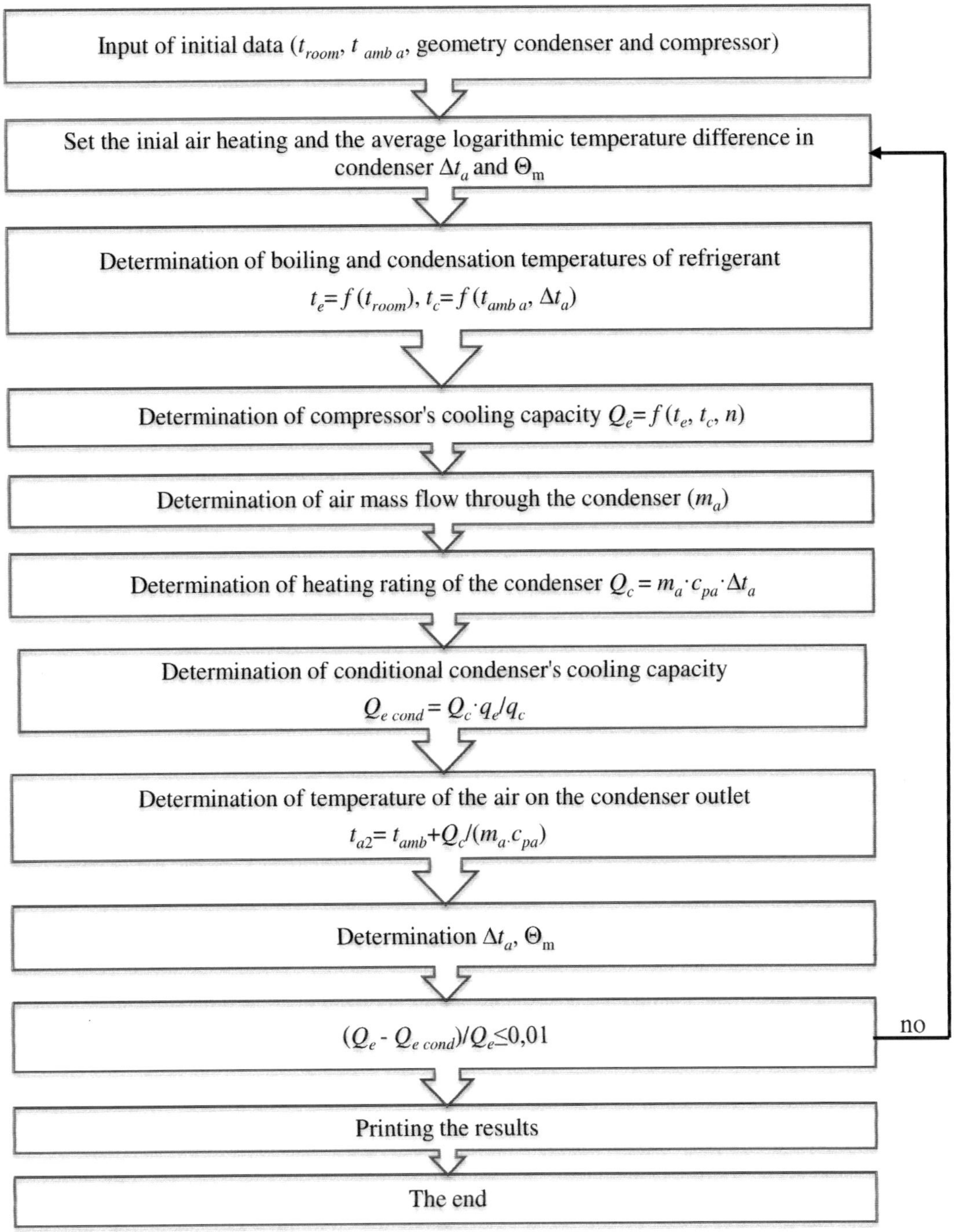

Figure 1. The flow chart of the program, implementing calculation of characteristics of the compressor-condenser unit

3. The results of the study

When considering the obtained data as a result of the program according to the algorithm presented in figure 1, it is possible to make a conclusion about the influence of the size of difference in air temperatures on an input and an output from the condenser of air cooling and the average logarithmic of a pressure in the device. Regarding the influence of relative humidity at the input to the condenser, it was found that it was insignificant. The maximum discrepancy of the temperature difference at the inlet and outlet of the air condenser is manifested only at ambient temperatures of more than 45°C and it is about 3 %. These characteristics are presented in figure 2 as the dependencies $\theta_m = f\left(t_{amb}, t_c\right)$ and $\Delta t_a = f\left(t_{amb}, t_c\right)$.

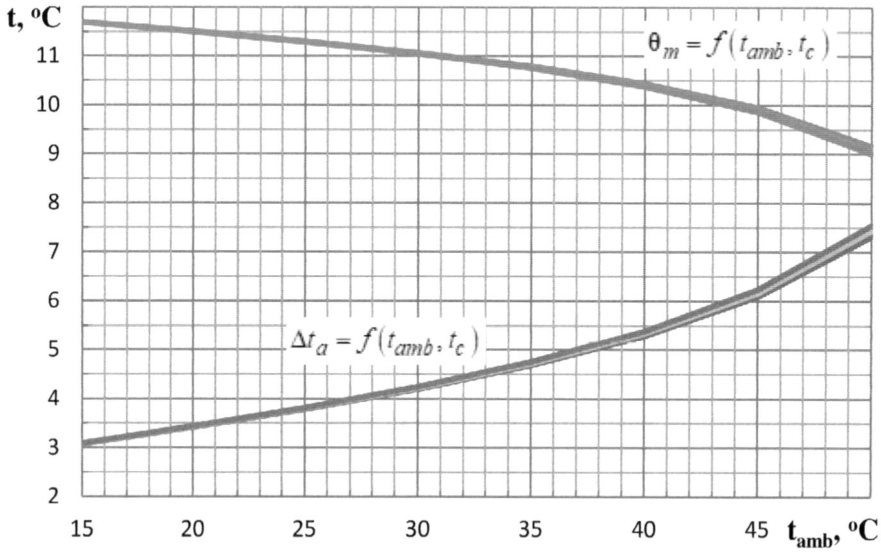

Figure 2. The characteristics of the air condenser

4. Conclusions

The mathematical model of the compressor-condenser unit was created. With its help, it is possible to obtain the parameters of the unit when working on any working substance and in operating mode.

In calculating the heat-transfer coefficient, it is possible to ignore the influence of the relative humidity of the environment. According to the calculations, it was found that its impact is insignificant.

It is necessary to continue this investigation including in the mathematical model of the subprogram of the evaporator to analyse the operation of the entire refrigeration machine as a whole.

References

[1] Montreal protocol on substances that deplete the ozone layer *Report of the Refrigeration* Air Conditioning and Heat Pumps Technical Options Committee 2014 Assessment.

[2] Tsvetkov O B and Laptev Yu A 2017 Working substances of low temperature engineering in the context of three climate agreements *Kholodilnaya Tekhnika* **5** 34–37

[3] Tsvetkov O B and Laptev Yu A The Kigali 2019 Amendment to the Montreal Protocol: evolution retrospection or the call of the future? *Kholodilnaya Tekhnika* **3** 2–8

[4] Tsvetkov O B, Zabodalova L A, Laptev Yu A, Mitropov V V and Pjatakov G L Choice of refrigerants 2018 *Dairy Industry* **9** 21-22

[5] Orth L A, Zietlow D C and Pedersen C O 1995 *ASHRAE Trans.* **23** 1367-1375

[6] Tatarenko I V 2015 *Introduction to mathematical modeling of the characteristics of steam compressor refrigeration machines* (St Petersburg: University ITMO)

[7] 2017 *ASHRAE Fundamentals* Handbook (SI)

[8] Tassou S A and Qureshi T Q 1997 Dynamic mode performance evaluation and energy analysis of mains and inverter driven refrigeration compressors *Proceedings of the Institution of Mechanical Engineers JPEE*A **211** (**4**) 339 - 346

[9] Ciconkov R 2016 *Refrigeration: Solved Examples* (Macedonia: University "Sv. Kiril i Metodij")

[10] Fotin B C, Pirumov I B, Priluckiy I K and Plastinin P I 1987 *Piston compressors* (Leningrad: Mashinostroenie)

[11] Danilova G N 1986 *Heat exchangers of refrigeration units* (Leningrad: Mashinostroenie)

[12] 2010 *CoolProp*, available at: http://www.coolprop.org/

IOP Conf. Series: Materials Science and Engineering **826** (2020) 012006 doi:10.1088/1757-899X/826/1/012006

Investigation of the influence of 3D printing modes with ceramics and sintering on the shrinkage process of thin-walled models

P Piterskov[1], S Egorov[1], E Boyko[2] and M Grigoriev[3]

[1] ITMO University, 49, Kronversky Ave., St. Petersburg, 197101, Russia
[2] Saint-Petersburg Mining University, 2, 21st Line, St Petersburg, 199106, Russia
[3] BSTU «VOENMEH» named after D.F. Ustinov, 1, 1st Krasnoarmeyskaya st., St. Petersburg, 190005, Russia

E-mail: piterskov@mail.ru

Abstract. The subject of this study focuses on the effect of 3D printing and heat treatment of thin-walled models made of ceramic paste. The plates with dimensions of 20*5*40 mm were used as the models. The methodology includes the creation of ceramic objects using 3D printing by the SLA method (stereolithography) and their further heat treatment. The surface quality and the shrinkage percentage of samples printed in different modes and thermally processed under different conditions are compared. It was found that with a decrease in layer thickness and particle size, as well as an increase in the duration of heat treatment, the surface quality of the samples improves and the percentage of shrinkage decreases approximately 2 times. Using new modes and improved heat treatment conditions, samples with the least number of cracks and a lower percentage of shrinkage were obtained, which allows using this process to create ceramic products of higher quality. The method of creating ceramic products used in the work in the future can be applied in such industries as aircraft manufacturing, medicine, biotechnology and other fields.

1. Introduction

3D printing of ceramics is a relatively new direction in the field of additive technologies, which is developing rapidly along with technologies for 3D printing with polymers and metals. It is believed that the technology of SLA (stereolithography) is the most famous and popular technology for 3D printing of ceramics and is widely used throughout the world [1-4].

One of the innovative and rapidly developing areas at the junction of biotechnology and materials science is the biological treatment of ceramic masses. This post-processing method improves the technological and consumer properties of ceramic products. Microorganisms in the composition of ceramic masses produce the formation of a number of metabolites and in combination have a biotechnological potential to change the properties of ceramic masses for the better. This technology can be used after 3D printing of ceramic materials [5-7].

The process of forming ceramic products using 3D printing has been widely developed and has reached high levels of accuracy and roughness of the product. However, since the 3D printing of ceramics involves the use of ceramic paste based on ceramic powder and a photosensitive polymer (the chemical composition, percentage, and chemical reagents used may be different depending on the manufacturer), the product was manufactured by means of additive technologies. [8-11]

At the moment, there are many foreign studies on the influence of 3D printing parameters on the surface quality of the product [12,13]; however, most of them relate to printing with polymers, where the resulting product is the final product and does not require subsequent heat treatment.

2. Experimental setup

To perform the study, the Russian 3D printer "AF200" for ceramics was used (manufacturer - Additive Manufacturing LLC). Ceramic plates were used as samples (Figure 1).

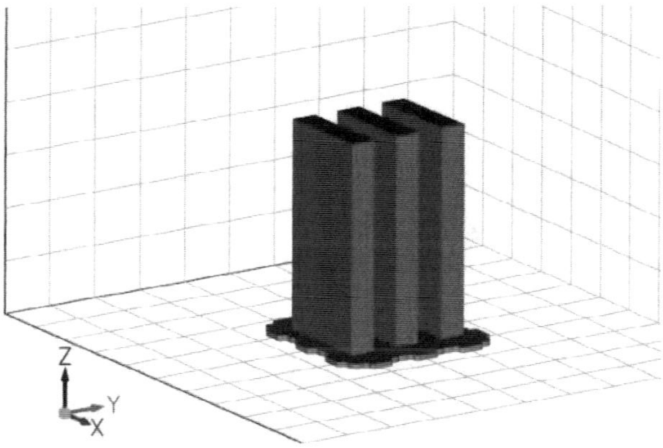

Figure 1. 3D model of ceramic plates

The Triangulatica program (in the public domain) was used to develop and select the parameters of the 3D printing, such as layer thickness, beam movement strategy, beam movement step, and others. For greater purity of the experiment, 9 plate samples (3 of each type) with different characteristics were printed. The main parameters of the samples and print modes are presented in Table 1.

Table 1. 3D printing parameters and plate characteristics

	Plates №1	Plates № 2	Plates № 3
Dimensions	20x5x40 mm		
Material	ceramic paste (porcelain in the form of a powder with a particle size of up to 5 microns in combination with a polymer binder) with a solid content of 44%		
Raft	solid, height - 1 mm		
Scan speed	11000 mm/s		
Layer thickness	100 µm	50 µm	25 µm
Beam movement strategy	Simple bidirectional		
Beam movement step	150 µm		
The distance between the wall and the internal movement of the beam	80 µm		
Print time	18 minutes	54 minutes	110 minutes

After separation from the substrate surface and washing with water, the obtained samples (Figure 2a) were placed in a furnace. The heat treatment in the furnace has 2 stages: the removal of the binder (debinding process) and the sintering of the samples. These processes were carried out in one furnace, but in different temperature conditions.

To study the influence of the environment in which the heat treatment is carried out, three groups of plates (3 plates in each group) were distributed in the working zone of the furnace (Figure 2b) [14, 15].

a)

Figure 2a. Samples obtained using 3D printing by the SLA method

b)

Figure 2b. Location of samples in the furnace

Heat treatment conditions at the stages of debinding and sintering of the products are presented in Table 2.

Table 2. Initial heat treatment conditions

	Temperature conditions at the stage of debinding	
t, °C	Time to reach, min	Duration, min
150	60	60
320	240	360
	Temperature conditions at the stage of sintering	
150	60	0
320	150	0
450	60	60
600	60	0
780	60	0
900	60	60
1050	60	0

As a result of heat treatment, 3 groups of plates with a large number of cracks, tears and leashes were obtained. The LEVENHUK DTX 90 training microscope was used to examine the structure of the samples. The appearance of the samples under the microscope is presented in Table 3. The samples were examined at a 60-fold increase. At this stage of the study, a larger increase is not required, since the samples have many visible cracks after sintering.

Based on the appearance of the samples under a microscope, it was concluded that the location of the samples in the additional medium (under a layer of quartz or under zirconium balls) increases the number of cracks and geometric deviations along the axes. Therefore, the location of ceramic objects at the stage of heat treatment in the free position in the furnace makes the heating more uniformed. In addition, it becomes easier to remove the polymer in the product composition.

Table 3. Appearance of samples at 60x magnification

	In the box	Under zirconium balls	On the surface of zirconium balls
100 μm	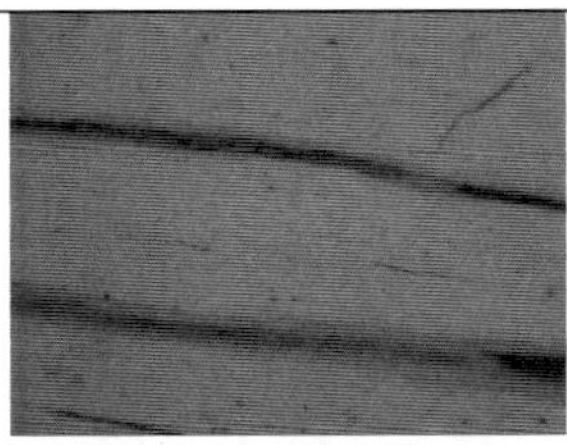	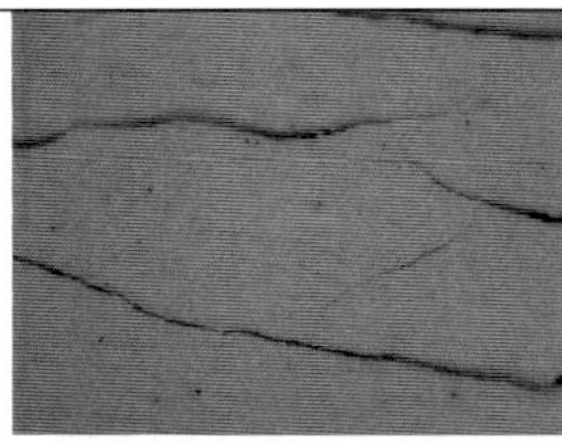	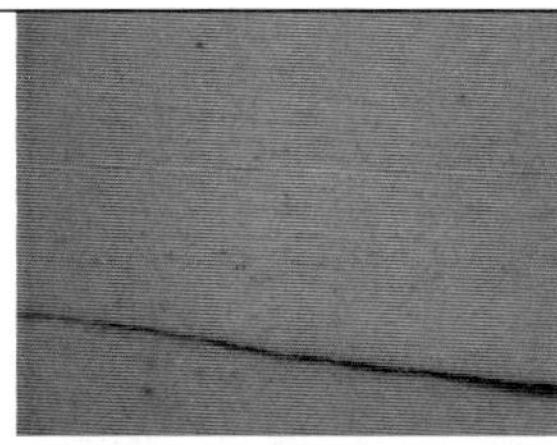
50 μm	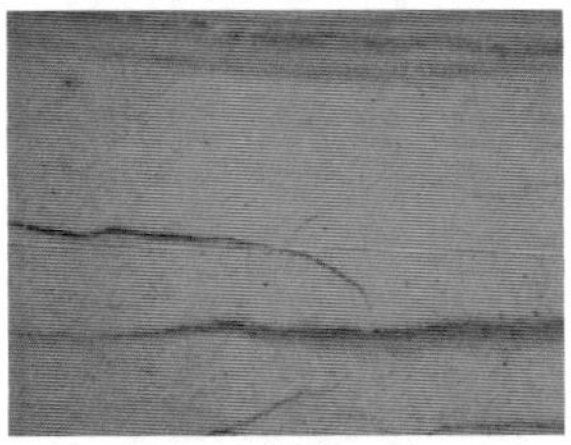	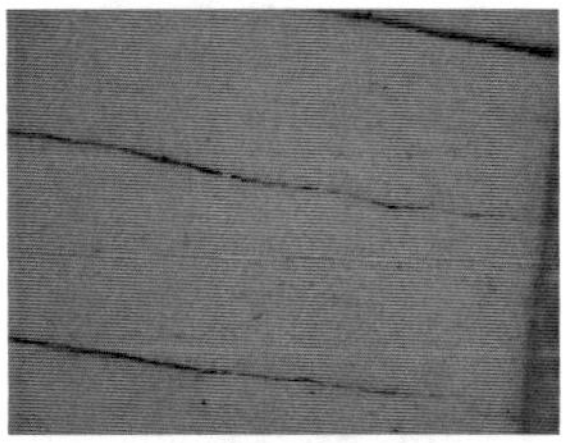	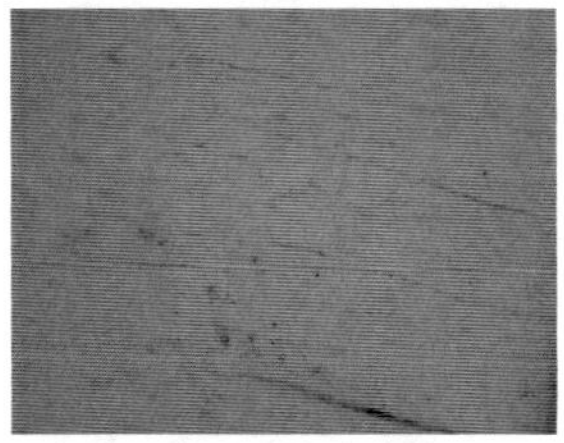
25 μm	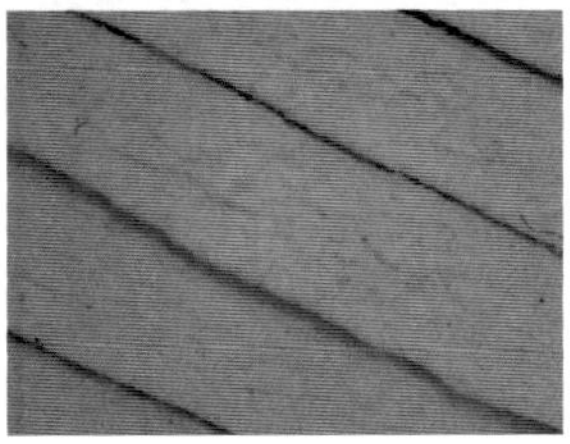	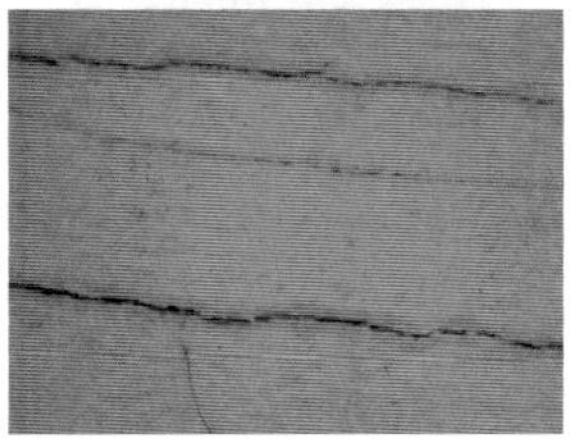	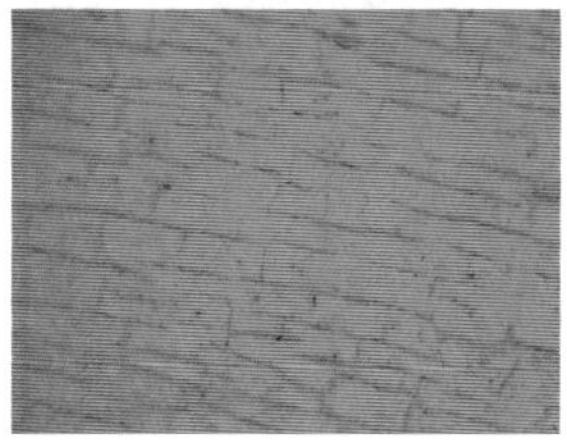

Based on the results, the following conclusions have been made:

1. The chemical composition of the polymer requires refinement for the more gradual and smooth debinding process.

2. It is essential to reduce the size of the powder fraction from 5 microns approximately to 0.01 - 1 microns.

3. It is necessary to increase the percentage of ceramic powder in the paste from 55% to 65% to reduce the percentage of shrinkage.

4. It is required to increase the duration of debinding and sintering processes to reduce the likelihood of leashes and cracks.

5. An adjustment of the beam movement strategy is required in order to increase the density of the product after printing. With the new strategy, internal shading (the filling pattern of the laser beam path along the inner area of the layer) starts at an angle of 45 ° and changes its direction with each layer by 90 °. In this connection, we receive a mesh filling structure and additional polymerization of each previous layer when the current is illuminated.

The above findings were applied to print a new plate. To prepare the ceramic paste, a smaller powder was used (particle size in the range 1-0.01 μm), the concentration of the powder in the paste was 50%, and the printing strategy was also changed. The printed plate went through a heat treatment process. To increase the speed of heat treatment, the debinding and sintering processes were performed sequentially and were not separated, the total time was 28 hours. Adjusted heat treatment conditions are presented in Table 4.

Table 4. Adjusted Heat Treatment Conditions

t, °C	Time to reach, min	Duration, min
150	60	0
320	150	330
450	60	60
600	120	60
780	150	60
900	90	60
1250	360	120

Thus, it was possible to obtain a sample with significantly smaller deviations in geometry, less shrinkage, and better surface quality. Figure 3a shows the appearance of the plate, and figure 3b shows the view of the plate under a training microscope.

a)

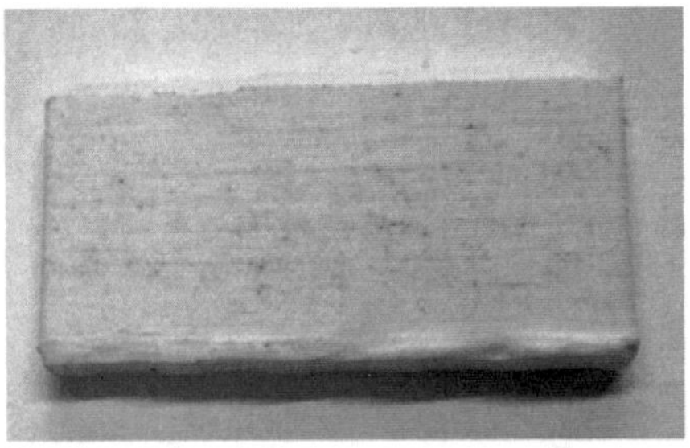

Figure 3a. The appearance of the plate.

b)

Figure 3b. View of the plate under the microscope.

Measurements were also made of the percentage of shrinkage of the sample along 3 axes. Thanks to the change in the composition of the paste, printing and heat treatment modes, it was possible to reduce the percentage of shrinkage almost 2 times (Figure 4).

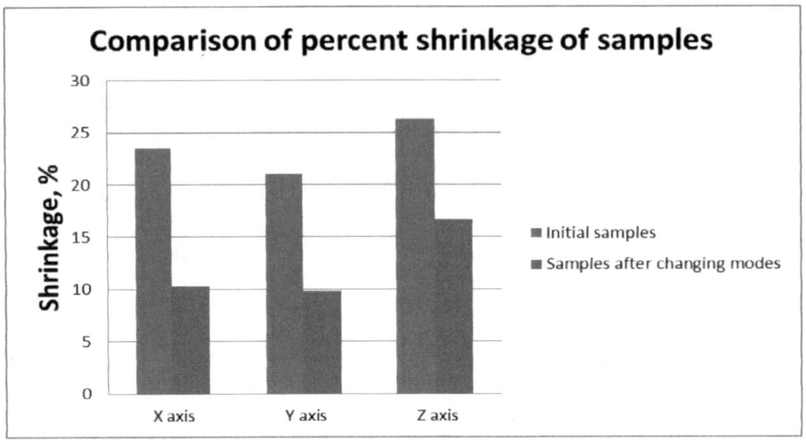

Figure 4. Comparison of percent shrinkage of samples

3. Conclusion

Based on the results, the following conclusions have been made:

- The location of the samples in the free position in the furnace allows achieving the most uniform heating, and also makes the debinding process more stable, which in turn leads to better surface quality with fewer cracks and leashes.
- Reducing the size of the fraction of ceramic powder and increasing its concentration in the paste can reduce the percentage of shrinkage of the product and make this process more uniform. It should be noted that these parameters also affect the viscosity of the paste, which in turn can lead to the impossibility of polymerization. The search for the optimal composition of ceramic paste requires the more detailed study.
- Changing the beam movement strategy for 3D printing allows you to increase the density of the final product, and, as a result, reduce the percentage of shrinkage.

The performed experiments have a positive result and require additional research. It is necessary to prepare the experimental series for 3D printing of samples based on ceramic paste with a powder percentage of 50% and higher, as well as strength testing of these samples.

References

[1] Chen Z, Li Z, Li J, Liu C, Liu C, Li Y, Wang P, Yi H, Lao C and Yuelong F 2019 *Journal of the European Ceramic Society* **39(4)** 661-687
[2] Griffith M L and Halloran J W 1996 *Journal of the American Ceramic Society* **79(10)** 2601–2608
[3] Badev A, Abouliatim Y, Chartier T, Lecamp L, Lebaudy P, Chaput C and Delage C 2011 *Journal of Photochemistry and Photobiology A: Chemistry* **222(1)** 117–122
[4] Xing H, Zou B, Li S and Fu X 2017 *Ceramics International* **43(18)** 16340–16347
[5] Yerezhep D, Umutbaeva N R, Piterskov P, Baranov A Y and Yerezhep A 2019 *J. Phys.: Conf. Ser.* **1333** 032091 DOI:10.1088/1742-6596/1333/3/032091
[6] Busuioc C, Stroescu M, Stoica-Guzun A, Voicu G and Jinga S-I 2016 *Ceramics International* **42(14)** 15449-15458 DOI:10.1016/j.ceramint.2016.06.196
[7] Yerezhep D, Yemelyanov A, Kuznetsov Y, Perminov A, Pronin V A, Minikaev A F and Yerezhep A 2019 *J. Phys.: Conf. Ser.* **1333** 032092 DOI:10.1088/1742-6596/1333/3/032092
[8] Chen Z, Li D, Zhou W and Wang L 2010 *Journal of Engineering Manufacture* **224(4)** 641–651
[9] Gibson I and Shi D 1997 *Rapid prototyping journal* **3(4)** 129–136
[10] Lombardo S J 2015 *Journal of the American Ceramic Society* **98(1)** 57–65
[11] Nigay P M, Cutard T and Nzihou A 2017 *Ceramics International* **43(2)** 1747-1754 DOI: 10.1016/j.ceramint.2016.10.084
[12] Raju B S, Chandrashekar U, Drakshayani D N and Chockalingam K 2010 *International Journal of Engineering Science and Technology* **2(7)** 3199-3205
[13] Chen Z, Li D and Zhou W 2012 *Proceedings of the Institution of Mechanical Engineers, Part B: Journal of Engineering Manufacture* **226 (7)** 1249–1258
[14] Özdemir H and Özdogan A 2017 *Dental Materials Journal* **37(1)** 24-32 DOI: 10.4012/dmj.2016-365
[15] Nigay Pierre-Marie, Cutard Thierry and Nzihou Ange 2016 *Ceramics International* **43(2)** 1747-1754 doi: 10.1016/j.ceramint.2016.10.084

IOP Conf. Series: Materials Science and Engineering **826** (2020) 012007 doi:10.1088/1757-899X/826/1/012007

Regulation of powder particle sizes during plasma spraying

B S Ermakov [1, 3], S B Ermakov [1], A Pavlenko[2] and S A Vologzhanina[3]

[1] Peter the Great St. Petersburg Polytechnic University, 29, Politekhnicheskaya str., Saint Petersburg, 195251, Russia
[2] Closed joint stock company Experimental plant «Micron», 5, Gogol' str., Velikie Luki, 182107, Russia
[3] ITMO University, 49, Kronverkskiy Ave., Saint Petersburg, 197101, Russia

E-mail: Ermakov55@bk.ru

Abstract. One of the factors constraining the development of 3D technologies is the shortage of powders for additive machines. St. Petersburg Polytechnic University (SPbPU) is developing the design and production of plasma nozzles to produce metal powders and alloys intended for additive machines. The article discusses the results of studies of the effect of pressure powders and flow rate generated in the spray chamber in the production of powders from 08Cr18Ni10Ti and CrNi60BWoTi alloys on the particle size distribution.

1. Introduction

The development of industry at the present stage can be characterized as a gradual departure from the creation and production of large-scale batches of machines and mechanisms with an average set of characteristics for the production of single and small series of equipment with the properties that best satisfy a particular customer. This way of development requires changes in a number of technological processes for the production of individual parts and assemblies, machines and mechanisms. Such changes, first of all, include the gradual replacement of subtractive processing technologies, for example, cutting the workpieces using additive manufacturing technology using the method of three-dimensional printing (3D). Wohlers Associates research showed that, in the world in 2014, the additive technology market amounted to about $ 3 billion with the average growth rate of about 20-30%. Probably in the near future, the market volume will be able to reach 16 billion dollars. The most active substitution of subtractors by 3D technology is observed in the actively developing aviation industry, shipbuilding, power engineering, including in dentistry and reconstructive surgery [1-3].

Today, Russia lags behind the leading countries of the world in the development of additive technologies due to the lack of domestic production of additive machines and the lack of a domestic base for the production of metal powders of small fractions. Today, most Russian companies use powders, which are mainly imported by foreign companies. According to academician E. N. According to Kablov, about 20 tons of powders is necessary per year for the Russian fleet of additive technologies equipment; it should be take into account that the volume of consumption of powders will constantly increase [1]. The production of powders is a major challenge for additive technologies. The quality of the powder depends primarily on the quality of the resulting parts. At present, especially with the introduction of sanctions, the problem of obtaining powders for additive technologies has become the most acute problem hindering the development of this technological direction in our country.

IOP Conf. Series: Materials Science and Engineering **826** (2020) 012007 doi:10.1088/1757-899X/826/1/012007

2. Materials and methods

Sputtering is the production of metal powders by dispersing a melt under the action of inertial forces by a jet of gas, liquid or plasma. Melting, spraying and solidification of the required metals and alloys are used in the production of powders. Melting furnaces of various types produce melting of metals and alloys under the action of an electric arc or plasma. Spraying is most often carried out using a stream of water at a high pressure, or in streams of inert and neutral gases. Spraying in various ways is currently the basis of high-quality metal production technologies. Such powders, subject to specific requirements, include a spherical shape of the particles, the particles pure from oxides and the volume of metal from foreign impurities, rapid crystallization, homogeneous microstructure [4]. Most often, when obtaining powders, liquid metal spraying is used. The undoubted advantage of this method can be consider the relative simplicity and cheapness of the technological process of obtaining metal powders with a wide temperature range of melting. The process of spraying a metal stream using a gas stream is either a gas stream that coaxially flows around the melt stream, or a stream of flowing gas directed at a given angle to the axis of the jet. Otherwise there is also a gas stream directed perpendicular to the axis of the jet. It is known that in one of the most efficient sputtering processes, the process takes place at a gas flow temperature that is equal to or greater than the melt temperature. However, if the temperature reaches 1500 °C or more, the creation of such conditions is difficult, because there are difficulties in heating the gas blast, and, in addition, become more complicated and expensive installations used for spraying. In such cases, it is justified to use the plasma method, when the spray jet of molten metal or wire is due to the kinetic energy of the jet of high-temperature plasma. In addition, the use of plasma atomization allows to obtain spheroidal powders. Wire or powder in this method can be as a source of raw materials. The plasma jet melts the initial billet, the droplets of the material are sprayed, after which the solidification of single-crystal particles occurs [3, 4]. By adjusting the energy parameters of the plasma flow [3], for example, increasing the temperature and kinetic energy of the gas flow, it is possible to produce a meltdown of the melt jet into smaller particles. The size and shape of the resulting powder particles can be adjusted, taking into account the physical properties of the sprayed material and the diameter of the sprayed wire. By increasing the diameter, the number of small particles in the powder can be reduced by increasing the mass of the melt entering the spray zone. Therefore, the solution of the problem of obtaining particles of the required size and increasing the yield of marketable products is one of the most important tasks in the development of technology and modes of plasma spraying.

The process of formation of a metal particle when sprayed in a low-temperature plasma can be conditionally divided into several successive stages, during which there is a formation of fine particles, the subsequent movement and heat exchange in the plasma flow and in the medium, and then the subsequent collection of particles. First, the feedstock is heated, and then it melts due to the heat of the plasma arc. Comparison of the residence time of the droplet during melting of the welding electrode or welding wire in the environment of protective gases with this method shows that at the end of the wire in this case it stays less time. The action of the arc pressure forces, as well as the reactive forces arising from the evaporation of metal and the release of gases, gravitational forces, surface tension forces, electrodynamic forces ensures the separation of the droplet from the end of the wire. At the moment of separation, the drop is characterized by certain dimensions, temperature and initial velocity [5]. The average diameter of the droplet can be estimated by the following formula:

$$d_d = \frac{\pi \sigma d_1 d_2}{M_1 V_1 + M_2 V_2}$$

where σ is the surface tension of the material feedstock in the contact zone of the metal and the plasma arc, d_1 – feedstock diameter, d_2 is the diameter of the nozzle of the plasma torch, M_1, and M_2 is the mass of the plasma stabilizing gas, V_1 and V_2 – speed of the plasma stabilizing gas.

After the drop leaves the plasma stream, it moves in a controlled environment of a protective gas, the temperature of which does not exceed several tens of degrees, gradually crystallizing and cooling. Depending on the spray conditions, the melting temperature of the sprayed material, the length of the free path of the particle inside the spray chamber (before impact with the bottom of the chamber), the protective medium may have a constant, normal or increased pressure. It can be either stationary or movable with the speed of movement directed towards the movement of the particle (flow). When certain counterflow parameters are reached, it is even possible to stop and change the direction of particle motion [6].

SPbPU is engaged in the development of equipment and technology for plasma spraying of metals and alloys and obtaining powders for additive machines. A number of plasma atomizers based on different atomization schemes with one or more plasma torches with horizontal and vertical solid feedstocks have been created.

3. Results and discussion

This paper presents the results of experiments on the melting of stainless steel wire 08Cr18Ni10Ti and alloy Woh98 (CrNi60BWoTi). (0.1%C, W-15.7%, Cr-25.2%, Fe -3.7%, Mo -1.4%< (Ti+V+Nb – 0.9%, nickel – base) with free run of particles and counterflow of protective gas-argon, the speed of which was determined through the volume of gas supply to the nozzles of the fluidized bed mechanism to create a counterflow. Also, the pressure in the spray chamber was regulated, which was $P_0 = 1$ ata - free flow of gas through the chamber and $P_1 = 1.25$ ata - work at high pressure. The mode of spraying was unchanged for both materials, spraying in an arc of indirect action: U-130 B, I-165 A, the volume of plasma-forming gas was 1800 l / h, the angle of wire feed – 90^0 to the direction of the gas-plasma flow. The volume of gas supply to the fluidized bed mechanism was $V_0 = 0$, $V_1 = 5$ and $V_2 = 10$ liters per minute.

Table 1. Particle size distribution of powders in the free run

Diameter of powder particles (not less than) mkm	The amount of powder, % Steel 08Cr18Ni10Ti		The amount of powder, % Alloy CrNi60BWoTi	
	P_0	P_1	P_0	P_1
315	14.3	9.1	12.7	9.0
200	21.1	12.8	16.3	13.4
140	17.8	17.0	18.1	15.2
100	12.8	17.6	13.2	15.5
80	11.3	15.7	12.0	15.4
63	10.5	12.2	11.7	13.0
40	6.2	8.0	9.1	10.1
20	5.9	6.1	7.3	7.0
Less than 20	1.0	1.5	0.6	1.4

Tables 1 and 2 present the results of the granulometric analysis of the obtained particles under the conditions of free run of particles (table 1) and table 2 – when creating a counterflow. The particle size and their percentage were determined by sieving the total mass of the sprayed powder on vibrating screens with sieves of appropriate cell sizes.

Increasing the pressure in the spray chamber allows increasing the share of commercial powder products, reducing the average diameter of powder particles.

Table 2. Effect of composition of powders at counterflow velocity (P_0 = 1 ata)

Fraction, mkm*	The amount of powder, %		The amount of powder, %	
	Less than steel 08Cr18Ni10Ti		Alloy CrNi60BWoTi	
	V_1	V_2	V_1	V_2
315	13.6	13.2	12.2	12.0
200	20.0	19.2	16.0	15.9
140	17.8	17.6	18.0	17.1
100	13.2	13.4	13.4	13.2
80	11.7	12.2	11.9	12.6
63	10.8	11.2	11.8	12.2
40	5.8	6.1	8.8	9.5
20	5.8	5.7	7.3	6,5
Less than 20	1.3	1.4	0.8	1,0

Creating a counterflow in the spray chamber has a very little effect on the percentage distribution of powder particles of both materials, but contributes to the production of powders of the best geometric shapes – the powders obtained when creating a counterflow have an almost ideal spherical shape. In the powder mass there are practically no satellites, fused with each of the powder particles (figure 1).

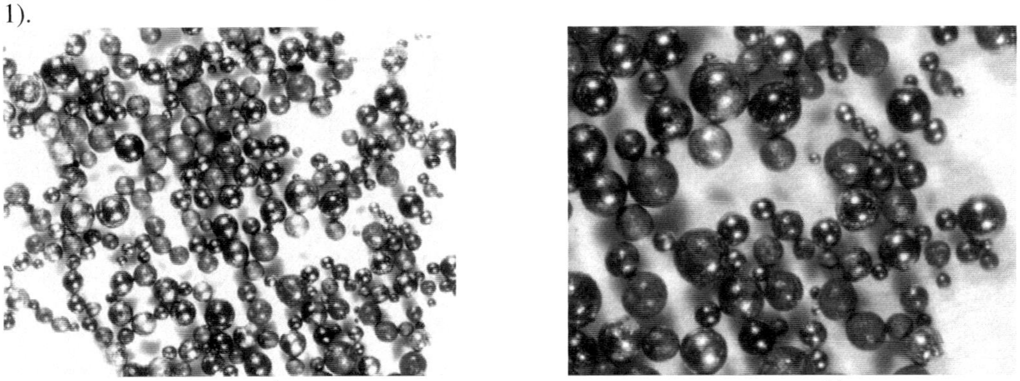

a) b)

Figure 1. The types of obtained powder: a) steel 08Cr18Ni10Ti; b) (CrNi60BWoTi)

4. Conclusion

On the basis of the conducted research, dependence of the sizes of the received powders on the size of excess pressure in the chamber of atomization was shown. The share of commodity production (powders with a diameter of 80-20 mkm) at normal pressure was 33.9% for steel 08Cr18Ni10Ti, 39.8% - for alloy CrNi60BWoTi increased by 1.25% at a pressure of 42 and 45.5%, respectively. Thus, the way of creating excess pressure in the spray chamber can be recognized as promising for further research.

References

[1] Kablov E N 2015 *Intellect and technologies* **2 (11)** 52-55
[2] Smurov I Yu, Konov S G and Kotoban D V 2015 *News of materials science. Science and technology* **2** 11-22
[3] Ermakov S B, Ermakov B S, Suleimenov E A, Protopopov A V and Abdaliev M A 1991 *Powder materials* (Alma-Ata: Gylym)
[4] Zlenko M A, Popovich A A and Mutiline I N 2013 *Additive technologies in mechanical*

engineering (St. Petersburg: Peter the Great St. Petersburg Polytechnic University Press)
[5] Strukov N N 2012 *Development of technology for plasma spraying of rod materials in a chamber with counterflow* Candidate technical sciences dissertation thesis, Perm national research Polytechnic University
[6] Dresvin S V and Ivanov D 2004 *Foundations of math. modeling of plasmatrons* (St. Petersburg: Peter the Great St. Petersburg Polytechnic University Press)

The use of sprayed powders to create coatings in the welds of oilfield pipelines

B S Ermakov [1, 3], A A Alkhimenko [1], N O Shaposhnikov [1], S B Ermakov [1], T E Shatsky [2] and A F Igolkin [3]

[1] Peter the Great St. Petersburg Polytechnic University, 29, Politekhnicheskaya str., Saint Petersburg, 195251, Russia
[2] Gazpromneft, Scientific and Technical Center, 75-79, letter D, Moika River emb., Saint Petersburg, 190000, Russia
[3] ITMO University, 49, Kronverkskiy Ave., Saint Petersburg, 197101, Russia

E-mail: ermakov55@bk.ru

Abstract. The most damaged elements of oil field pipelines are welded joints, the life of which often does not exceed one or two years. The problem of anticorrosive protection of internal surfaces of welded pipe joints has not been solved to date. In this paper, the possibility of applying powder protective stainless coatings to the inner edges of the pipe adjacent to the annular weld is considered. The research of corrosion resistance and mechanical properties of pipe welded joints is presented.

1. Introduction

Field pipelines can be distinguished among the facilities most susceptible to oil production. Although the oil itself is not an aggressive medium, the presence of salts, chlorine compounds, oxygen, hydrogen sulfide, highly mineralized water in commercial oil products reduce the corrosion resistance of pipes, and the presence of highly abrasive particles in this substrate leads to the emergence and development of abrasive damage. Analysis of the actual service life of oil field pipelines in Western Siberia shows that corrosion and erosion defects reduce the service life of pipelines many times. Sometimes this service life does not reach even one year [1], and the main damage is concentrated in the areas of welded pipe joints. To reduce the effect of corrosion damage, several methods of protecting the metal of welded joints have been developed, aimed both at deactivating the corrosive environment and at creating protective devices that prevent contact between the corrosive environment and the metal surface. Corrosion inhibitors are used when carrying out measures to deactivate the transported product, but their use is not always technically justified, which is due to both the uneven distribution of the inhibitor along the length of the pipe and the need to periodically stop pumping during inhibition operations. Another way is to reduce the aggressiveness of corrosive products directly in the well, for example, by creating high rates of transportation of petroleum products in the "emulsion" mode to eliminate the product separation and isolation of aggressive components from it. The method of protecting the tread is widely used. There is also the method of using pipes with increased wall thickness, taking into account the increase in thickness "for corrosion", the method of creating the facing coatings of the inner surfaces of the pipes. Such coatings are made of relatively inexpensive and easily applied organic and inorganic materials, mainly based on thermosetting plastics and elastomers. Coatings can solve the problem of protecting the base metal of the pipe from corrosion and erosion effects of the environment; however, having low heat resistance, they burn out or evaporate in the places of welding of pipes. At the same time, there are zones of increased corrosion defects in the pipeline - zones of welded joints, the number of which reaches one hundred pieces per kilometer of the pipeline.

The analysis of operability and damage of pipeline systems showed that there are several main mechanisms of damage of welded joints of pipe wires. The main danger to pipelines is general and selective corrosion in neutral aqueous media with pH 6-8, containing hydrogen sulfide and carbon dioxide, also contaminated with solid inclusions. The occurrence of damages manifested through the mechanism of corrosion-erosion wear and the development of corrosion defects according to the classical mechanism of electrochemical corrosion, which is enhanced by abrasive particles that constantly destroy the passive oxide layer on the metal surface, preventing the attenuation of corrosion processes. It is also possible to develop defects by the mechanism of sulfide stress cracking and hydrogen cracking.

2. Materials and methods

To date, almost the only method of protection of welded joints is the method of installation of protective sleeves installed during the assembly of the pipeline. The protective sleeve is a metal ring with anti-corrosion coating. Installing a protective sleeve allows you to achieve isolation of the weld zone from the pumped medium, to slow down corrosion processes. However, this method has a number of significant drawbacks, which include a decrease in the diameter of the pipeline, complicating the hydrodynamics of pumping products, complicating the technology of repair work, increasing the cost of installation work and the pipeline as a whole. Thus, we can state the fact that the problem of protecting the internal surfaces of welded joints of oil field pipelines has not yet been completely solved and remains one of the most pressing problems facing large oil companies. In St. Petersburg Polytechnic University together with STC developed a methodology and conducted research on the application of combined protective coatings consisting of a polymer coating of the base metal of the pipe and a protective ring of stainless steel in the weld zone. The width of the metal coating ring shall be guaranteed to exceed the width of the heat-affected zone of the annular weld, and the polymer coating shall be applied to the inner surface of the pipes in such a way that it overlaps the metallization zone outside the heat-affected zone of the weld. In this work, we studied the creation of protective metal coatings for welded pipe joints for field pipelines. In the work, we studied model objects-samples on pipes with a diameter of 109 mm and a length of 300 mm, made of steel 09Mn2Si.

It is known that metal coatings on the inner surface of pipes can be applied in various ways, by surfacing stainless wire in an inert gas medium, laser surfacing, plasma spraying of metal powder [2]. At the first stage of work, a preliminary assessment of each of the methods was carried out and the main advantages and disadvantages of each technology were studied. The advantage of surfacing in the argon environment was the technological simplicity of the process, the possibility of wide automation and robotization of the surfacing plant, the use of standard filler wires. The disadvantages include low quality (high roughness) of the surfacing surface-the inability to obtain an equidistant coating without additional machining, the complexity of quality control of the surfacing for the detection of pores and other micro-surface defects. These shortcomings, identified in the course of research, did not allow recommending the technology of surfacing in the application of metallization coatings, so the main attention was paid to the method of laser surfacing and plasma spraying. The advantages of the methods of plasma spraying and laser surfacing is that the resulting coatings provide a uniform layer thickness, the ability to adjust this thickness from the maximum at the end of the pipe and gradually decreasing as it moves away from the end, the relative ease of quality control of the coating. All this makes the methods of applying powder metallization coatings extremely attractive and can minimize the change in the internal diameter of the pipes in the transition layer zone, where a polymer coating will be applied to the metallization surface. A common disadvantage of these methods is the need to use metallic powders, usually used in additive machines, which have special requirements for the shape (spherical or approaching spherical shape of the particles, the absence or minimum number of

IOP Conf. Series: Materials Science and Engineering **826** (2020) 012008 doi:10.1088/1757-899X/826/1/012008

satellites, particles fused together when spraying the powder) and sizes-from 60 to 160 microns, the absence of oxide films on the surface of powder particles, etc. The main drawback of these methods is the scarcity and high cost of the metal powder used for coating. Such powders are produced in very small volumes on the territory of the Russian Federation, and the purchase of imported powder is associated with a number of difficulties in its delivery and storage - due to the high chemical activity and the tendency to oxidation of fine powders. This leads to certain restrictions on the timing and volume of its use and storage. To reduce the cost of the metallized surfacing process, it is necessary to apply an integrated approach, when the surfacing complex includes not only a powder-coating unit, but also a unit for the production of such a powder. The presence of its own atomization system allows you to reduce the cost of powder, the cost of which is usually two to three times lower than the cost of small wholesale batches on the Russian market; to apply powders of different compositions at spraying, selecting own protective composition optimum for each aggressive environment; reduces to zero terms of delivery of powders, allows, on the one hand, to have, necessary technological reserve of the put material, with another not to create difficult warehouse systems on storage of powders. Ensuring the necessary requirements for the metal powder can be achieved by obtaining the powder by the melt dispersion method. Dispersion is the most productive, economical and efficient method of obtaining small and medium-sized metal powders-up to 70% of the volume of all industrial powders is obtained by this method [3]. It is widely used in the production of powders of multicomponent alloys-dispersion allows to achieve a uniform chemical composition of the composition, even when the content of alloying components above their solubility limit in the main component of the alloy. The powders obtained using the methods of dispersing the melts are spherical or close to spherical shape particles. For the production of metal powder from stainless steel, the method of plasma spraying of wire in a plasma atomizer developed and manufactured at St. Petersburg Polytechnic University of Peter the Great was used (figure 1).

a)

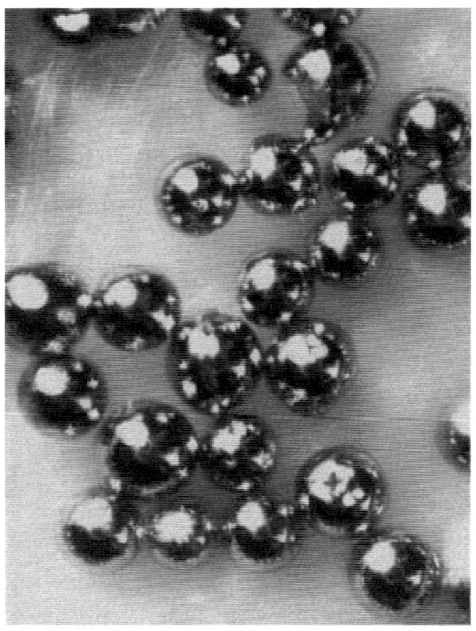

b)

Figure 1. The atomization chamber of the plasma atomizer – a; the shape of the powder particles in the analysis using an optical microscope (D_{av} = 50 microns) – b.

Plasma atomization refers to gas atomization, when a metal drop melted in the arc of a plasma torch is sprayed by a jet of plasma-forming gas. As a plasma generator in a plasma atomizer, plasma torches with arc fixation by means of interelectrode inserts were used, which allow one to obtain an ascending voltage characteristic of the arc, providing stable operation of the plasma torches without ballast resistance in the power supply circuit from a source with a rigid characteristic. The atomizer makes it possible to obtain powders of spherical shape with sizes from 10 to 500 microns, the performance of the atomizer depending on the chemical composition of the material, the required particle size and other requirements can vary from 1 to 5 kg of powder per hour.

3. Results and discussion

In the course of the research, the task of working out the technology of obtaining powder from steel 08Cr18Ni10Ti of spherical shape with particle sizes from 40 to 80 microns was set and solved. In the course of the research, the current strength from 150 to 250A, the velocity (flow rate) of the plasma-forming gas from 30 to 40 liters per minute, the wire feed rate from 0,5 to 5 meters per minute were taken as factors of variation.

The mathematical quasi-d-optimal plan of the second order was used in the work. Optimization of technological parameters of the spraying process allowed obtaining a spherical stainless steel powder (figure 1) with the retention of particles of specified sizes (in the total mass of the sprayed powder) of about 30%. The resulting powder was applied to the inner surfaces of samples of 109x7 mm pipes made of steel 9Mn2Si with a length of 150 mm. The surface for coating was prepared in advance – mechanically cleaned from the oxide film and degreased with organic solvents. The width of the coating was 50 mm, the thickness of the coating was -1 mm at the end of the pipe and 0,6 mm at a distance of 50 m from the end. After coating, the ends of the pipes were separated according to the standard technology according to GOST 16037-80 and welded according to the technology and modes used in the installation of oil field pipelines of Gazpromneft. In the process of welding, the root pass from the side of the protective coating brew stainless steel electrode brand Sv-04Cr19Ni9 (ER 308 LSi) and filling cutting was carried out with welding wire Sv-08Mn2Si (CITOFIL 1). The resulting compound was cut into samples for metallographic investigations, mechanical and corrosion tests. Mechanical tests performed in accordance with GOST 6996-66, confirmed the quality of the welded joint-the destruction of samples after all tests occurred outside the weld zone, showing results that meet the requirements of GOST 32678-2014.

During visual inspection and metallographic studies, it was found that after welding of stratifications of the basic metal and the covering put by laser surfacing and plasma spraying, it is not revealed. Samples destroyed during mechanical tests did not have a clear reference of the origin and development of the crack. The most important tests of coated materials are bend tests [4]. Such tests were carried out on a breaking machine with the installation of a coated sample on supports and bending it around mandrels having a diameter equal to two, three and four times the thickness of the pipe of 14, 21 and 28 mm. The width of the sample was equal to twice the thickness of 14 mm, and the length of 200 mm. The bend was carried out at an angle of 90^0. Two series of tests were carried out. In the first tests, the samples were bent so that the coating experienced tensile stresses, in the second – the coating experienced compressive stresses. The test results are shown in the table.

Table 1. The results of the test bend

Type of coating	Corner bend 90^0					
	tension on mandrel D, mm			compression on mandrel D, mm		
	14	21	28	14	21	28
plasma spraying	M, O	M	-	M	-	-
laser cladding	MM	M	-	M	-	-

* M – single microcracks, MM – numerous microcracks, O – detachment.

Thus, only in the conditions of a rigid bend on the mandrel with a diameter of 14 mm we found significant defects. On the mandrel with a diameter of 28 mm – no defects were found. On the mandrel with a diameter of 21 mm two microcracks were found on two of the five samples after plasma spraying and one of the five after laser surfacing. Such results may be considered satisfactory.

The tendency of the welded joint protected by metallization coatings to general corrosion was determined by conducting corrosion tests while holding samples in a corrosive medium (in a solution of 3,5% NaCl for 240 h). To assess corrosion resistance, samples were produced with a surface covering the base metal, weld and coating. The estimated characteristics were appearance and weight loss. The ends of the samples were insulated with a paint coating. Coating and welded joint were without insulation. It is established that after corrosion tests in the zone of protective metallization coatings, corrosion products and corrosion defects are completely absent. Extended loose corrosion products, ulcers up to 1,5 mm deep and up to 5 mm in diameter were found in the areas of the welded joint unprotected by the metallization coating (control samples).

4. Conclusion

According to the results of the tests, it was found that the protection of a welded joint by the coating applied by plasma spraying and laser surfacing fulfils its function in full. The powders obtained on the plasma atomizer provide the necessary shape and size of the particles. The proposed technology, after comprehensive tests on the experimental sections of pipelines, can be recommended for use in the assembly of field pipelines.

References

[1] Rodionova I G, Baklanova O N and Shapovalov E T 2005 *Metallurgist* 44-50
[2] Ermakov B S 2005 *Technology of structural materials* (SPb.: Institute of refrigeration and biotechnology Press)
[3] Ermakov S S and Ermakov B S 1996 *Powder and composite* (SPb.: SPbSTU Press)
[4] Yudin P E, Petov S M, Fedotova A V, Elistratova N V and Markov Yu M 2014 *Pipeline transport* 16-18

Features of the use of light alloys in the marine climate of the arctic zone

S B Ermakov[1], B S Ermakov[1, 2] and A K Andreev[2]

[1] Peter the Great St. Petersburg Polytechnic University, 29, Politekhnicheskaya str., Saint Petersburg, 195251, Russia
[2] ITMO University, 49, Kronverkskiy Ave., Saint Petersburg, 197101, Russia

E-mail: s.b.ermakov@bk.ru

Abstract. The movement of mining and processing industries to the North poses the task of analyzing the performance of materials traditionally used in structures and products. The paper presents the results of comparative studies of magnesium alloy MA2-1 and aluminum alloy AMg6 in the marine climate of Northern latitudes. The studies were carried out after simulating the long-term operation and storage of alloys in northern latitudes, namely after long 1000-hour exposures at a temperature of -600 ° C, aging in a salt fog chamber. We analysed the impact of external climatic factors, typical of northern latitudes, on the mechanical properties and a comparative assessment of the corrosion resistance of alloys in a marine climate.

1. Introduction

The primary task facing the Russian Federation is the development of the shelf zone of the Arctic seas and adjacent coastal areas, in the depths of which there are significant reserves of hydrocarbons, ore deposits of such scarce and expensive metals as gold, platinum, tungsten, copper, cobalt, nickel, tin. The movement of the mining and processing industries to the north sets the task to re-examine the possibility of using traditional structural materials, such as aluminum and magnesium alloys, in the marine climate of the Arctic Ocean.

2. Materials and methods

Light alloys are increasingly used in structures and products for various purposes. Their use becomes especially relevant in equipment with strict requirements for weight and overall characteristics. Such alloys have a relatively high specific strength and several times lower density than the density of traditional steels. They have been used in aviation and shipbuilding, equipment for the extraction and processing of hydrocarbon raw materials, radio engineering, etc. These materials include widely used light alloys aluminum and magnesium [1].

In this work, comparative tests were carried out on two alloys – MA2-1. They were deformable magnesium alloy containing 0,2-0,6% of manganese, 3,8-5,0% of aluminum, 0,6-1,5% of zinc, and deformable aluminum alloy AMg6 (5,8-6,8% of magnesium, 0,5-0,8% of manganese, not more than 1% of iron, silicon, titanium, copper and zinc). The MA2-1 alloy is mainly used in the form of sheet blanks, bars, plates and is widely used in instrument housings, in the housing elements of mobile equipment, electronic systems, for welded parts of complex configuration, fittings, gasoline and oil systems. Alloy AMg6 is used for the manufacture of welded parts of tanks and pipelines: shells, bot-

IOP Conf. Series: Materials Science and Engineering **826** (2020) 012009 doi:10.1088/1757-899X/826/1/012009

toms, flanges, pipe grids, etc. Semi-finished products of alloy AMg6 have been used in shipbuilding [2].

In the work we used rods of both alloys with a diameter of 100 mm according to GOST 21488-97. This was the AMg6 alloy and GOST 18351-73 and the MA2-1 alloy. Features of the equipment used in the marine climate of the Northern latitudes is prolonged (storage and operation) of equipment outdoors in extremely low temperatures, icing of the working surfaces (frost); when defrosting and high humidity conditions, the formation, in the working of ultrasonic crystals water drops (dew). In the warm part of the year, the equipment is exposed to aggressive corrosive effects of sea air [3, 4]. The influence of these factors should be investigated when determining the possibility of using alloys AMg6 and MA2-1 as hulls and other components of systems, long-term operation in the coastal zones of the Northern seas. The influence of the above factors on the mechanical properties of alloys has been studied. The effects of prolonged exposure in the open air were simulated by cooling the tensile and impact-bending test specimens in a climate chamber at minus 60°C. The cooling time varied from 200 to 1000 hours.

The influence of external climatic factors such as frost and dew was also simulated in the climate chamber. The same samples were placed in a climatic chamber and kept at a temperature of minus 60 °C for 100 hours with periodic purging of the chamber with wet water vapor to form a stable ice crust on the working surfaces of the samples. The samples processed in this way were placed in the grips of the test machines. The transfer time was not more than 60 seconds, which allowed preserving the ice crust during the tests. Dew-the temperature in the chamber was 20 °C, humidity 95%, exposure time 100 hours. The influence of sea air was simulated in two ways: the first – by exposure of samples to mechanical tests in a salt mist chamber at a temperature of 35 ^{0}C for 100 hours. The second tested the resistance of alloys to corrosion in accordance with the requirements of ESZKS GOST 9.913-90. For testing according to GOST 9.913-90, two types of samples were prepared - disks with a diameter of 100 mm and a thickness of 2,5 mm and samples for mechanical tests. Samples attached to nylon threads were completely immersed in a reactive corrosion medium heated to a temperature of 23 °C and representing a 3% solution of sodium chloride with the addition of 1% hydrogen peroxide. Such reagent composition and test conditions were selected in accordance with the requirements for the category of equipment performance for this climatic region, the conditions of its operation, storage and transportation in terms of the impact of climatic factors of the environment.

During the tests, the 0,1% hydrogen peroxide solution was added to the test medium once every five days; the solution itself was replaced every 10 days of the test. When replacing solution we photographed the actual state of the surface of the disc-shaped samples.

3. Results and discussion

The results of mechanical tests of alloys are given in table 1. As follows from the table, alloy AMg6 was practically not affected by external climatic factors, its properties after prolonged exposure. Simulating the impact of external climatic conditions of operation of equipment in coastal zones of Northern territories remains virtually unchanged. The properties of the alloy MA2-1 also do not undergo changes under the influence of temperature and ice action. At the same time, studies of alloy samples after exposure to salt mist showed a sharp decrease in time resistance, ductility and toughness. Visual inspection of the surface of the samples after exposure in the salt mist chamber showed that their surface is covered with a layer of white deposits. After removing a layer of deposits on the surface of impact samples and samples intended for uniaxial static tensile tests, corrosion defects such as flat ulcers with a diameter of up to 2-4 mm and a depth of more than 1 mm were found on the surfaces of the working parts of the samples.

To clarify the corrosion resistance of the alloys, additional corrosion tests of the samples were carried out. The tests were carried out by completely immersing the samples in a solution containing 3% sodium chloride and 0,1% hydrogen peroxide.

Table 1. Mechanical properties of alloys AMg6 and MA2-1 after prolonged exposure to external climatic factors*

Climatic impact	T	Alloy AMg6				Alloy MA2-1			
		σ_B	$\sigma_{0,2}$	δ_5	KCV	σ_B	$\sigma_{0,2}$	δ_5	KCV
	°C	MPa		%	J/cm^2	MPa		%	J/cm^2
Delivery condition	20	370	173	24	31.6	255	155	11	22.7
	-60	410	180	21	27.4	310	215	8	16.4
Freezing at minus 60 °C - 200 h	20	375	176	24	32.0	260	160	12	22.1
	-60	408	179	20	27.2	310	210	8	16.2
Freezing at minus 60 °C - 500 h	20	371	175	23	31.3	257	157	11	21.0
	-60	412	181	21	27.0	305	211	8	15.8
Freezing at minus 60 °C - 1000 h	20	370	177	24	31.3	260	160	10	20.8
	-60	415	182	21	27.5	305	155	9	15.4
Exposure in salt mist at 35 °C - 100 hours	20	369	171	23	31.0	210	150	4	11.9
	-60	404	179	20	27.3	235	205	1	8.7
Frost	20	372	170	23	31.2	252	155	10	22.0
	-60	410	183	21	27.5	310	200	8	16.2
Dew	20	370	174	24	31.6	256	160	11	22.3
	-60	407	181	21	27.0	311	205	9	16.5

* Note. The sizes of working parts of all samples were measured before imitation of external climatic influences

The evaluation of the test results was carried out by changing the appearance of the samples; the presence, depth and amount of corrosion damage, as well as by changing the mechanical properties of the samples after aggressive environmental exposure. The test results are given in table 2 and in figure 1.

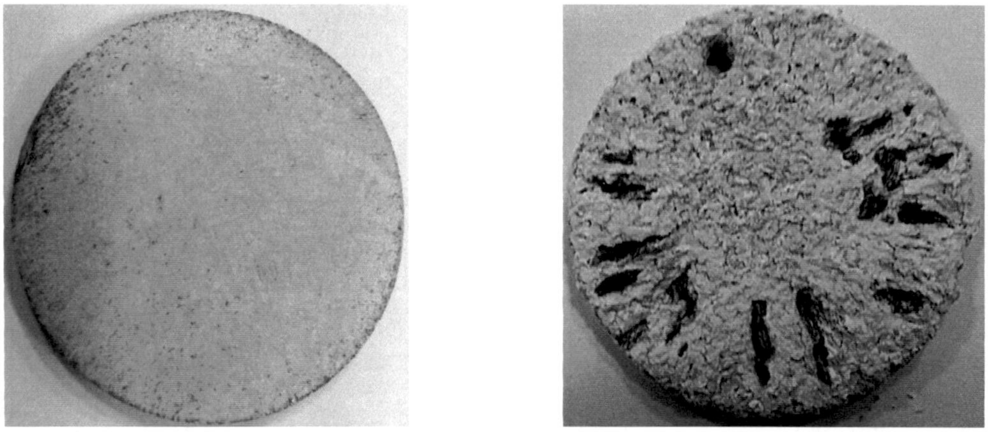

a) b)

Figure 1. Samples after corrosion tests within 1000 hours: a) alloy AMg6, b) alloy MA2-1

The results of mechanical tests of the AMg6 alloy after a 100-hour exposure in a medium of 3% sodium chloride and 0,1% hydrogen peroxide are given in table 3. The samples for mechanical tests of the alloy MA2-1, sustained for 1000 hours in a solution of 3% sodium chloride and 0,1% hydrogen peroxide, were completely damaged by corrosion and it was not possible to carry out tests.

Table 2. Description of corrosion damage of AMg6 and MA2-1 alloy samples

Time, hour before testing	Description of sample surface condition, the samples are machined, Ra 2.0 mkm
	Alloy AMg6
240 hours	The surface of the sample is covered with a thin translucent layer of deposits.
720 hours	The surface of the sample is completely covered with a layer of white deposits, on the surface of the sample marked individual ulcers with a diameter of 0,2 mm.
1000 hours	The surface of the sample is completely covered with a layer of white deposits. Chains of small corrosion ulcers are visible on the surface of the sample. The diameter of the ulcers is not more than 0,3 mm. Single ulcers with a diameter of 1,0 mm and a depth of 0,5 mm are noted.
2500 hours	The surface of the sample is completely covered with a layer of white deposits. On the surface there are visible corrosion ulcers at a depth of more than 0,5 mm, diameter of up to 2,5 mm.
	Alloy MA2-1
24 hours	The surface of the sample has lost its metallic luster, is completely covered with thin oxide films, individual corrosion ulcers are visible, mainly concentrated in the peripheral part of the sample.
48 hours	The surface of the sample is completely covered with a layer of deposits of gray and gray-green colors. Numerous corrosion ulcers were found. After removal of surface deposits, flat ulcers of a corrosive nature were found on the surface of the sample. Depth is up to 0,25 m, length is up to 1,5 mm. Density of defects is 2-3 ulcers per 1 cm^2.
500 hours	The surface of the sample is completely covered with a layer of loose corrosion deposits of gray and white-gray colors. Deposits are easily exfoliated from the over-surface of the sample. Large corrosion defects up to 1 mm deep were found in the peripheral part of the sample.
1000 hours	The surface of the sample is completely covered with a layer of loose corrosion deposits of gray and white-gray colors. Deposits are easily exfoliated from the over-surface of the sample. The composition is deposition-oxides and chlorides of magnesium and aluminum chloride. Numerous large corrosion ulcers, including through ulcers, were found in the peripheral part of the sample.

Table 3. Mechanical properties of alloy AMg6 exposed for 1000 hours in corrosive environment

T test, ^{0}C	σ_B , MPa	$\sigma_{0,2}$, MPa	δ_5 , %	KCV, J/cm^2
20	365	171	11	31.2
-60	412	181	8	27.0

Based on the corrosion resistance tests of the AMg6 and MA2-1 alloys, it was established that both alloys retain the stability of mechanical properties during long-term low-temperature operation. This is confirmed by the results of mechanical testing of samples after such external climatic influences as prolonged exposure to low climatic temperatures, icing, loss condensed moisture (dew) and can be used to manufacture casings and other components and parts of equipment operated on the Far North and the Arctic. Particular attention should be paid to the conditions in which the equipment will be operated, in the functions of the material for the manufacture of low-temperature equipment. Magnesium alloy MA2-1 is subject to corrosion under the action of chlorine-ion medium, the impact of which leads to the emergence and development of corrosion defects such as ulcers, developed by the mechanism of general corrosion. Even within 2 days of testing, the depth of corrosion ulcers was 0,25 mm, the diameter was up to 1,5 mm, the density of defects was 2-3 ulcers per 1 cm^2. With further exposure of samples in a corrosive environment, the density of corrosion defects increases and after 1000 hours of exposure grows into an almost continuous field; corrosion damage was detected. Thus, it is shown that the use of MA2-1 alloy can lead to accelerated destruction of products, operated and stored in the sea climate. Analysis of the surfaces of the samples of the AMg6 alloy after corrosion showed that the alloy has a significantly greater resistance to aggressive chlorine-ion environment. However, in the course of studies of the surface of the samples after a 1000- and 2500-hour exposure in an aggressive environment, damage was detected, developed by the mechanism of ulcerative corrosion and corrosion deposits of chlorides and oxides.

4. Conclusion

It is shown that alloys AMg6 and MA2-1 can be recommended for the manufacture of equipment operated outdoors in northern latitudes. However, when selecting the material for the manufacture of equipment, it is necessary to take into account the additional climatic conditions of operation and the area of the equipment. If in the areas, remote from the seas, these alloys can be used without restriction, then in the marine climate the use of magnesium alloy is possible only with the use of protective coatings [5].

Also on the basis of the conducted research, it was established that the relatively short-term use of the equipment made from the AMg6 alloy in the conditions of sea climate of northern latitudes (about one hour) does not need additional measures on anticorrosive protection. Long-term corrosion effects of the environment leads to the appearance of corrosion defects on the surface of the alloy AMg6 in the form of corrosion ulcers, looseness, chains of small corrosion defects. The layer of corrosion deposits consists of two groups of deposits. The first is a dense passivating layer of aluminum oxide with good adhesion to the surface of the sample. The second is separate conglomerates of loose deposits of magnesium oxide and chloride and aluminum chloride, easily removed from the surface of the sample. Therefore, despite the fact that the results of mechanical tests did not reveal the influence of the corrosive environment on the mechanical properties of the alloy. In order to prevent corrosion damage and loss of performance of equipment intended for long-term operation in a marine climate, it is recommended to protect the surface of the equipment using protective anticorrosive coatings.

References

[1] Kornysheva I S, Volkova E F, Goncharenko E S and Mukhina I Yu 2012 *Aviation materials and technologies* **8** 7-17
[2] Solntsev Yu P, Ermakov B S and Sleptsov O I 2008 *Materials for low and cryogenic temperatures: Encyclopedic reference* (SPb.: KHIMIZDAT)
[3] Ermakov B S, Ermakov S B, Vasiliev A V and Karginova V V 2011 *Dairy industry* **7** 47-49
[4] Sleptsov O I, Petrov P P, Ermakov B S *Modern problems of science and education* **9** 51-58
[5] Kozlov I A, Karimova S A 2014 *Aviation materials and technologies* **2** 14-20

Lanthanide doping of $A^{III}B^{V}$ crystals

V V Romanov[1], B S Ermakov[1, 3], V A Kozhevnikov[1], K F Stelmakh[2] and S A Vologzhanina[3]

[1] Peter the Great St. Petersburg Polytechnic University, 29, Politekhnicheskaya str., Saint Petersburg, 195251, Russia
[2] FT. A. F. Ioffe, 26, Politekhnicheskaya str., Saint Petersburg, 194021, Russia
[3] ITMO University, 49, Kronverskij Ave., Saint Petersburg, 197101, Russia

E-mail: ermakov55@bk.ru

Abstract. The state of rare earth impurities at concentrations of about 10^{18} cm^{-3} in volume-doped $A^{III}B^{V}$ semiconductor crystals can be described in the framework of the model of quasi-molecular rare earth centers of Ln_2O_3 type and spontaneous polarization regions-coupled spin-polarons localized on small donors.

1. Introduction

The interest in lanthanides as an alloying admixture of semiconductor crystals is due to their bright physical and chemical properties. For example, some studies have shown that due to the high chemical activity of rare earth elements, it is possible to form metal clusters in Germany and silicon, which are clusters that serve as effective effluents for radiation defects [1, 2]. In addition, the introduction of rare earth impurities can be used to "purify" crystals by binding uncontrolled impurities. However, this does not exhaust the possibility of using rare earth impurities in crystal alloying. Apparently, no less interesting is the possibility of using lanthanides to create new optical materials based on semiconductors doped with rare earths, which are characterized by unique properties due to the structure of the electron shell of rare earth atoms.

Due to the high chemical activity of the elements of the lanthanide group, the question of the state of atoms of rare earth elements introduced into semiconductor crystals and their distribution over the crystal is relevant.

The electrical and optical properties of semiconductor materials, due to impurity and intrinsic defects and of greatest interest in the creation of solid-state electronics devices, are inextricably linked to their magnetic properties. The latter are most adequately reflected in the behavior of the magnetic moment, which can be reliably controlled by the temperature dependences of the magnetization of the studied samples.

In this article, we discuss the behavior of rare earth impurities in volume-doped $A^{III}B^{V}$ crystals grown by the Chochralski, Bridgeman, and solution-melt methods. Alloying occurred during the growing process by introducing a rare earth metal into the melt. The concentration of the introduced impurity was about 10^{18} cm^{-3} and was controlled by spectral and neutron activation analysis [3].

IOP Conf. Series: Materials Science and Engineering **826** (2020) 012010 doi:10.1088/1757-899X/826/1/012010

2. Materials and methods

Measurements of temperature dependences of magnetization in the temperature range from 3,2 to 300 K at the intensity of the external magnetic field up to 15 ke were carried out by the Faraday method on a special installation, which was created on the basis of the spectrometer MGD 312 FG. The samples were placed in an external magnetic field in a quartz Cup suspended from a thin quartz thread. Registration of intensity of interaction with a magnetic field was made in conditions when the sample reached a state of thermodynamic equilibrium. Control of this state was carried out by means of the automated control system of measurements with the specially developed software. Calibration of the installation was carried out using a reference sample, which was used as a single crystal of magnetically pure indium phosphide with a susceptibility of $\chi = -313 \cdot 10^{-9}$ cm^3/g. The mass of the sample was determined on the scales BP 211 D with an accuracy of 10^{-5} g [1].

3. Results and discussion

Studies of rare-earth impurity states carried out in [4,5] showed the fundamental possibility of the formation of substitutional solid solutions with single lanthanides statistically distributed over the crystal in the Ln (III) charge state. However, the success was very limited – the total number of lanthanides has managed to register the EPR spectra of only two lands – ytterbium and gadolinium. In addition, the erbium spectrum on crystals doped by ion implantation was observed in [6], which was interpreted as tetrahedral with weak axial distortion.

All attempts to detect single impurity centers of other rare earth elements from studies of EPR spectra and temperature dependences of magnetization have not been successful.

The most obvious explanation of the obtained results of the research, apparently, is a significant excess of the ionic radii of lanthanide atoms of the radius of the substituted cation [7]. The known effect of lanthanide compression [7] allows explaining why it was possible to achieve success with the use of "heavy" rare earth elements – their size due to the mentioned effect is close to the ionic radius of the cation of the semiconductor lattice.

However, further studies conducted with crystals doped with "heavy" lanthanides revealed the instability of the resulting solid solutions – their decay. As expected, the most "long-lived" was doped with an ytterbium sample.

For example, the EPR spectrum that identified the tetragonal center of ytterbium registered at temperatures of $T < 4{,}5K$ immediately after the synthesis of the InP <Yb> sample [4] could not be reproduced after six months.

An even shorter "lifetime" was given to a solid solution of gadolinium in indium phosphide, the EPR spectrum of which was recorded on the ER-220D unit in the temperature range from 3 to 300 K and showed a close to ideal angular dependence of a single gadolinium center in the $^8S_{7/2}$ state. Disintegration of the InP<Gd> solid solution with a concentration of rare earth impurity of $4 \cdot 10^{16}$ cm^{-3} occurred within a month.

Thus, solid solutions of lanthanides containing InP <Gd> rare earth elements as substitutional impurities in the cationic lattice sites of III – V crystals completely lost their initial properties in a relatively short period of time.

All attempts to detect single impurity centers of lanthanides in $A^{III}B^{V}$ crystals doped with rare earth elements, with the exception of Yb, Er and Gd, by measuring the temperature dependence of static magnetic susceptibility and by the EPR method, failed [3].

Further studies showed the temperature dependence of the static magnetic susceptibility of $A^{III}B^{V}$ crystals containing lanthanides with the possible exception of europium in thermodynamical equilibrium. States of the materials under study receive an adequate description if we assume that rare earth impurities form in the crystal a system of weakly interacting quasimolecules such as Ln_2O_3 - quasimolecular centers, in which exchange interactions occur between atoms of rare earth elements [3]. The proposed model corresponds to the nature of temperature dependences of static magnetic susceptibility in a wide temperature range, demonstrating almost complete absence of dependence on the strength of the external magnetic field.

IOP Conf. Series: Materials Science and Engineering **826** (2020) 012010 doi:10.1088/1757-899X/826/1/012010

On a competitive basis, along with weakly interacting quasimolecules, "regions of spontaneous polarization – coupled spin-polarons" can be formed [1, 8-10]. The formation of spin-polarons occurs when ten or more quasimolecules fall into the orbit of a small donor. This becomes possible at low temperatures and in the region of high local concentrations [1, 8, 14] of rare earth elements (Figure 1).

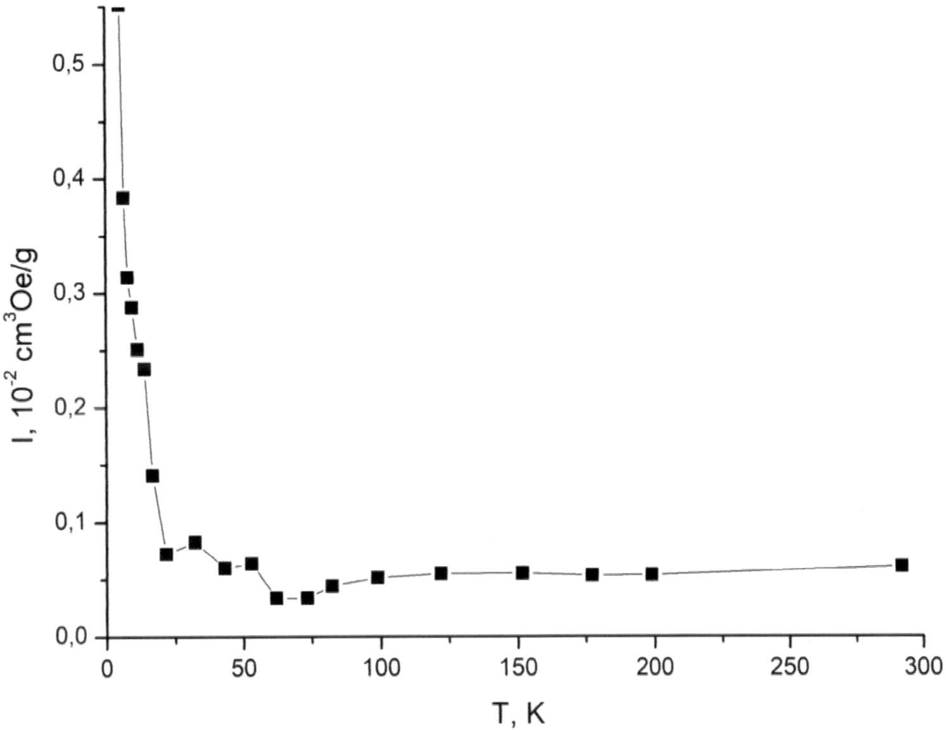

Figure 1. Temperature dependence of magnetization of InP< Er > crystals with high erbium concentration, demonstrating magnetic anomalies caused by ferro-ordering of quasi-molecular centers

However, even in the composition of quasimolecular rare earth centers, due to the unique structure of the magnetic 4f-shell of the rare earth ion, shielded by 5s and 5p–electrons, is slightly distorted (split) by the electric field of ligands. This makes it possible to observe the processes of intracenter luminescence due to the presence of an effective excitation and absorption channel in almost any matrix, since the influence of the latter will be relatively weak [2, 8]. In this case, the spectra of intracenter luminescence will be formed by a set of narrow lines (~10 Å) due to the fact that the basic states of rare earth impurities will be practically not hybridized with the zone states of the crystal, and the excited ones will be weakly hybridized [2, 4, 6,11-13].

At the same time, a further increase in the concentration of the introduced rare earth impurity significantly changes the situation, demonstrating the chemical activity of rare earths and leading to the formation of "second phase inclusions" in the studied crystals, which were identified by the characteristic temperatures of the phase magnetic transitions.

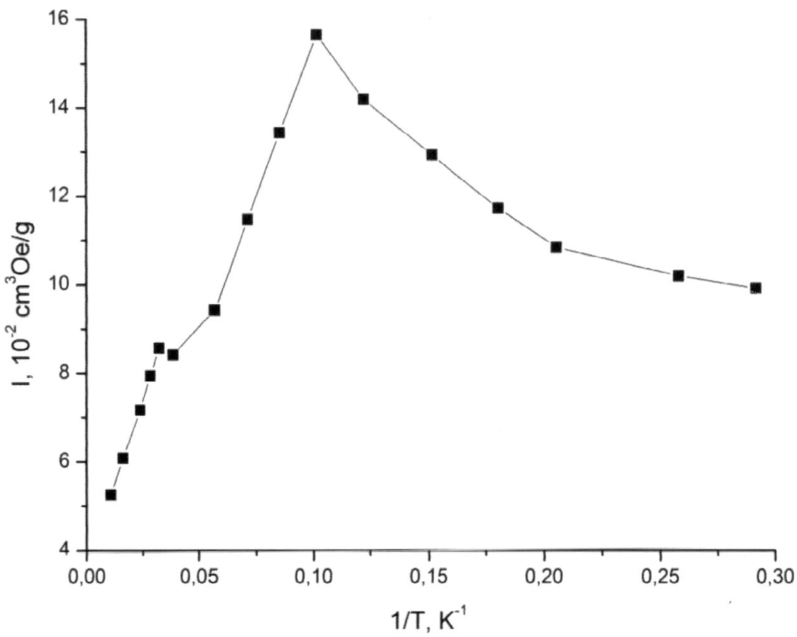

Figure 2. Temperature dependence of magnetization of crystals InP<Nd>, «second phase switch-on».

4. Conclusion

Using the mentioned temperature dependences of magnetization of crystals of indium phosphides doped with neodymium (Figure 2), characteristic features are clearly observed. This may be due to strong magnetic "inclusions of the second phase" with residual impurities, which are part of the main lattice, in this case – with phosphorus [1].

The behavior of rare earth impurities in low-dimensional semiconductor structures is the subject of separate studies.

References

[1] Bagraev N, Romanov V V 2005 *FTP* **39(10)** 1173
[2] Andrianov D G, Bochkarev E P, Grishin V P, Karpov Yu A 1978 *FTP* **12** 511
[3] Romanov V V 2002 *Magnetic properties of metastable defects in semiconductor glasses, crystals and nanostructures* Doctor of physical and mathematical sciences dissertation, Saint Petersburg
[4] Kurdyukov D A, Pevtsov A B, Smirnov A N et al. 2016 *FTT* **58(6)** 1176-1182
[5] Masterov V F, Romanov V V and Shtelmakh K F 1983 *FTT* **25(5)** 1435
[6] Masterov V F, Romanov V V, Samorukov B E and Stelmakh K F 1983 *FTP* **17(5)** 948
[7] Ennen H, Schneider J, Pomrenke G and Axmann A *Appl. Phys. Lett.* **43** 943
[8] Taylor K, Darby M 1974 *Physics of rare earth compounds* (Moscow: Mir)
[9] Bagraev N T and Romanov V V 2002 *Proceedings of SPIE* **4627** 84
[10] Bagraev N T 1983 *USSR Academy of sciences. Ser. Phys.* **47** 2331
[11] Nagaev E L 1979 *Physics of magnetic semiconductors* (Moscow: Nauka)
[12] Ennen H, Kaufmann U, Pomrenke G, Schneider J, Windscheif J and Axmann A 1983 *J. of Crystal Growth* **64** 165
[13] Kasatkin V A and Savelyev V P *FTP* **18(9)** 1634
[14] Wagner J, Ennen H and Muller H D 1986 *J. Appl. Phys* **59(4)** 1202

Investigation of structure and properties of welded joints of pipe steels

B S Ermakov[1, 2], A R Khasanov[1] and E O Khasanova[1]

[1] ITMO University, 49, Kronverksky ave., Saint-Petersburg, 197101, Russia
[2] Peter the Great St. Petersburg Polytechnic University, 29, Politekhnicheskaya str., Saint Petersburg, 195251, Russia

E-mail: hasanoffalick@yandex.ru

Abstract. The work is devoted to the study of microstructure and crack resistance of welded joints of pipe materials of main gas and oil pipelines for the Far North regions. Reliable operation of pipelines in harsh climatic conditions of such regions can be provided by the use of materials that have a combination of high values of strength, toughness, cold resistance and weld ability. The results of studies of samples from large diameter pipes with an operating pressure of up to 7.4 MPa at test temperatures of -10 and -20 °C. are presented. These pipes are made of hot-rolled low-carbon low-alloy steel. Various welding methods were used to produce welded joints: high-frequency induction welding, multi-arc automatic welding, laser and hybrid welding. This allowed us to evaluate the influence of the heating and cooling rate during welding on the characteristics of the structural States of the weld, the fusion line, and the overheating area in the zone of thermal influence. It was revealed that a rough needle structure of Vidmanshtett is formed in the zone of overheating of the longitudinal seam at a high cooling rate. Using a low cooling rate allows you to get the Vidmanshtett structure within 2-3 points. It is shown that the crack resistance of the base metal of the pipes practically does not change with a decrease in the test temperature. The values of crack resistance for the metal of the zone of thermal influence have a large spread, which may be due to the influence of technological features of cooling the pipe billet after welding.

1. Introduction

The development of pipeline transport in Russia is associated with the construction of large Russian, Intercontinental and transnational pipeline systems. In the course of improving the methods of construction of gas and oil pipelines, higher requirements for materials for the manufacture of pipes are put forward [1, 2].

The increase in diameters and pressure, pipe wall thicknesses, improvement of welding technologies cause tightening not only in the requirements for mechanical parameters, chemical composition of steels, but also to the production technology of sheet metal, pipe forming processes, as well as physical means of product quality control.

The microstructure in the heat affected zone (HAZ) of welded joints of pipe steels of the line of fusion of the weld with the base metal varies continuously and is determined by the cooling rate, chemical composition and hardenability of steel, grain size and degree of homogenization of austenite by carbon content, alloying elements and microeconomic before transformation [3, 4]. The main structural changes occur in the following zones of HAZ:

- large grain area directly adjacent to the confluence line;
- areas of complete and partial recrystallization;
- heated area near AC1.

2. Materials and methods

The objects of research were materials of various zones of welded pipe joints, which can be used in the installation of designed oil pipelines. For the study, samples of pipe material were obtained, the chemical composition of which is presented in table 1.

Table 1. Results of studies of chemical analysis of base metal patterns

Pattern number		C	Mn	Si	∑(V+Nb+Ti)	N
		Mass. %				
The average values of the chemical composition of the base metal of pipes are given	1	0.12	1.32	0.22	0.12	≤0.02
	2	0.11	1.44	0.40	0.08	≤0.02
	3	0.13	1.55	0.24	0.15	≤0.02
	4	0.09	1.66	0.39	0.08	≤0.02
	5	0.11	1.47	0.28	0.13	≤0.02
	6	0.12	1.33	0.23	0.12	≤0.02
	7	0.12	1.46	0.41	0.09	≤0.02
	8	0.08	1.68	0.37	0.07	≤0.02
	9	0.10	1.65	0.44	0.08	≤0.02
	10	0.12	1.45	0.23	0.13	≤0.02
In accordance with table.5 clause. 9.2.2. GOST R ISO 3183-2009-section "Welded pipes" L-L450M or X65M (no more)		0,12	1.60	0.45	≤0.15	
		Steel grade of control sample 12Mn2VNb (12Mn2V)				

In accordance with GOST R ISO 3183-2009, the grade of material is defined as steel 12Mn2VNb.

Studies of the microstructure of the zones of welded joints were carried out using an optical metallographic microscope. The analysis of the microstructure of the base metal and weld metal of all pipes was carried out in the non-etched and etched states. The plates were manufactured in accordance with the requirements of GOST 5639 and GOST 5640.

Crack resistance tests were performed using INSTRON 8801 test equipment. Characteristics of crack resistance δc were determined by the results of tests of standard samples. Test samples made with incisions in the base metal zones (BM) and heat-affected zones (HAZ) were cut from the templets in such a way that the longitudinal axes of the samples were directed in the circumferential directions of the pipes, since their welds had longitudinal directions with respect to the pipes.

3. Results and discussion

The speed of heating and cooling during welding depends on several factors: the thickness of the metal, the welding current while maintaining the welding speed, the temperature of the metal billet, the number of passes during welding, heat input welding. Each of the above factors and their totality can significantly affect the characteristics of the structural States in each of the four selected zones in steel of different alloying.

The structure of low-alloy pipe steel is not resistant to high-temperature welding effects, so during welding in the zone of thermal influence of the weld, a number of structures with different physical and mechanical properties are formed.

During cooling after welding as a result of $\gamma=>\alpha$ transformation in low-carbon low-alloy tube steel, the structure change begins with the release of free ferrite, and an unfavorable structure of coarse-grained needle ferrite with low impact values is formed at the fusion boundary.

The most unfavorable zones in the welding of ferrite-perlite pipe steels are the fusion boundary and the zone of softening at the site of full recrystallization of the HAZ due to the formation of a significant proportion of large polygonal ferrite due to the decrease in the cooling rate associated with a high level of heat investment in high-performance multi-arc welding.

In areas of partial recrystallization (since the structure of the base metal contains in addition to ferrite and cementite), when heated in the region of $\alpha+\gamma$, the origin of austenite occurs first in the form of separate areas mainly along the grain boundaries, and then their growth. Upon subsequent cooling, the resulting austenite is completely or partially converted into a ferrite-carbide (ferrite-perlite) mixture. The areas of complete and partial recrystallization formed at a lower temperature compared to the HAZ areas located closer to the fusion boundary have the highest viscosity.

In this regard, in order to obtain satisfactory viscous properties of the pipe metal, especially in the area of negative temperatures, it is necessary to limit the linear welding energy. This limitation is caused to the fact that with increasing linear energy, the total length of HAZ increases, and the sizes of the newly formed phases and structural components increase in the process of recrystallization.

The study of the metal longitudinal welded joints of pipes for oil pipelines, aimed at assessing the microstructure and its impact on the complex properties of the material. The microstructures presented in the figure are typical Widmanstett structures with different degrees of severity and sizes of needle ferrite, which primarily depends on the cooling rate of the zone of thermal influence of the weld. It is known that as the cooling rate increases (with a constant carbon content in steels), the shape of ferrite grains gradually changes from polygonal to needle-like.

As indicated in [5], when cooled in the heat-affected zones of both steels (in the areas of overheating adjacent to the fusion boundary) ferrite needles grow from the borders to the center of the austenite grains and they have a distinct orientation: the plane of the dodecahedron ferritic plates (110) and located parallel to octahedron planes of the austenitic grain (111) and close-packed direction (111) body-centered cubic structure of ferrite parallel close-packed directions the (110) face-centered cubic lattice of austenite. This mutual arrangement of the two phases provides the best coupling between the two structures and the lowest surface energy. The figure shows typical structures of needle ferrite formed by the above mechanism – with a clear orientation of the needles in the specified crystallographic directions, the space between the needles is filled with ferrite-cementite aggregates of pearlite origin.

During the analysis of microstructures the probability of preservation of some quantity of residual austenite in Widmannstetter structures of lines of fusion of steels as it is known that in case of partial formation of bainite in mezhigolchatom space in structures of steels the residual austenite has to remain was investigated. However, the studies did not reveal traces of bainite and residual austenite in the structures of both steels.

The high cooling rate of the weld in the steel pipe No. 1 leads to the formation of a rough needle structure (see Figure 1, a), on average 4-5 points on a scale of 4 series A according to GOST 5640; while in steel pipe No. 2, due to more favorable conditions for the formation of a weld, the beam of the

IOP Conf. Series: Materials Science and Engineering **826** (2020) 012011 doi:10.1088/1757-899X/826/1/012011

Widmanstett structure (see Figure 1, b) is within 2-3 points on a scale of 4 according to GOST 5640.

Based on the studies of the microstructure of the zones adjacent to the border of fusion welds in steels (lines of fusion), we can conclude that the value of fracture toughness of fusion line of the two steels is directly dependent on technological parameters of welding, increasing the heat input the welding causes the growth of austenite grains in the line of fusion of the weld and during accelerated cooling to the formation of ferritic structures dominating in the group with increased fragility, especially in the area of negative temperatures.

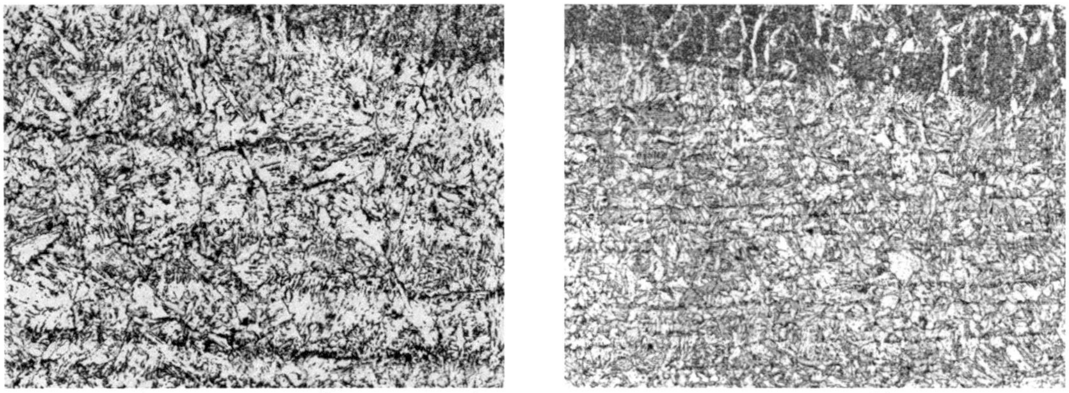

a) b)

Figure 1. Weld, fusion line and overheating area in the heat affected zone weld: a-pipe 1; b-pipe 2

The numerical values of fracture toughness of steels is a function of the size and shape of needles of ferrite grain in steel pipe № 2 with the best combination of fracture toughness – Widmannstetter structure of the line of fusion level 2-3 on a scale of 4, A GOST 5640; in the case of steel pipe № 1 with the worst indicators of fracture toughness – as a 4-5 score structure Widmannstetter.

In work tests on crack resistance of the pipes welded with different speed of movement of a welding source on technology of high-frequency induction welding in a zone of the basic metal and a zone of overheating of welded connection were carried out. The parameter δc (CTODmin) was used as a crack resistance parameter - critical opening of the crack banks in the dead-end part during its straining. A total of 5 clippings (templets) from two pipes were tested (see table).

Table 2. The results of the tests fracture toughness CTOD min, mm metal pipes and overheating zones of welded joints

№ of sample	Temperature, °C / place of incision			
	Base metal		Overheating zone	
	-10	-20	-10	-20
1-1	0.62	0.68	0.07	0.28
1-2	0.72	0.69	0.11	0.10
1-3	0.77	0.78	0.09	0.14
1-4	0.76	0.99	0.12	0.14
1-5	0.61	0.61	0.12	0.15
2-1	0.68	0.76	0.51	0.47
2-2	0.79	0.75	0.44	0.38
2-3	0.79	1.01	0.37	0.35
2-4	0.67	0.66	0.36	0.37
2-5	0.60	0.59	0.33	0.41

The results of the studies allowed us to establish that the values of crack resistance of the main material of the pipes are close in values. For the metal of the superheat zone there is a significant spread of data, which may be the result of the peculiarities of the formation of the microstructure in the zone of thermal influence due to different modes of welding processes.

4. Conclusion

The paper analyzes the influence of the metal microstructure of the longitudinal weld seam of the main gas and oil pipelines. The most common welding technology used for oil and gas pipes is the so-called "high frequency induction technology", which is the application of induction current on the outer surface of the pipe, able to create a strong weld and tightly connect the two sides of the steel billet. Multi-arc automatic welding, laser and hybrid welding are also used. By adjusting the speed of the welding source, it is possible to achieve the formation of various microstructures in the weld zone. It was established that:

1. the high cooling rate of the longitudinal seam (pipe № 1) leads to the formation in the overheating zone of the coarse needle structure Widmannstetter, estimated 4-5 points GOST 5640; while in steel pipes with low cooling rate (pipe № 2) due to more favorable conditions for the formation of the weld structure Widmannstetter score is within 2-3 points GOST 5640;
2. the fracture toughness of the base metal of both pipes are almost identical, and the difference in the magnitude of fracture toughness of metal in the zone of overheating of welded joints of pipes can only be explained by differences in the technological characteristics of the cooling tubular workpieces after welding and is connected with the peculiarities of the formation of the microstructure of the heat affected zone of a welded joint.

References

[1] Podolsky Yu 2008 *Petroleum geology. Theory and practice* **3 (4)** 84-96
[2] Kozhukhova O S 2011 *Oil and Gas complex of Russia: state and directions of development, Economic and legal issues* **7** 174-177
[3] Solntsev Yu P, Ermakov B S and others 2005 *Metals and alloys. Handbook* (SPb: Professional)
[4] Cochrane R K and Morrison W B 1985 Influence of production technology on the transformation characteristics and microstructure of some niobium-molybdenum-vanadium tube steels *Steel for gas pipes and fittings. Proceedings of the conference.* (Moscow: Metallurgy)
[5] Lyakishev N P, Shamray V F and Egiz I V 2000 *Metalls* **2** 68-72

Specificities of control over multifactorial objects in food production

V A Balyubash, S E Alyoshichev, E A Pashkova and V L Ivanov

ITMO University, 49, Kronverksky ave., St. Petersburg, 197101, Russia

E-mail: vniig-audit@mail.ru

Abstract. The research deals with control methods of multifactorial objects of food production, representing diverse technological processes occurring in continuous, discrete-continuous and discrete modes. This is especially relevant for systems in food production having the series connection of technological units, separated by apparatus-technological capacities and resistances. For such units, a transient delay value is significantly greater than that of single-capacity objects. A solution proposed: the use of two-loop automatic control systems capable of improving system quality parameters in facilities of chemical and technological processes. The two-loop automatic control system is relevant when the object is subjected to frequent disturbing influences from the regulator and has a low inertia parameter. The internal control loop suppresses arising internal disturbance caused by the specifics of the process itself, and the external loop is responsive to disturbing influences coming from outside, e.g., with changing properties of the incoming raw materials.

1. Introduction

Food production as a complex system for control represents different categories of processes occurring in discrete, continuous, and discrete-continuous modes. For example, serial connection of separate devices with differing quality parameters of raw materials, depending on the process of preparation of a biotechnological product, leads to the appearance of abrupt perturbation representing one of the most severe disturbances in the control system. It should also be noted that the series connection of technological units separated by technological capacities and resistances generates a capacitive or transient type of delay, which, together with the "net" (transport) delay, significantly exceeds the delay value in single-capacity objects.

When selecting the control system structure for the technological complex (TC) in food production, it is necessary to consider specificity of technological and equipment organization of TC which is dependent on such aspects as normalized processing time of certain types of food raw materials, its variable properties and composition, requirements for quality parameters of the finished product [1]. At the same time, modern highly efficient TC bear an increased risk of significant losses due to possible rework of the defect products.

A single-loop control system used in modern TCs, manufacturing food, is not always able to provide the required quality parameters of the finished product.

Selection of the multi-loop control system for multi-factor food production TC is preferred, taking into account the equipment and technological combination of processes that perform a full cycle of operations affecting the quality indicators of the finished product [2]. Increasing quality indicators in the transient regulation process, in this case, is provided by the reduction of the control system delay time and increased levels of regulatory effects when connecting control channels affecting the output parameters of the object [3].

2. Research methods

Figure 1 is a structural-parametric diagram of the TC producing butter, comprising of external and internal factors [4]. The external factors include cream which is the raw material for production of butter. Cream undergoes preliminary technological preparation operations, potentially providing an impact on the process of butter churning and its moisture content. Internal factors include technological operations of the transfer of incoming cream into butter with a given level of moisture content [5]. One of the main sources of disturbances is the cream coming in the butter-churning machine. This is due to the fact that at the stage of cream ripening, technological factors affecting the final moisture content in butter – a ripening temperature of the cream and a fat content in the cream – are usually uncontrollable. When the next cream-ripening tank is connected, their values may differ since these disturbances of a stepped nature.

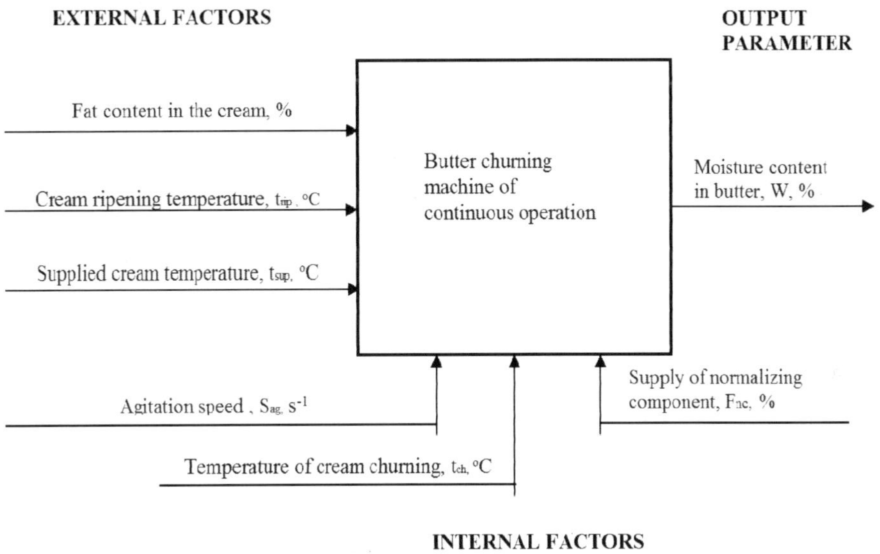

Figure 1. Structural scheme of butter manufacturing in butter churning machine of continuous operation

Factors, affecting moisture content of the final product at the stage of churning, are supplied cream temperature, agitation speed of churning and supply of a normalizing component. The control of these factors is achieved by using relevant equipment. The use of various combinations of control factors can be applied to provide synchronization of modes in the manufacturing process.

To determine the level of influence of the technological factor on 1.0% change of the moisture content, the coefficient of humidity formation (transition factor) is used. The transition factor is calculated according to the formula:

$$K_h = \frac{\Delta W}{\Delta S} \tag{1}$$

where K_h – coefficient of humidity formation, % humidity / factor measuring units;
ΔW - change in the moisture content in butter, %;

ΔS - change of the factor, factor-measuring units.

3. Research results

The results of analysis of technological factors influence on 1.0% change in the moisture content in butter within technological limits are given in table 1.

Analysis of technological factors in the production of butter determines the cumulative effect of uncontrollable factors as a disturbance. At the same time, the simultaneous deviation of the fat content in the cream and the cream ripening temperature within the technically permitted range may lead to an output deviation in moisture content as high as ± 4,0%. This situation is characterized by the transient mode butter churner work associated with the connection of another reservoir and caused by occurrence of disturbances of stepwise nature, which greatly complicates the stabilization of the moisture content of the finished product [7].

As the results of the analysis, in this case, the total control action of three factors – supplied cream temperature entering the churning, agitator rotational speed and supply of the normalizing component, potentially can be used to stabilize the butter humidity, since the level of their total influence exceeds the impact of uncontrollable factors

Table 1. Influence of factors on 1% change in the moisture content in butter within technologically permitted limits

Factors forming humidity parameter		Technologically allowed range of variation of the factor	Possible deviation in butter humidity,%	Coefficient of moisture formation (transition factor), $K_в$
External	Fat content in the cream, %	38 - 45	5.6	0.8
	Cream ripening temperature, °C	4 - 7	2.6	0.9
	Supplied cream temperature, °C	8 - 14	15.0	2.5
Internal	Agitation speed of butter churn, s^{-1}	17.4 - 30.0	13.6	1.0
	Supply of normalizing component, %	0 - 2	2.0	1.0

The proposed solution allows us to quickly compensate the deviation of moisture in the finished product caused by the disturbing influence in operational and transient operation modes of butter churn.

Analysis of equipment and technological solutions allows justifying the choice of controlled and uncontrolled process factors and setting the level of their influence on the formation of moisture of the finished product, as well as the type and nature of the disturbances at the stages of cream ripening and butter production. Considering the complexity of the process of formation of butter humidity, presence of intermittent external disturbances and delay of the equipment-technological system "cream-ripening reservoir – butter churn", we proposed a structure of multi-channel control, including realization of control actions depending on the current value of the moisture content of output butter with computing potentially expected humidity deviation of the finished product. The structure of the multi-channel control of the butter production process is represented in Figure 2.

In the operational control mode, when connecting the next reservoir to the churn, cream flows through the heat exchanger which stabilizes the predetermined temperature level before cream enters the churn. Thus, an allowable temperature drop of 3-4 °C inside the churn is additionally provided. For the mentioned two modes of operation, two separate independent control loops for temperature control of cream are used, respectively.

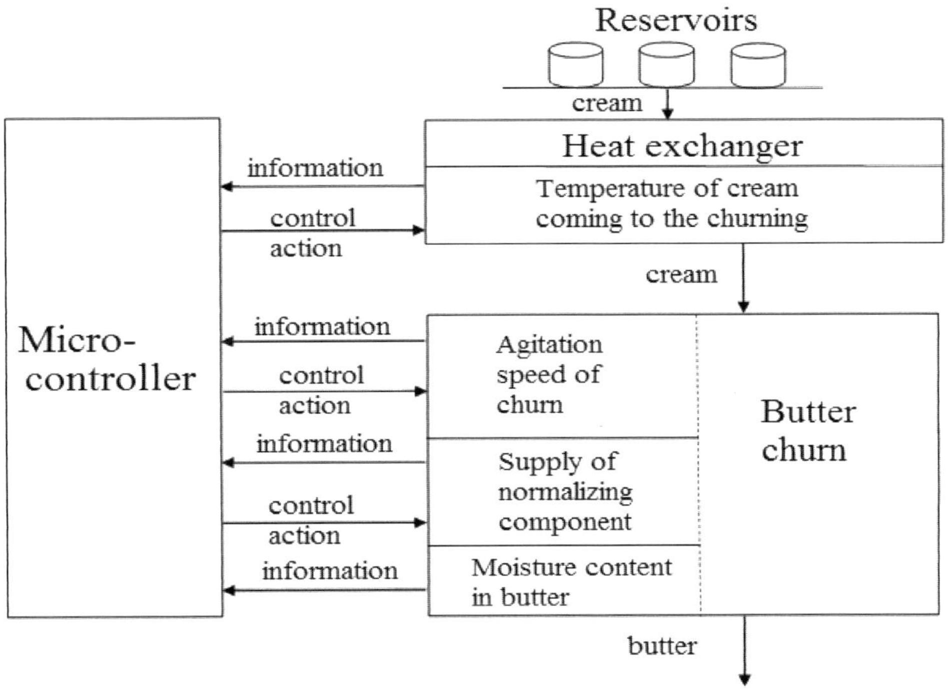

Figure 2. Structure of multi-loop control system for butter production process

In this structure of the multi-loop control system, the control actions are realized through equipment and technological channels of the butter production process. In the operational mode of butter churning, deviations in the temperature of supplied cream in the heat exchanger before entering the churn avoided because of the decreased time delay achieved by preliminary computation of expected disturbance. To increase the transition factor, the cumulative effect of the equipment and technological channels of systems interconnected with the moisture content of butter is used.

In case of uncontrolled intermittent disturbances, when connecting the reservoir with the deviation in the content of fat in cream or in ripening temperature, the cumulative effect of parallel-sequential introduction of control actions is used. In such case increased influence of the control action from the agitation of churning on stabilization of moisture content replaces control action from normalizing the component channel. Such a method of making control actions, considering the complex shape of the perturbation, provides rapid stabilization of the moisture content in butter in the final stage of production.

4. Conclusion

A structural basis of this multi-loop control scheme is a microprocessor providing formation of operations stabilizing the moisture content of butter produced in the multifactor equipment-technological complex. An adopted structural scheme in a process of performing operations stabilizing the moisture content of butter provides operating mode of control channels within technologically and hardware-recommended limits. Economic efficiency of the considered technical solutions is proved in [9].

IOP Conf. Series: Materials Science and Engineering **826** (2020) 012012 doi:10.1088/1757-899X/826/1/012012

References

[1] Tikhomirova N A 2011 *Technology of milk and dairy products. Butter technology (technological notebooks)* (SPb.: GIORD)

[2] Brusilovsky L P and Weinberg A Ya 1993 *Automated process control system for whole milk and milk-canning productions: handbook* (Moscow: Kolos)

[3] Blagoveshchenskaya M M and Zlobin L A 2005 *Information technology of process control systems* (Moscow: "Higher School")

[4] Stegalichev Yu G, Balyubash V A and Zamarashkina V N 2006 *Technological processes of food production. Textbook for high schools* (Rostov / D-SPb.: Phoenix)

[5] Grishchenko A D 1983 *Butter* (Moscow: Light and food industry)

[6] Vyshemirsky F A and Ivanova N V 2005 Cow's milk butter: standardization and terminology *Cheese making and butter making* **1** 2-24

[7] Balyubash V A and Alyoshichev S E 2007 Formation of a multi-channel system for stabilizing the humidity of butter *Cheese making and butter making* **2** 45–46

[8] Ivanov V L, Polyakov R I and Travina E A 2019 Problems and economic efficiency of automation of product quality control at food industry enterprises *Economics: yesterday, today, tomorrow* **9 (1A)** 121-134

The structure of the manipulators of industrial robots

T L Ten[1], V L Ivanov[2], G D Kogay[3], V G Drozd[1], A L Te[4]

[1] Karaganda Economic University, 9, Akademicheskaya str., Karaganda, 100009, Kazakhstan Republic
[2] ITMO University, 49, Kronverksky ave., St. Petersburg, 197101, Russian Federation
[3] Karaganda State Technical University, 56, Nursultan Nazarbaev ave., Karaganda, 100027, Kazakhstan Republic
[4] Almaty Academy of Economics and Statistics, 59, Zhandosova str., Almaty, 050035, Kazakhstan Republic

E-mail: tentl@mail.ru, vlivanov@itmo.ru

Abstract. The paper discusses the specifics of using the technology of applying the force action of servo-drives to manipulator components in accordance with the control signals from the robot control device. Consideration is given to solving the inverse kinematics of manipulators of the robot manipulator in continuous path control. Solving the inverse problem description of dependencies is formed in an analytical and tabular form.

1. Introduction
A robot is a complex technical device created on the basis of the mechatronic method, consisting of a manipulator and a control device and intended for the transport of working tools in a three-dimensional space. In robotic manipulators, single rotational or rectilinear sliding kinematic pairs of the fifth class are used [1].

The power effects of the servos on the components of the manipulator are formed in accordance with the control signals coming from the robot control device which are formed according to a predetermined capture motion as well as the position of the robot and the surrounding environment. Under the influence of the servo, the robot arm and the gripper will make very definite movement in space.

2. Materials and methods
In a robot, to perform the motor function, the manipulator is used which is a series of kinematic links coupled together by kinematic pairs (Figure 1).

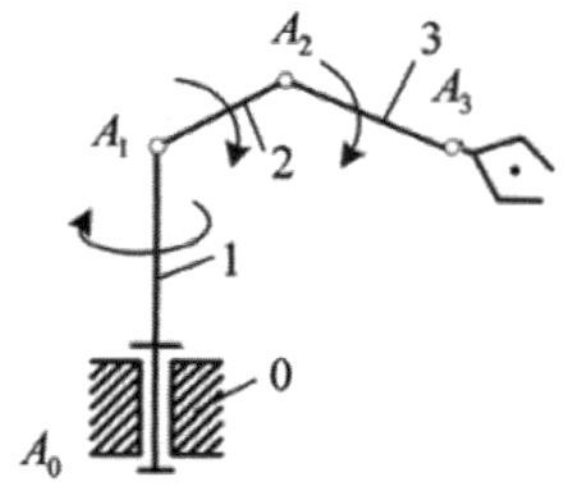

Figure 1. Manipulator based on the open chain

One of the links is stable and fixed, while others may make controlled movements due to the influence of drives.

The robot manipulator can be formed according to the principle of the open kinematic chain (Figure 2) with consecutive links (anthropomorphic humanoid robots) or on the basis of a closed kinematic chain (Figure 3) with a series-parallel connection (the robot with parallel kinematics, specifically hexapods on Stewart's platform) [2].

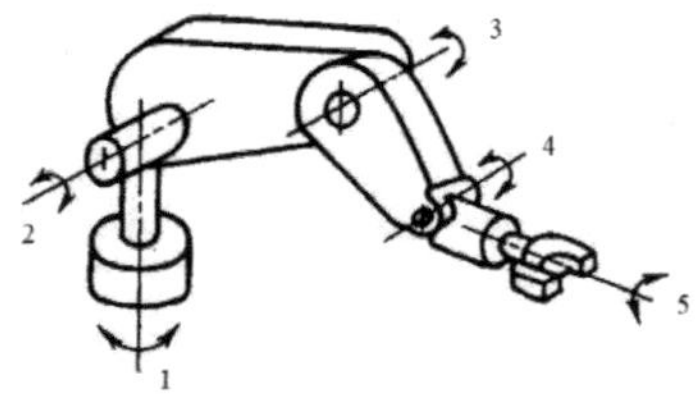

Figure 2. The scheme of the manipulator based on the open chain

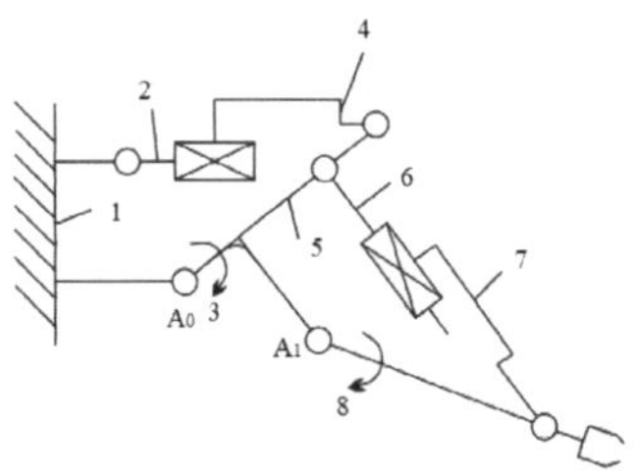

Figure 3. The manipulator with parallel kinematics

The manipulator is a mechanism that itself is a closed or open kinematic chain that is used to obtain the desired grip in the space.

Transformation laws depend on the nature of kinematic pairs and their relative positions in the manipulator.

The kinematic pairs allow some movement of the constituting parts relative to each other. This certainty is due to constraints of the pairs depending of the geometry of their elements.

Pairs are classified into classes, the number of which equals the number of constraints (the number of communication conditions) imposed by this pair:

– 1 limitation (1 connection condition) - a pair of class I;

– 2 limitations (2 connection conditions) - a pair of classes II;

– 3 limitations (3 connection conditions) - a pair of classes III;

– 4 limitations (4 connection conditions) - a pair of classes IV;

– 5 limitations (5 connection conditions) - the pair of classes V (pair of 5-th class can be translational and rectilinear sliding) [3].

One of the important characteristics of the manipulator is the number of degrees of mobility, the number of degrees of freedom.

In an arbitrary kinematic chain the formula of Luba-Malyshev can be applied: Σ

$$W = 6n - 5p_5 - 4p_4 - 3p_3 - 2p_2 - p_1 = 6n - \sum_{i=1}^{i=5} i p_i, \quad (1)$$

where p_i - the number of kinematic pairs of the i-th class.

For robotic manipulators with kinematic pairs of the 5-th class [4]:

$$W = 6n - 5p_5. \quad (2)$$

Solution of the inverse kinematics problem of manipulators is one of the main tasks of the kinematic and dynamic analysis and synthesis of manipulators.

The inverse problem is solved by a continuous path control of the robot. In case where the gripper must move along a path defined in time and space, we need to find values of the generalized coordinates of the manipulator in accordance with a predetermined position of the gripper.

Given there is a law of gripper motion, i.e. the rules of changing its coordinates and orientation angles:

$$\left.\begin{array}{lll} x_{0n} = x_{0n}(t); & y_{0n} = y_{0n}(t); & z_{0n} = z_{0n}(t) \\ \widehat{x_0 z_n} = f_1(t); & \widehat{y_0 z_n} = f_2(t); & \widehat{x_0 y_n} = f_3(t) \end{array}\right\}. \quad (3)$$

Consequently, we can define a matrix, setting positions and orientations of the gripper:

$$T_{0n}^{\Pi} = \begin{vmatrix} a_{11}^{\Pi}(t) & a_{12}^{\Pi}(t) a_{13}^{\Pi}(t) & a_{14}^{\Pi}(t) \\ a_{21}^{\Pi}(t) & a_{22}^{\Pi}(t) a_{23}^{\Pi}(t) & a_{24}^{\Pi}(t) \\ a_{31}^{\Pi}(t) & a_{32}^{\Pi}(t) a_{33}^{\Pi}(t) & a_{34}^{\Pi}(t) \\ 0 & 0 \quad 0 & 1 \end{vmatrix}, \quad (4)$$

where

$$a_{12}^{\Pi}(t) = cos[\widehat{x_0 y_n}];\ a_{13}^{\Pi}(t) = cos[\widehat{x_0 z_n}];\ a_{23}^{\Pi}(t) = cos[\widehat{y_0 z_n}]; \quad (5)$$

$$a_{14}^{\Pi}(t) = x_{0n}(t);\ a_{24}^{\Pi}(t) = y_{0n}(t);\ a_{34}^{\Pi}(t) = z_{0n}(t). \quad (6)$$

We suppose that at some k-th point of time t_k (k = 0, ..., K), the predetermined gripper position: $a_{12}^{\Pi,k}(t_k), a_{13}^{\Pi,k}, \dots, a_{34}^{\Pi,k}(t_k)$ coincides with the actual position supported by the current values q_i^k of generalized coordinates [5].

Then:

$$T_{0n}^{\Pi,k}(t_k) - T_{0n}^{Q,k}(q_1^k, \dots, q_n^k; t_k) = 0, \quad (7)$$

whence:

$$a_{12}^{\Pi,k}(t) - a_{12}^{\Pi,k} = 0,\ a_{13}^{\Pi,k}(t) - a_{13}^{\Pi,k} = 0, \dots, a_{34}^{\Pi,k}(t) - a_{34}^{\Pi,k} = 0. \quad (8)$$

Generalized coordinates, at which the equivalence of the matrices T_{0n}^{Π} and T_{0n}^{Q} is achieved, are taken as the value q_i^{k+1}tk + 1, corresponding to time t_{k+1}.

Thereafter position (k +1) is taken as the k-th and the next new position of the manipulator, i.e. a mismatch again introduced into the resulting equation, which should be removed after the determination of the following values of the generalized coordinates. This continues until the moment when the values are defined for each of generalized coordinates required by positions K of the manipulator at a predetermined trajectory.

Let us define the criterion function. Since the aim of the solution is to find such generalized coordinate values, for which the difference between this position of gripper and its actual position is equal to zero, the parameter reflecting this difference should be taken as a criterion.

In this regard, for the automatic execution of the probable range of changes of generalized coordinates, the penalty function is introduced in the form of the of the following restrictions on the values q_i:

$$F_i^1 = \begin{cases} 0, if\ q_i - q_i^{min} \geq 0; \\ W_i^1(q_i^{min} - q_i), if\ q_i - q_i^{min} < 0, \end{cases} \quad (9)$$

$$F_i^2 = \begin{cases} 0, if\ q_i^{max} - q_i \geq 0; \\ W_i^2(q_i - q_i^{max}), if\ q_i^{max} - q_i < 0, \end{cases} \quad (10)$$

where W_i^1. W_i^2 - the weight of the penalty function, which can adjust its slope [6].

In addition to these, there can be other limitations.

In the role of the objective function, one can use a function that is the sum of test functions and penalty function:

$$K(q_1, \dots, q_n) = \Delta + \sum_{i=1}^{i=n} F_i^1 + \sum_{i=1}^{i=n} F_i^2. \quad (11)$$

We will use as a method of finding new values of generalized coordinates, the q method of the gradient which is expressed by the following relation:

$$q_i^{m+1} = q_i^m - h_i \cdot \frac{\Delta(q_1^m,\dots,q_i^m+\Delta q_i,\dots,q_n^m) - \Delta(q_1^m,\dots,q_i^m,\dots,q_n^m)}{\Delta q_i}, \quad (12)$$

where h_i - a step for the i-th generalized coordinate; Δq_i - a small increment of the i-th generalized coordinate, used for finding the partial derivative of q_i.

It should be noted that the old value q_i^k is used as a first approximation $q_i^m (m = 1)$, i.e. at the initial moment $q_i^1 = q_i^k$ is supposed.

The computational process of approaching the values of q_i^{k+1} may be interrupted by any attribute used in the nonlinear programming. In particular, it can be supplemented here by condition $|q_i^{m+1} - q_i^m| \leq [\delta_i]$, here δ_i is the value of the deviation of the i-th generalized coordinate permitted by tolerance accuracy of the trajectory.

Given there is the nature of the function (11), the final value should be taken as:

$$q_i^{k+1} = 0{,}5 \cdot (q_i^{m+1} + q_i^m). \quad (13)$$

Let us note that the (m + 1) th step to the point (k + 1) in the calculation process may be performed using various algorithms typical of nonlinear programming techniques. Using a method of the gradient step (m + 1) is performed simultaneously on all the coordinates after calculating the orientation anti-gradient movement.

To do this, we will use the central subtraction:

$$\dot{q}_i^k = \frac{q_i^{k+1} - q_i^{k-1}}{2 \cdot \Delta t}; \ \ddot{q}_i^k = \frac{q_i^{k+1} - 2 \cdot q_i^k + q_i^{k-1}}{\Delta t^2}. \quad (14)$$

Thus, functions $\dot{q}_i^k$. $\ddot{q}_i^k$ will be also obtained in a tabular form. To form the control action as continuous functions, rather than a table, it is important to bring the table values of the generalized coordinates, velocities and accelerations. To do this we can use, for example, Lagrange interpolation equation:

$$q_i(t) = q_i^0 \cdot \frac{(t - t_1) \cdot (t - t_2) \cdot \ldots \cdot (t - t_K)}{(t_0 - t_1) \cdot (t_0 - t_2) \cdot \ldots \cdot (t_0 - t_K)} + +q_i^1 \cdot \frac{(t - t_0) \cdot (t - t_2) \cdot \ldots \cdot (t - t_K)}{(t_1 - t_0) \cdot (t_1 - t_2) \cdot \ldots \cdot (t_1 - t_K)} +$$

$$+q_i^2 \cdot \frac{(t - t_0) \cdot (t - t_2) \cdot \ldots \cdot (t - t_K)}{(t_2 - t_0) \cdot (t_2 - t_1) \cdot \ldots \cdot (t_2 - t_K)} + \cdots + +q_i^K \cdot \frac{(t - t_0) \cdot (t - t_2) \cdot \ldots \cdot (t - t_{K-1})}{(t_K - t_0) \cdot (t_K - t_1) \cdot \ldots \cdot (t_K - t_{K-1})}. \quad (15)$$

3. Conclusion

As a result, a continuous function $q_i(t)$ was found, which will pass through the points q_i^k at fixed moments in time k.

References

[1] Novikov V A, Savva S V and Tatarintsev N I 2014 *Electric drive in modern technologies* (Moscow: Academy)
[2] Ivanov A A 2011 *Fundamentals of Robotics: Textbook.* (Nizhny Novgorod: NSTU named after R.E. Alekseeva)
[3] Pavlov A N and Nikiforov S O 2006 Requirements for the movements of manipulated objects and methods for implementing devices for their movements *Vestnik mashinostroeniya* **3** 14-18
[4] Borisov O I, Gromov V S and Pyrkin A A 2016 *Methods for managing robotic applications Textbook* (Spb: ITMO University)
[5] Featherstone R and Orin D 20002 *Robot Dynamics: Equations and Algorithms, Proceedings of the 2000 IEEE International Conference on Robotics & Automation* (San Francisco, CA)
[6] Biagiotti L and Melchiorri C 2008 Trajectory Planning for Automatic Machines and Robots (Berlin, Heidelberg: Springer-Verlag)

The cost of liquid nitrogen for WBC sessions

A V Vasilenok, A Y Baranov, T A Malysheva, O A Filatova, L R Kutyanova, A M Andreev and Y A Solonina

Information Technologies, Mechanics and Optics University, 9, Lomonosova str., St. Petersburg, 191002, Russia

E-mail: vasilenok_anna@mail.ru

Abstract. The first unit for the whole-body cryotherapy (WBC) used liquid nitrogen (LN). Liquid nitrogen has a low boiling point (-196 °C) and a high heat sink capacity (HSC). The HSC of LN is 260 kJ/kg at a temperature level of -130°C. Approximately 600 kJ of heat should be removed from the body surface of a patient during a WBC session. The average heat flux from the body surface makes up 3.4 kW. The use of liquid nitrogen makes it possible to remove heat without complicated equipment. Refusal to use LN is caused by the temperature in the WBC zone (-110 °C). This fact reduces the efficiency of WBC procedures. Cryoagent consumption is 5 times less than the demand in some units with LN coolings.

1. Introduction

The apparatus for the whole-body cryotherapy (WBC) is the most expensive type of the cryomedical equipment. High costs for the acquisition and operation of WBC units limit the availability of cryotherapy procedures for the public. In 1985, the approximate price of the first Japanese cryotherapy unit was $ 2 million. This was approximately 6 million dollars in 2018 [1]. The high cost of original equipment has become an insurmountable barrier for European consumers. Due to the high price of Japanese WBC units their purchase by European clinics was not even considered. Instead, European companies tried to make similar devices on their own. A lot of current WBC problems were initiated by this decision since European developers had only the superficial information about the device and the operating principle of the original equipment. As shown in the 21st century studies, the algorithm of the nitrogen cooling system of the units used by T.Yamauchi to implement the technology was not taken into account in the design. The change in the basic operating principles of the nitrogen cooling system (NCS) did not allow reproducing the temperature regime of the original WBC units. Thus, the European WBC devices resembled the original units but did not maintain the optimum temperature in the main treatment cabin.

In addition, the price of European WBC equipment remained too high. During operation of these units with NCS a lot of liquid nitrogen was consumed. The LN costs significantly increased the cost of procedures. The desire to reduce the operating costs and make the procedures more affordable resulted in appearance of cryotherapy units with “nitrogen-free cooling systems” (NFCS) in Europe in 1990 - 1995. The NFCS is based on refrigeration machines. The independence from LN supply has greatly simplified the operation of the WBC units. The operating costs have decreased significantly. However, the gas temperature in the main cabin does not drop below 170 K in the WBC units with the NFCS.

IOP Conf. Series: Materials Science and Engineering **826** (2020) 012014 doi:10.1088/1757-899X/826/1/012014

This is considerably higher than the air temperature in the original Japanese unit (100 K). The fact that the air temperature in the WBC zone has increased by a factor of 1.7 has fundamentally changed the effect of the procedure on a patient's body. The effectiveness of the WBC procedures has become 10 times lower [2].

At the current stage of WBC development the NCS technology is the only workable version of heat removal from the WBC zone. To substantiate the advantages of nitrogen cooling systems, the operating principle of the original Japanese WBC unit is to be considered in detail.

2. Nitrogen cooling system of multi-seat WBC units

The NCS was used on the first units for the WBC procedures. These devices were designed to carry out WBC group procedures. There were from 5 to 10 patients at the same time in the cabin. The design of a multi-seat WBC unit is identical to the design of a cold room for long-term food storage. To maintain a cryogenic temperature level, liquid nitrogen was supplied to the heat exchangers of multi-site WBC units instead of freon. The use of the NCS made it possible to abandon energy-intensive equipment. The energy required to conduct the WBC procedures is expended in liquid nitrogen plants. To hold WBC exposure for one patient, it is necessary to remove about 600 kDj of heat within 3 minutes at a temperature level no higher than 140 K. Under such conditions the average heat flux from the patient body is 3.4 kWt. For removal of such heat flux at the temperature level of 140 K a refrigerator with an electric drive power of 13 kW is required. The use of LN eliminates the need for the WBC owner to connect the powerful cooling systems to the electric power network and to operate them. This is particularly relevant for the multi-seat WBC systems [3].

The multi-seat WBC units with the NCS provided first data on the therapeutic possibilities of cryotherapy. The information about the possibility of successful treatment of rheumatoid arthritis (RA) was of special interest to doctors. The WBC procedures relieved patients from pain and stiffness in joints for 6 hours. T. Yamauchi, the author of the WBC method, used up to three cryotherapy procedures per day to treat the RA. Thanks to the effect of the WBC procedure the patients could do remedial exercises which in turn provided RA treatment [4].

Today the multi-site WBC units with the NCS are used only in Poland. Most of the multi-site units are cooled with the NFCS. The refusal to use the NCS took place for economic reasons. The use of the NFCS led to a sharp increase in the air temperatures in the WBC zone which fundamentally changed the cold air effect on the patient's body. The recent research of the WBC with the NFCS does not show a lot of positive WBC effects described when using the units with the NCS. Thus, the WBC method is discredited.

The NCS is widely used for the WBC procedures in single units. These units were developed in Russia at the end of the twentieth century. The single WBC units with the NCS are known under the "cryosauna" trade name. The NCS is the basis for the competitive and therapeutic benefits of the cryosaunas.

Twenty-year clinical application of the cryosaunas has shown that their therapeutic effectiveness is similar to that of the original Japanese units. This is related to the high heat dissipation capacity of the NCS. The current WBC units with the NCS made in Poland are 2 to 2.5 times less effective as compared to the original Japanese units. This loss of efficiency is connected with modification of the operating principle of the NCS and inadequate consumption of LN.

With the similar operating principle of the NCS the WBC technology of the cryosaunas differs significantly from that of the multi-seat devices. It is useful to determine these differences for further development of the WBC technology.

2.1. Cooling system for Japanese WBC units

Published works about the Japanese WBC units operating mode contain conflicting information concerning the temperature in the main low-temperature chamber.

There are different data on the temperature in the publications: -186°C (87 K), -175°C (98 K) and -160°C (113 K) [5]. This uncertainty confuses the issue of the temperature conditions of the device. Taking into account the complex algorithm of the device, it is unclear at which procedural stage the temperature is measured. An inaccurate description of the temperature conditions has caused the errors in the assessment of the main factor of cryotherapy – the gas temperature in the WBC zone. Cryogenics specialists suppose that it is impossible to maintain the temperature of 87 K in the treatment cabin. The surface temperature of the cooling coil of the heat exchanger should be at least 10 K lower than the air temperature (77 K). However, the surface temperature of the heat exchanger shall be not less than 82 K to prevent air condensation on the heat exchanger surface. When designing the European cryotherapy systems, the nominal air temperature in the WBC zone was assumed to be -160°C (113 K). This decision led to a number of negative effects which significantly reduced the effectiveness of the WBC units manufactured in Germany and Poland.

A universal mathematical model of the WBC unit was developed at ITMO University which allowed simulating the full cycle of operation of the multi-site WBC unit with the NCS. It is found that the air temperature in the main (low-temperature) cabin of the WBC device can be reduced to the level of 85-87 K. The reduction can be achieved using the minimum thermal load on the heat exchanger with the NCS. The minimum thermal load is observed when patients are absent [6].

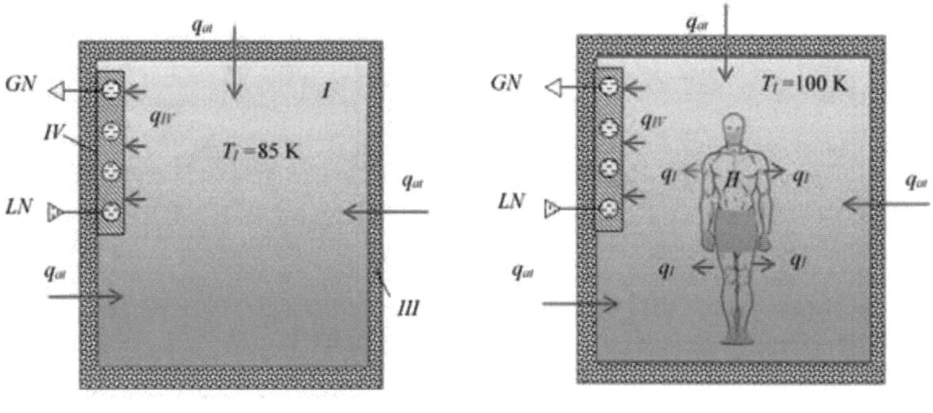

Figure 1. Temperature conditions of the main cabin of the Japanese unit for the WBC in the absence of patients and in the presence of patients.

The figure (Figure 1) explains the operating principle of the NCS under changing conditions of the thermal load. Cold air I is located inside the insulated cabin II. Heat flow with a density of q_{at} = 20 W/m^2 penetrates from the atmosphere through the thermal insulation of the cabin II. Considering the standard dimensions of the multi-cabins for the WBC (2x2x2 m), no more than 500 W of heat penetrates through all walls of the cabin II. This heat warms up the air I. Natural convection arises in the cabin. Air movement transfers heat supplied from the atmosphere to the surface of the heat exchanger IV. The heat exchanger IV consists of a large number of metal plates cooled by pipes passed through them. Liquid nitrogen is inside the pipes. The LN boiling point depends on the pressure in the heat exchanger pipes.

To ensure that the air does not condense on the heat exchanger plates IV, the temperature of the plates shall be higher than the air condensation temperature $T_{HE}>T'_{Ai} = 81$ K. To fulfil this condition, the nitrogen vapour pressure in the pipes of the heat exchanger IV is to be more than 0.05 MPa. The boiling point of nitrogen is $T''_{LN}> 81$ K. The surface area of the heat exchanger IV is calculated taking into account the maximum thermal load which increases 20-30 times in the presence of the patients. With a low thermal load, the air temperature drops to the values close to the surface temperature of the heat exchanger IV which tends to equilibrate with LN boiling in the pipes $T_{Ai} \rightarrow T_{HE} \rightarrow T''_{LN}> 81$ K. Under such conditions, the air temperature inside the main cabin of the multi-seat WBC unit can drop to 85 - 87 K. The operation of these devices starts with a cooling mode which takes 2-3 hours. During this mode the NCS cools down the cabin walls II. At the end of the cooling mode the air temperature reaches its minimum value and the temperature of the inner surface of the cabin walls is close to the air temperature $T_S \rightarrow T_{Ai} \rightarrow T_{HE}> 81$ K. After completion of the cooling process the unit goes into the procedural mode in which the thermal load on the NCS depends on the presence of the patients in the cabin. As a rule, the duration of patients' stay in the cabin does not exceed 50% of the total duration of the procedural cycle. The thermal load is reduced and the NCS restores its temperature equilibrium reached in the cooling mode.

The thermal equilibrium of the cabin is disturbed at the moment of patient II entry (Fig. 1). The warm air from the intermediate chamber ($T_{A1} = 210$ K) penetrates through the open doors of the cabin when the patients enter. As a result of air movement in the main cabin, the average temperature is set $T_{Ai}\approx 148$K.

Owing to the high temperature difference between skin and air ($\Delta T_{Sk\text{-}Ai}\approx 150$ K) a significant amount of heat is released from the surface of the patients' bodies. The heat intake from one patient is more than 3 kW. Due to the penetration of the warmer air from the pre-chamber and the heat supply from the patients' bodies the air temperature in the main cabin and the heat load on the heat exchanger IV are sharply increased. There is a reserve of liquid nitrogen in the heat exchanger pipes which maintains a constant temperature of the metal pipes. Owing to this the temperature of the heat exchanger plates IV does not rise significantly and the temperature difference between the heat transfer plates and air increases to $\Delta T_{Ai\text{-}HE}\approx 75$ K. With such temperature gradient, the intensity of heat removal from the air in the cabin increases 20 to 30 times which results in gradual lowering of the temperature. The thermal equilibrium can be established within 30-60 s at the sufficiently large area of the heat transfer surface of the heat sink surface. This process is somewhat facilitated by the fact that the temperature of the cabin walls III is lower than the air temperature at this stage. In the empty cabin the air temperature is about 87 K, so the inner surface of the walls has a temperature lower than 95 K. The heat transfer to the surface of the heat exchanger IV and the cabin walls III reduces the air temperature from 150 K to 100 K [7].

The air temperature in the main cabin of the WBC unit is set at 87 K when the cabin is empty and at 100 K when the patients are in the cabin. The algorithm of the air temperature change in the WBC zone of Japanese units can be reproduced in practice, so the doubts about the reliability of the air temperature data in the zone given by T.Yamauchi are unreasonable.

Due to the fact that the air temperature reaches a minimum permissible level in the empty cabin, it is possible to compensate for the sharp increase of the thermal load at the moment when the patients enter the cabin. The heat exchanger pipes IV are constantly filled with liquid nitrogen, so the increase of the thermal load is compensated by intensive boiling of the cryoagent.

The use of the NCS in the first WBC units was critical for achievement of the therapeutic effect. Due to the large heat removal capacity the NCS compensated for all design and manufacturing imperfections of the multi-site WBC systems.

The thoughtless reconstruction of the nitrogen cooling systems has led to deterioration of European multi-seat systems.

2.2. European multi-seat units with NCS for WBC
In the design of the European units with the NCS the nominal temperature in the WBC zone was assumed to be -160°C (≈ 115 K), i.e. above the temperature level in the Japanese units at the time of the procedure. The increased nominal temperature has significantly changed the operating principle of the NCS. To maintain the air temperature at 115 K in the empty cabin, LN should be fed into the heat exchanger IV in portions. The heat transfer from the air to liquid nitrogen is carried out only in some part of the heat exchanger IV. It ensures thermal equilibrium at a given temperature level. LN supply is controlled by the temperature control system (TCS) which was unavailable in the Japanese units. When LN is fed in portions the heat exchanger pipes are periodically cooled to a level of 81 K. Pipe cooling leads to cooling of the heat exchanger plates and to the increased removal of heat from the air in the cabin. The air temperature drops below 115 K and the TCS stops supplying LN to the heat exchanger. In the empty cabin, when the heat input to the heat exchanger IV is small, LN is virtually not supplied. As a consequence, the temperature of the plates and pipes of the heat exchanger tends to the nominal air temperature. The temperature difference between the inner surface of the pipes and boiling LN can exceed 16 K [8].

When the patients enter the cabin the air temperature sharply increases to 160 K. The TCS resumes the supply of LN to the heat exchanger. Due to the fact that the temperature of the heat exchanger pipes exceeds the boiling point of LN more than 16 K, cryoagent boiling goes to the ineffective "film" mode. A vapour layer is formed between LN and the inner surface of the pipe wall which poorly transfers heat. The efficiency of the heat transfer to LN is 10-20 times lower and the cryoagent has no time to evaporate in the heat exchanger pipes. A combination of the above-mentioned negative factors may result in the air temperature increase in spite of the supply of LN to the heat exchanger.

Due to the low-efficiency NCS operation with nitrogen control using the TCS, the air temperature does not fall below 160 K in the presence of the patients. The WBC procedure is performed at the air temperature 55 K higher than that in the original Japanese units. The inefficient boiling mode results in the large LN consumption which requires additional financial costs.

The shortcomings of the NCS operation described above led to the development in the 1990s of "nitrogen-free" cooling systems which could maintain the nominal temperature not exceeding 160 K. The NCS continued to use only in WBC units made in Poland where the nominal temperature was 170 K. Nowadays the researchers note that some WBC effects described by T. Yamauchi are reproduced with the Polish NCS units, while application of the units with the NFCS makes it possible to achieve the WBC efficiency comparable to that of hypothermic procedures [9].

2.3. Liquid nitrogen consumption estimation for multi-seat WBC units with NCS
Another reason for the low efficiency of the multi-site WBC units with the NCS is insufficient supply of the cryoagent. NCS developers underestimate the intensity of heat transfer from the patient's body surface. When calculating the LN consumption, only the heat is taken into account which a living organism produces by means of metabolic reactions. At a daily calorie intake of 3000 kcal the heat flow from the body to the environment is only 150 W. It appears from the NCS characteristics of the modern WBC units that the designers increase this value by half when evaluating the heat production by the patients. In fact the metabolic heat is less than 5% of the total heat release from the patient's body surface. The main heat supply is provided by heat capacity of skin. One patient emits up to 600 kJ of heat within 200 seconds, which creates a thermal load of 3 kW.

It is not hard to calculate that during the WBC 5 patients release 3000 kJ of heat to the treatment cabin volume. 15 kg of LN are required so that the NCS can compensate for this thermal load. In the multi-seat WBC units the heat emission by the patients provides only 50% of the thermal load on the NCS, so each procedure requires 30 kg of LN. This condition is not satisfied in practice.

The Polish unit KR-2005N designed for 5 patients carries out 10 WBC procedures per hour. As reported by the manufacturer, the LN consumption is only 100 kg / h (10 kg per a procedure). Inadequate LN supply reduces the NCS efficiency. When the patients enter the cabin the temperature in the WBC zone rises by 40-50 K which reduce the procedure efficiency [10].

To restore the operating efficiency of the WBC units with the NCS, the LN consumption shall correspond to the heat load.

3. Conclusion

The multi-seat units are mostly used for the WBC procedures in Western Europe. In recent years a growing number of articles has been published claiming that the WBC provides no curative effects. The reason for these statements is that the WBC devices with the NFCS with a nominal temperature of not less than 160 K are used for the studies.

The researchers who apply the devices with the NCS find some WBC effects but the duration of such effects is 2 to 3 times lower than expected.

The reason for the low efficiency of the WBC units with the NCS is misuse of LN.

To restore the efficiency of the multi-seat WBC units, the NCS is to be used which maintains the air temperature of 90 K in the empty treatment cabin.

It is necessary to develop LN logistics technologies, to automate the cryoagent supply systems and to use large volume tanks for the cryoagent.

The use of the secondary sources of cooling capacity for cheap cryoagent production would be the right decision so as to reduce the costs for liquid nitrogen purchase.

References

[1] Baranov A, Filatova O, Vasilenok A, Safonov A and Sokolova E 2019 *J. of International Academy of Refrigeration* **1 (70)** 92-97

[2] Yerezhep D, Tukmakova A, Fomin V, Masalimov A, Asach A, Novotelnova A and Baranov A 2018 *J. Phys.: Conf. Ser.* **1015** 032151 DOI: 10.1088/1742-6596/1015/3/032151

[3] Cholewka A, Stanek A, Sieroń and Drzazga Z 2012 *Skin Res Technol* 180–187

[4] Greenwald E, Christman M, Penn L, Brinster N and Liebman T 2018 *JAAD Case Rep.* **4(4)** 344-345

[5] Rymaszewska J, Urbańska K, Szcześniak D, Stańczykiewicz B, Trypka E and Zabłocka A 2018 *CryoLetters* **39 (3)** 166-176

[6] Savica M, Fonda B and Sarabona N 2013 *J. of Thermal Biology* **38 (4)** 186-191

[7] Yamauchi Y, Yamauchi T and Miura K 1987*curable/ IASP Pain* **30 (1)** 261

[8] Carrard J, Lambert A and Genné D 2017*BMJ Case Rep* DOI: 10.1136/bcr-2017-221431

[9] Engel P, Fricke R, Taghawinejad M and Hildebrandt G 1989 *Phys Med Baln Med Klim.* **18** 37-43

[10] Krueger M, Costello J T, Achtzehn S, Dittmar K H and Mester J 2018 *Cytokine* **10** 751-58

Strength of structural elements with cracks

A A Peregudov, S A Vologzhanina and A F Igolkin

ITMO University, 49, Kronverksky ave., Saint-Petersburg, 197101, Russia

E-mail: aaperegudov@itmo.ru

Abstract. In almost every detail, despite careful control at all stages of production, there may be cracks. The reasons for their occurrence can be very different, including the shape and size of the part, the conditions and nature of loading, the nature of the environment, defects of a crystal lattice, the possible presence of non-metallic inclusions adverse forms, structure and grain size, depending on the technology upstream processing, fatigue, etc. The reliability of the equipment is determined by the correct choice of material and taking into account the operating conditions, including the level and nature of loading. The paper provides a comparative review of the main approaches for selecting strength criteria for structural elements with cracks. The energy and power strength criteria are considered. The energy criterion of A. A. Griffiths considered a perfectly brittle material without taking into account the plasticity reserve of real metal materials. The force criterion allows us to estimate the value of the stress taking into account the critical moment of loading, at which unstable destruction begins due to the reserve of elastic energy. It is shown that the transition of steel from a plastic state to a brittle one is influenced by the test temperature, the type and speed of deformation, and changes in the chemical composition and mode of heat treatment. Thus, when choosing a certain strength criterion, you should take into account the operating conditions and the degree of responsibility of the structure.

1. Introduction

The influence of cracks on the strength of structural elements is an urgent issue, as currently recorded a large number of cases of sudden destruction of parts, and the destruction took place at stresses below the yield strength of the material. It is impossible to guarantee the complete absence of cracks in the structures, despite the control at all stages of manufacture [1-3].

The reasons for the appearance of cracks can be very diverse: defects in the crystal lattice, technological origin, fatigue, etc. Since there is a high probability of the presence of cracks in the products, the reliability of the product can be solved in two ways: either by choosing a material that will "hold" the cracks - such that the cracks will not spread under operating loads, or by determining the level and nature of the loading. A special danger is the spread of cracks in those cases when it is committed as if spontaneously, without additional external load. This pattern of crack spread is called unstable [4, 5]. During the research, the strength criteria of structural elements with stress concentrators were considered. In developing the strength criteria, the greatest attention was paid to brittle fractures, which are the most dangerous.

One of the first attempts to obtain such a criterion belongs to A. A. Griffiths [6], who considered the question of the strength of a plate of unlimited dimensions and brittle material containing an elliptical crack.

Considering the balances of energy consumption and arrival at the appearance of a crack in a stretched and fixed plate allowed Griffiths to obtain an expression for the critical stress σ_{kr}, above which the unstable crack spread begins.

In Griffiths' solution, the material was considered to be perfectly brittle, so the crack propagation work equated to surface energy. In real metal materials, plastic deformation occurs during the propagation of cracks, which greatly increases the work of crack propagation, or, as they say, the fracture toughness.

In this regard, E. Orovan [7] proposed to unify the expression for σ_{cr}. According to Orovan, the critical stress formula has the form:

$$\sigma_{cr} = \sqrt{\frac{2\,EG_c}{\pi\, l}}\,, \tag{1}$$

where l is the crack length. The formula (1) can be used to determine the critical length of the l_{cr} crack at which its unstable propagation occurs:

$$l_{cr} = \frac{2\,EG_c}{\sigma_{cr}^2\,\pi}\,. \tag{2}$$

The idea of creating a force criterion is to determine the stress at the critical moment of loading, when unstable destruction begins due to the reserve of elastic energy [3-5]. Expressions for stresses in the vicinity of the dead-end of the crack were obtained on the basis of works when considering a plate of an elastic material of infinite width with a through crack subjected to uniform stretching. In the case of a plane deformation at a triaxial stress state the stress on the crack extension will be equal to:

$$\sigma_r = K_I\sqrt{2\pi r}, \tag{3}$$

where: K_I - stress intensity factor,

r – the distance from the top of the crack to the point in question on the continuation of the crack.

The stress σ_r determined at the critical moment of loading (i.e. at failure) could play the role of a force criterion. However, research have shown that when changing the size of the crack, the constant is not σ_r, but the stress intensity coefficient, experimentally determined at the critical moment of loading and is therefore a characteristic of the strength of the material, in the case when the structural element has cracks. This coefficient is denoted by K_{Ic}.

Expression (3) was obtained from the fracture condition at the separation displacement of the crack banks. Apart from the aforementioned case of shifting can be implemented well as transverse shear, in which the surface of the crack slide over one another in a direction perpendicular to the front edge cracks and longitudinal shear, where the crack surface slide over each other parallel to the front edge of the crack. For transverse and longitudinal shear, the stress intensity coefficient is denoted by respectively K_{IIc} and K_{IIIc}.

For a linear-elastic material, a fairly strict theory describing the stress state at the crack apex and is a good foundation for the analysis of brittle fractures is currently applied. Unfortunately, there is no such theory for describing the stress and strain state at the top of a crack with a developed plastic zone, since the solution of this problem for a material with hardening encounters insurmountable obstacles. Meanwhile, the vast majority of practically important and dangerous fractures are accompanied by the constant presence of a plastically deformed region at the top of the crack. This plastic deformation affects the shape of the tip of the crack, leads to smoothing of the stress peak, and the properties of the material being destroyed. When a sample with a crack is deformed, a plastic zone is first formed at the end of the crack, the length of which grows to some relatively small size d, after which a local rupture occurs at the beginning of this region coinciding with the end of the crack. The crack begins to move, and the plastic region moves with the moving end of the crack without changing its size and shape. At a distance l > d from the end of the crack, the deformations will only be elastic.

Evaluation of the effective surface energy in the destruction of metals and its comparison with the true surface energy shows that the first can exceed the second by several orders of magnitude, thus determining the exceptional importance of taking into account the role of local plastic deformation at the crack tip.

The fact that the destruction of the sample in determining the intensity coefficients in most cases is plastic is the reason for the sensitivity of these coefficients to a number of factors that contribute to the transition of steel from a plastic state to a brittle one. Such factors include the test temperature, the type of deformation, changes in the chemical composition and mode of heat treatment, the rate of deformation.

Increasing the rate of application of the load affects the value of K_{ID} differently in different materials (K_{ID} -the value K_{Ic}, obtained by the dynamic application of forces). In some cold-breaking steels, the K_{ID} is lower thanK_{Ic}, i.e., the dynamic coefficient is lower than the static one. In other steels, especially non-cold-breaking steels, the stress intensity coefficient K_{ID} increases with increasing loading speed. In the first case lowering K_{ID} – the result of brittleness, in the second case, the increase in the K_{ID} is the result of increasing the resistance of plastic deformation of a plastic area. In this case, the dependence $\sigma = f(\dot{\varepsilon})$ is reflected in the K_{ID} value, since:

$$K_{ID} = \sigma * \sqrt{\pi e} = f(\dot{\varepsilon}) * \sqrt{\pi e}.$$

Based on the research, it was concluded that the choice of a certain strength criterion depends on the operating conditions and the degree of responsibility of the structure.

2. Materials and methods

Metastable austenitic steels of grades 10Cr14Mn14Ni4Ti and 10Cr14NMn20 used for the manufacture of equipment operated including at low temperatures were chosen as the object of the study [8]. The chemical composition of the studied steels is given in table 1.

Table 1. Chemical composition of low-temperature steels

№	Grade of steel	Content of elements, %										
		C	Mn	Si	P	S	Cr	Ni	Cu	Ti	N	B
1	10Cr14Mn14Ni4Ti	0.10	14.9	0.70	0.020	0.019	14.6	4.6	-	0.23	-	-
2	10Cr14NMn20	0.10	20.3	0.50	0.012	0.011	14.8	-	-	-	0.30	0.1

To influence stress concentrators on mechanical properties, static tensile tests were performed in a wide temperature range on samples with an annular incision. A deep annular incision don't complicates the development of plastic deformation, beginning at its apex, as the proportion of tangential stresses decreases sharply from the incision to the center of the sample. In addition, it is known that the effect of acute deep incision is manifested for all steels, regardless of their strength level, lattice type, viscosity and plasticity [8-11].

3. Results and discussion

In research practice, when testing for static tension, the coefficients are used as structural strength criteria:

$$\alpha = {\sigma_B^H}/{\sigma_B};\ \alpha' = {\sigma_B^H}/{\sigma_{0,2}};\ \beta = {\psi^H}/{\psi},$$

where σ_B – tensile strength, σ_B^H – tensile strength of the sample with an incision, $\sigma_{0,2}$ – conditional yield strength, ψ – relative contraction, ψ^H – relative contraction of the sample with an incision.

In table 2 the values of the coefficients α, α' *and* β for the studied steels are given.

Table 2. Variation of the sensitivity coefficient to the stress concentration α, α′ and β depending on the test temperature

№	Grade of steel	The diameter of the samples	$\alpha = \sigma_B^H / \sigma_B$		$\alpha' = \sigma_B^H / \sigma_{0,2}$		$\beta = \psi^H / \psi$	
			20°	-196°C	20°	-196°C	20°	-196°C
1	10Cr14Mn14Ni4Ti	6	1.24	1.40	2.98	3.25	0.54	0.43
		10	1.24	1.11	3.53	3.01	0.32	0.29
2	10Cr14NMn20	6	1.31	1.51	2.45	2.05	0.38	0.41
		10	1.49	1.21	2.43	1.93	0.27	0.35

The obtained results allow us to conclude that these coefficients are not sufficiently informative from the point of view of the performance of steels, especially at low temperatures. For example, the coefficient α allows, on the one hand, to compare the studied materials on the resistance of plastic deformation in the incision. The more the strength characteristics increase (especially at low temperatures), the less plastic the material is. But, on the other hand, the coefficient of sensitivity to stress concentrators $\alpha = \sigma_B^H / \sigma_B$ characterizes only the change in the average stress of destruction in the presence of stress concentration, regardless of at what stage of deformation this destruction begins.

The coefficient α' gives more objective information about the performance of the metal in the presence of stress concentration, because it answers two questions: 1) does the stress concentration lead to destruction at a voltage below the calculated one, and if it does not, then 2) what is the permissible margin of safety in case of excess of the calculated stresses due to possible operational loads.

Before proceeding to the analysis of the obtained values α', it should be noted that in the presence of a stress concentrator (sharp incision or crack), the yield strength of the incised sample in plastic steel, regardless of the test temperature, should be close to the yield strength for smooth samples. In this case, the stress corresponding to the elastic deformation does not change its value regardless of whether the crack first appears in the material during deformation, or whether it already existed in the material before the deformation began. The transition of the material from the viscous to the brittle fracture region will be accompanied by a drop in the time resistance, and the ratio $\sigma_B^H / \sigma_{0,2}$ will decrease. Therefore, the drop in strength should predetermine the beginning of brittle fracture. The disadvantages of the criterion α^' include the fact that during the destruction we do not get any information about the nature of the destruction, i.e. it is not known whether there was a noticeable plastic deformation or destruction occurred in the elastic region.

The coefficient β characterizes the degree of influence of the incision on the reduction of steel ductility at both normal and low temperatures. However, without analyzing the numerical values of ψ^H, it is impossible to judge the reliability of steel only by the values of β. For example, the same value of the coefficient β =0,25 for steel 10Cr14NMn20 on samples of the same type at the same test temperature, means only the degree of reduction of plasticity. While the true values of ψ^H at -196°C for steel 10Cr14NMn20 are respectively 12 and 4%.

For a more complete assessment of the performance of steels at the operating temperature, the following coefficients can be additionally used:

1) $K_1^t = \sigma_B^{Ht} / \sigma_B^{20}$, the ratio of the tensile strength of the sample with the incision, to the tensile strength of a smooth specimen determined at room temperature.

2) $K_2^t = \sigma_B^{Ht} / \sigma_{0,2}^{20}$, the ratio of the tensile strength of the sample with the incision, at operating temperature, to the yield point a smooth sample determined at room temperature.

3) $K_3^t = \sigma_{0,2}^{Ht} / \sigma_{0,2}^{20°C}$, the ratio of the yield strength of samples with incision or crack determined at operating temperature, to the yield strength of a smooth sample determined at room temperature.

The values of the coefficients K_1^t,K_2^t and K_3^t for the studied steels are given in table 3.

Table 3. Variation of stress concentration sensitivity coefficients K_1^t, K_2^t, K_3^t depending on test temperature

№	Grade of steel	Diameter, mm	$K_1^t = \sigma_B^{Ht} / \sigma_B^{20}$				$K_2^t = \sigma_B^{Ht} / \sigma_{0,2}^{20}$				$K_3^t = \sigma_{0,2}^{Ht} / \sigma_{0,2}^{20}$			
			20°C	20°C - 50°C	20°C - 100°C	20°C - 196°C	20°C	20°C - 50°C	20°C - 100°C	20°C - 196°C	20°C	20°C - 50°C	20°C - 100°C	20°C - 196°C
1	10Cr14Mn14Ni4Ti	6	1.25	1.77	2.08	2.43	4.00	5.77	6.81	7.95	1.73	2.16	2.46	3.16
		10	1.20	1.58	1.76	1.81	2.73	3.56	3.96	4.33	1.66	1.93	2.06	2.60
2	10Cr14NMn20	6	1.32	1.80	2.07	2.34	2.40	3.26	3.77	4.08	1.91	2.48	3.11	4.26
		10	1.52	1.84	2.07	2.33	2.45	3.02	3.20	3.80	1.67	2.06	2.44	3.48

As follows from the obtained data, the total average fracture stress at the operating temperature in comparison with the fracture stress at room temperature for all studied steels increases, as evidenced by K_1^t.

The coefficient K_2^t characterizes the safety margin at the operating temperature if $K_2^t > 1$ or the absence of safety margin if $K_2^t > 1$, and thus gives information about the permissible level of operational loads. For all studied steels in the entire operating temperature range $K_2^t > 1$. The coefficient K_3^t allows us to judge the possible application of the calculated permissible voltages at the operating temperature equal to room temperature. As follows from the data (see table. 3), for steel 10Cr14NMn20 coefficient value K_3^t at all test temperatures is higher than that for steel 10Cr14Mn14Ni4Ti.

The characteristics of the strength indicators obtained on cylindrical smooth samples with a diameter of 10 and 6 mm are approximately at the same level. Testing of non-standard samples with a diameter of 10 mm with an annular incision did not result in a more complex stress state compared to non-standard samples with a diameter of 6 mm with an annular incision. It follows that the use of samples of increased cross-section does not give advantages in determining the mechanical properties under static tension.

4. Conclusion

An important practical task for assessing the reliability of the performance of the material is to determine the influence of stress concentration on their mechanical properties. It is shown that the presence of stress concentration at low temperatures leads to a decrease in strength during the transition of the material to the brittle state. There is a continuous increasing ratio of yield point to ultimate strength with decreasing temperature, that is, the ability of hardening under elastic-plastic deformation decreases the transition temperature of the material in fragile condition disappears completely.

References

[1] Mockvichev V V, Mahutov N A and Chernyaev A P 2002 *Crack resistance and mechanical properties of structural materials of technical systems* (Science, Novosibirsk)
[2] Soltsev Yu P and Titova T I 2008 *Steels for the North and Siberia (*Himizdat, St. Petersburg)
[3] Ermakov B S, Vologzhanina S A, Bobrovskij I N, Bobrovskij N M and Erisov Y 2018 Resistance to brittle fracture and availability of austenitic steels *IOP Conference Series: Materials Science and Engineering* **3 (450)** 032041
[4] Vologjanina S A, Ermakov B S, Bobrovsky S, Lukyanov A and Lee R 2019 Cast Austenitic Steels for Cryogenic Technology *Key Engineering Materials* **822** 60-65
[5] Soltsev Yu P, Ermakov B S, Vologzhanina S A and Krutikov N V 2005 The problem of resource exhaustion of cryogenic objects *Journal International Academy of Refrigeration* **2** 31-33
[6] Griffith A A 1924 The theory of rupture. *Proc. Ict. Int. Congr. Appl. Mech. Delft*
[7] Orowan E 1948 Fracture and strength of solids *Repts. Progn. Phis.* **49 (12)** 185-232
[8] Bannykh O A, Bannykh I O, Lukin E I and Sorokin A M 2018 Structure and mechanical properties of high-strength structural steels *Russian metallurgy (Metals)* **6** 528-532
[9] Vologjanina S A and Ermakov B S 2019 Decrease in mechanical properties, transient brittleness temperature, fatigue strength and corrosion resistance of Fe-C alloys *IOP Conference Series: Materials Science and Engineering* **2 (537)** 022021
[10] Li, Wang Y L, Zhang H L, Li S X, Zheng K, Xue F and Wang X T 2013 *Nucl. Mater.* **41** 433
[11] Sagaradze V V and Uvarov A I 2013 *Hardening and properties of austenitic steels* (Editorial and publishing Department of the Ural branch of the Russian Academy of Sciences, Ekaterinburg)

IOP Conf. Series: Materials Science and Engineering **826** (2020) 012016 doi:10.1088/1757-899X/826/1/012016

Study of the structural state of the metal welded joint 2.25Cr-1Mo-V steel depending on the parameters of post-weld heat treatment

Titova T I, Shulgan N A and Borovskoy A S

«TK «OMZ-Izhora» Ltd, 13/BM Finlyandskaya st., 196650, St. Petersburg

E-mail: tatyana.titova@omzglobal.com

Abstract. The results of studies of the structural state and properties of welded joints of steel 2.25Cr-1Mo-V after various modes of post-weld heat treatment are presented. The study of all zones of the welded joint using an optical microscope at magnifications up to 1000 times was conducted. It was established that the structure of the weld metal and HAZ of the welded joint 2.25Cr-1Mo-V steel in all thermal conditions is represented by bainite with different contents and distribution of the carbide phase. The level of hardness of the welded joint in Vickers units according to ASTM E92 is studied. The following critical values of the Larsen-Miller parameter were determined: for which the weld metal and HAZ ensure the absence of brittle structures with hardness higher than 350 HV10, exposed to hydrogen and the formation of cold cracks, at which the required level of hardness for trouble-free operation is not more than 248 HV10 and at which the coagulation of carbides is too intense and leads to a decrease strength properties.

1. Introduction

As is known, modern petrochemical reactors are highly loaded pressure vessels, the materials of which are subject to very strict requirements for the level of service properties – a high level of "cold" and "hot" strength, cold resistance, taking into account the operation at low climatic temperatures, and others [1-3]. For the production of such reactors, materials of the 2.25Cr-1Mo-V alloying system are used, which are prone to cracking during welding, so the technology of their welding and post-welding heat treatment is complex and requires knowledge of the processes of structure formation.

In the manufacture of the petrochemical reactor vessel from 2.25Cr-1Mo-V steel, all pressure welds are subjected to immediate post-weld heat treatment, which is divided into low-temperature dehydrogenation heat treatment (DHT) and intermediate stress relieved heat treatment (ISR, high tempering), and then a final post-weld heat treatment (PWHT, high tempering) is performed to form operational properties. As a rule, the welded joint of the 2.25Cr-1Mo-V alloying system passes the following manufacturing sequence: welding → DHT → technological operations → ISR-technological operations → PWHT. In this case, the first seam of the reactor vessel Assembly may be subjected to more than ten high tempering, and the closing seam of the reactor vessel to only one high tempering of PWHT.

Information about the structural state of the weld metal depending on the parameters of post-weld heat treatment is necessary to determine the most optimal modes of its conduct, allowing avoiding the formation of cold cracks and providing the required performance properties.

2. Materials and methods of research

Experimental welds made of 2.25Cr-1Mo-V steel were made by automatic submerged welding method for the research. The welding materials used for this purpose had a type of alloying close to the base metal (table1). After welding, the welded joints were subjected to heat treatment according to the following modes: DHT (P_{LM} = 13.0), ISR (P_{LM} = 19.0-19.6), PWHT (P_{LM} = 20.3-21.6).

Table 1. Chemical composition of metal of experimental welded joints

The weld joint area	The content of elements, wt. %											
	C	Cr	Mo	V	Mn	Si	Nb	P	S	As	Sn	Sb
Base metal SA-336V F22V	0.13	2.38	1.01	0.28	0.48	0.07	0.01	0.006	0.002	0.004	0.003	0.001
Weld metal 2,25Cr-1Mo-V	0.10	2.49	1.08	0.35	0.86	0.17	0.02	0.007	0.003	0.003	0.007	0.0006

3. Findings and discussion of results

The study of the structural state of the metal of different zones of the welded joint by direct method using an optical microscope at magnifications up to 1000 times and indirect method, namely determination of the level of hardness of the welded joint by Vickers method according to ASTM E92. The effect of thermal treatments on the structure and properties of the weld metal was expressed through the release parameter-Larsen-Miller parameter (P_{LM}). The Larsen-Miller parameter (P_{LM}) is calculated by the formula [4]:

$$P_{LM} = TK(20+\lg\tau)\times10^{-3},$$

where TK - the temperature of exposure in degrees Kelvin, τ - the duration of the release, hour.

The criterion of acceptability of the metal structure of the welded joint 2.25Cr-1Mo-V steel was considered the absence of pronounced coagulation of carbides, which leads to a sharp decrease in the strength and stress-rupture of the steel.

As is known, the bainite structure characteristic of the weld metal of 2.25Cr-1Mo-V steel is optimal for ensuring operational properties, including a high level of low-temperature impact work, tensile strength and fluidity in the PWHT state [5]. However, the bainite structure formed during welding is characterized by a high hardness - more than 350 HV_{10}, which indicates its high sensitivity to hydrogen, and therefore to the formation of cold cracks [5, 6]. Obviously, the parameters of the ISR in the process of manufacturing the product should provide a hardness of not more than 350 HV_{10} in all areas of the welded joint.

In order to reduce the tendency of petrochemical reactor vessel metal to crack during operation under the influence of hydrogen sulfide-containing medium at high temperature and pressure, it is necessary to ensure the metal hardness of welded joints not more than 248 HV. This requirement is included in all standard specifications for the manufacture of modern petrochemical reactor housings.

In the present study, it was found that the structure of the weld metal and HAZ of the welded joint of 2.25Cr-1Mo-V steel in all thermal states is represented by bainite with different content, distribution of the carbide phase. HAZ has a fine-grained structure, the weld metal consists of sections of large-crystal and fine-grained structure. Figure 1 shows the change in the structure of the weld metal as the value of the P_{LM} increases. After heat treatment with the tempering parameter P_{LM} = 13.0 (350°C), the structure of the weld metal and HAZ is close to the original (after welding), and therefore has similar properties. An increase in temperature-time parameters of tempering to P_{LM} = 19.3 and above is accompanied by the release of carbides, an increase in their size and quantity, and coagulation

IOP Conf. Series: Materials Science and Engineering **826** (2020) 012016 doi:10.1088/1757-899X/826/1/012016

along the grain boundaries (Figure 1 b, c), which should lead to a decrease in the tensile strength of the metal and its stress-rupture properties [7]. Tempering with P_{LM} much higher than 21.0 is impractical, as it leads to intensive coagulation of carbides, and with P_{LM} more than 21,6 to the dissolution of small and the formation of large carbides (Figure 1 d).

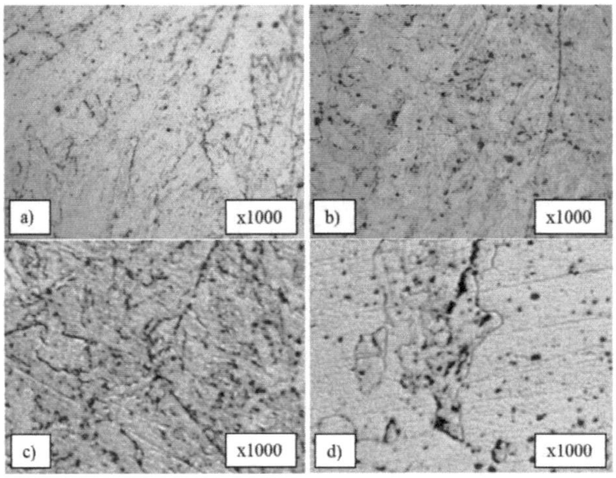

Figure 1. Microstructure of weld metal after various post welding thermal treatments: a) 13.0 P_{LM}(350°C), b) 19.3 P_{LM}, c) 20.3P_{LM}, d) 21.6 P_{LM}. Etching is performed with 4% alcohol solution HNO_3

Changes in the level of hardness of welded joints of 2.25Cr-1Mo-V steel were evaluated in the state after welding and after tempering with P_{LM} = 13.0-20.6. The results of determining the hardness depending on the tempering parameter P_{LM} are shown in figure 2.

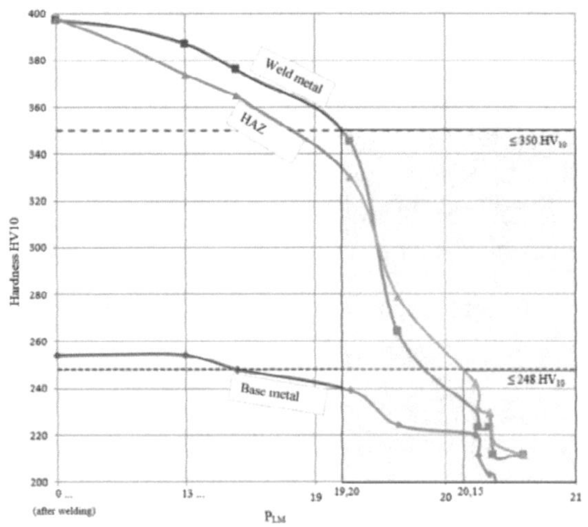

Figure 2. Hardness of various zones of welded joints 2,25Cr-1Mo-V steel in various thermal states

As a result of researches it is established that in a condition after welding (before heat treatment) the level of hardness of welded connection made in metal of a seam - to 398 HV, in metal of HAZ - to 397 HV. High values of hardness in the weld metal and HAZ immediately after welding indicate the presence in these areas of the weld brittle structures prone to cracking. At the same time, the initial hardness level of the thermally improved base metal is quite low, it does not exceed 250 HV_{10} and subsequently, when increasing the P_{LM} to 20.2, it decreases to 220 HV_{10}. After DHT with P_{LM} = 13.0, performed at 350°C, immediately after welding, the level of hardness of the welded joint is close to the initial (post-welding) level and is in the weld metal-up to 374 HV_{10}, in the metal of HAZ - up to 387 HV_{10}, which corresponds to the presence of brittle structures prone to cracking in the presence of hydrogen. These data suggest a high probability of cracking is in the weld metal or HAZ, where the hardness exceeds the critical level of 350 HV_{10}. Thus, the DHT of welded joints of 2.25Cr-1Mo-V steel applied directly after welding, to remove hydrogen, does not significantly affect the reduction of hardness and can only be used when providing a low initial level of hydrogen in the weld metal and HAZ. To reduce the hydrogen content in the welded joint, a lower temperature mode DHT can be used [8].

According to the data presented in figure 2, to reduce the hardness below the critical value of 350 HV_{10} and confirm the absence of brittle structures in the metal of welded joints of 2.25Cr-1Mo-V steel, the P_{LM} tempering parameter must be at least 19.2 (Figure 2). An increase in the P_{LM} tempering parameter to 19.6 leads to a more significant decrease in the hardness level: in the weld metal-to 279 HV_{10}, in the HAZ - to 264 HV_{10}.

However, tempering with the P_{LM} parameter up to 19.6 does not allow achieving the required level of weld hardness for defect-free operation-no more than 248 HV_{10}. In accordance with the graphical representation of the dependence "P_{LM} tempering parameter - hardness" to ensure the hardness of welded joints of 2.25Cr-1Mo-V steel below 248 HV_{10}, it is necessary to temper with the P_{LM} parameter > 20.15. The implementation of the experimental mode of post-weld heat treatment with the tempering parameter P_{LM} = 20.2 at a temperature of 695°C led to a decrease in the hardness of the weld metal to 242 HV_{10}, which is close to the maximum permissible values. However, after tempering with the parameter P_{LM} = 20.2, the hardness drop over the cross-section of the welded joint reaches 50 HV_{10}, which indicates the possibility of structural stresses and uneven strength characteristics. Obviously, the PWHT temperature should be increased. The use of tempering with a holding temperature of 700°C (P_{LM} = 20.3) showed that it is not only more effective than tempering at 695°C (P_{LM} = 20.3), reduces the level of hardness of the welded joint (up to no more than 236 HV_{10}), but also provides a smaller hardness drop across the cross section of the welded joint -38 HV_{10}. Thus, a PWHT regime with a temperature of 700°C corresponding to a P_{LM} of not less than 20.3 should be considered sufficient to ensure the required level of hardness of welded joints of 2.25Cr-1Mo-V steel of not more than 248 HV_{10}.

4. Conclusion

As a result of the performed studies of welded joints of 2.25Cr-1Mo-V steel, it was established that the absence of brittle structures with hardness above 350 HV_{10} in the weld metal and HAZ, prone to hydrogen influence and formation of cold cracks, is provided at $P_{LM} \geq 19.20$. Guaranteed obtaining required for operation of a uniform level of hardness of not more than 248 HV_{10} is possible during highpost-welding tempering, the temperature and duration of which exceed 20.3 P_{LM}. It was found that tempering with P_{LM} being significantly higher than 21.0 is impractical, as it leads to intensive coagulation of carbides, and therefore to a decrease in strength properties. The established limits of the P_{LM} tempering parameter values can serve as a basis for the development of temperature-time parameters of post-weld heat treatment of welded joints of 2.25Cr-1Mo-V steel for guaranteed absence of cold cracks in them in order to improve the safety and reliability of materials of pressure vessels such as petrochemical reactor housings operated in regions with low climatic temperatures.

References

[1] ASME BPVC-VIII-2: 2017 *Alternative rules - rules for construction of pressure vessels* section VIII division 2 867
[2] API RP 934-A 2012 *Materials and Fabrication of 2-1/4Cr-1Mo, 2 1/4Cr-1Mo-1/4V, 3Cr-1Mo, and 3Cr-1Mo-1/4V Steel Heavy Wall Pressure Vessels for High-temperature, High Pressure Hydrogen Service*

[3] Titova T I, Shulgan N A and Borovskoy A S 2012 *St. Petersburg International Scientific and Technical Conference Welding materials* (St. Petersburg: Publishing House of the Polytechnic University) 193-201
[4] Larson F and Miller J 1952 *Transactions of ASME* **74** 765-771
[5] Hrivnak I 1984 *Weldability of steels* (Moscow: Mashinostroenie)
[6] Smirnov I V 2012 *Welding Special Steels and Alloys: A Training Manual* (St. Petersburg: Publishing House "Lan")
[7] Livshits L S and Khakimov A N 1989 *Welding metal science and heat treatment of welded joints* (Moskow: Mechanical Engineering)
[8] Abe M, Nakatani M, Namatame N and Terasaki T 2012 *Welding in the world in multi-layer welds of Cr-Mo-V steel* **05/06(56)** 114-123

IOP Conf. Series: Materials Science and Engineering **826** (2020) 012017 doi:10.1088/1757-899X/826/1/012017

Energy and environment options of working fluid alternative for different refrigeration configurations

O B Tsvetkov, Yu A Laptev, A A Nikitin, V V Mitropov, A V Sharkov and Yu A Rakhmanov

ITMO University, 49, Kronverksky pr., St. Petersburg, 197101, Russia

E-mail: tsvetkov@itmo.ru

Abstract. Hydrofluorocarbons, currently used as refrigerants, in air-conditioning systems, for high-temperature heat pump systems in industrial applications are potent greenhouse gases. Their contribution to climate change is to increase. Future use of the hydrofluorocarbons as the hydrochlorofluorocarbons will be phased down. There are very different options to replace them and only a few pure fluids possess the combination of chemical, environmental, thermodynamic, and safety properties necessary for a working fluid. Low global warming potential refrigerants are anticipated to be the refrigerants of choice in industrial applications.

1. Introduction

Refrigeration is an essential technology for our present-day life. It is used in air-conditioning, manufacturing of different materials, medical applications freezing, cooling, food processing and preservation. The amount of energy used for refrigerant engineering is a substantial part of the total energy consumption in the modem world. The refrigeration industry today is in a state of intensive challenge to meet various mandatory regulations and deadlines. Early refrigerants were toxic, flammable, or both.

2. Chlorofluoro- (CFC) and hydrochlorofluorocarbons (HCFC)

The use of chlorofluorocarbons as refrigerants was proposed in 1930. As refrigerants, the CFCs have important advantages: high molecular mass, nonflammability and extremely stability. Since 1930 they were manufactured using large-scale production processes. These refrigerants have changed the world of refrigeration. Later HCFCs substances were developed. During the following 50 years, the CFCs and HCFCs refrigerants used CO_2 and ammonia in large-scale applications.

The chlorofluorocarbons are a class of compounds containing the halogens chlorine and/or fluorine on a carbon skeleton. For many years once emitted in the atmosphere the CFC and HCFC refrigerants persist for many years in the atmosphere. Diffusing to the upper atmosphere the molecule finally breaks down releasing the chlorine, which is responsible for catalyzing the destruction of stratospheric ozone. Stratospheric chorine from CFCs and HCFCs is believed at least partly responsible for peak ozone concentrations occurring in the stratosphere and an ozone deficit at the poles [1]. Manufacture or import of CFCs has been ceased. A phase out date of 2030 was established for HCFCs.

Nowadays, the environmental problems associated with refrigerants in the light of their impact on the environment are the ozone depletion potentials (ODP) and global warming potential (GWP) [2]. The ozone depletion potential (ODP) is the ration of ozone destroyed by 1 kg of the substance emitted to the atmosphere that is destroyed by 1 kg of dichlorodifluoromethane (CFC-12). The global warming potential due to 1 kg of substance emitted instantaneously to the atmosphere from 1 kg of carbon dioxide. Table 1 gives some values.

Table 1. ODP and GWP potentials of CFCs and HCFCs refrigerants to be phased down by Montreal Protocol

Refrigerant	Molecular formula	ODP	GWP	Flammability
R11	CCl_3F	1	4750	A1
R12	CCl_2F_2	1	10890	A1
R113	$C_2Cl_3F_3$	1	6130	A1
R502	R22/R115	0.25	4700	A1
R22	$CHCl_2F$	0.055	1760	A1
R142b	$C_2H_3ClF_2$	0.07	2310	A2

According to the internationally agreed amendments, the HFC refrigerants such as HFC-134a, HFC-410A, HFC-407C and HFC-404A, entirely harmless to the ozone layer, are considered as alternative refrigerants to replace CFC and HCFCs. The criteria often applied to the assessment of alternative refrigerants in the light of their impact of the environment have been the ODP and GWP. However, one should consider both for the direct contribution of the alternatives as greenhouse gases and the indirect contribution of the CO_2 emissions resulting from the energy required to operate the refrigeration system. It should be noted that among the chlorine-free halocarbons to be considered from the environmental viewpoint, there is no single possibility to replace HCFC-22 as well as CFC-502.

The chlorine-free alternatives offered by the chemical industry are HFC-134a as a pure fluid instead of CFC-12, and mainly non-azeotropic mixtures are based on the HFC-32, HFC-125, HFC-134a and HFC-143a. It is well known that these mixtures require changes in the manufacture of components, new sealing materials, new lubricants and decomposition problems of the non-azeotropic mixtures. One of the major disadvantages of zeotropes concerned by the industry is a problem of fractionation of the mixtures during the charging and leaking situations. It is clear that the composition of the refrigerant mixture can change dramatically if a leak occurs in the system under a vapor-liquid coexistence and is recharged with the original composition a number of times. In accord with an urgent demand, an attention was paid to zeotropes named as near-azeotropes or zeotrope-like blends, which exhibit minimal temperature and composition changes or «glides» during the vapor-liquid phase change. Table II summarizes some examples of HFC-mixture alternatives being proposed. The chlorine free alternatives offered by the chemical industry are HFCs as pure fluids and mainly non-azeotropic mixtures based on the HFCs as HFC-32, HFC-125, HFC-134a and HFC-143a. However, one of the major disadvantages of zeotropes is a problem of a large global warming potential (GWP) of the these synthetic refrigerants.

3. Global warming impact

An amendment to the Montreal Protocol, signed in 1987 and ratified today by more than 190 countries, asked for a reduction in the consumption and production of CFCs ad HCFCs. It was adopted in October 2016 [3]. It requires a significant reduction of the weighted value of global warming potential (GWP) of fluids used in the refrigeration industry. Consequently, hydrofluorocarbon (HFC) working fluids having a high GWP will be eliminated [4].

Another choice is to change the old natural refrigerants like ammonia (R744), water (R718), the hydrocarbons propane (R290) and isobutane (R600a) or CO_2 (R744) [5, 6].

Ammonia is an old refrigerant, which has even survived the CFCs and HCFCs in large refrigeration plants, where ammonia has always been the first choice. Ammonia has excellent thermodynamic properties similar to those of CFCs and HCFCs working fluids. It has neither an ozone depletion nor a global warming potential. But ammonia is toxic, and its characteristic smell can cause panic. Copper is not compatible with ammonia, and open compressors are usually used.

Table 2. HFC substances as working fluids

Working fluids	Molecular formula/ composition	ODP	GWP	Flammability
R134a	$C_2H_2F_4$	-	1430	A1
R125	C_2HF_5	-	3500	A1
R32	CH_2F_2	-	674	A2
R143a	$C_2H_3F_3$	-	4470	A2
R152a	$C_2H_4F_2$	-	138	A2
R404A	R125/R143a/R134a	-	3900	A1
R407C	R32/R125/R134a	-	1600	A1
R410A	R32/R125	-	1924	A1
R507A	R125/R143a	-	4000	A1
R438A	R32/R125/R134a/ R600/R601a	-	2265	A1

Hydrocarbons becomes interesting again, even although they are flammable. Hydrocarbons are non-toxic, chemically stable, low cost and widely available. Propane in contrast to the majority of HFC mixtures is presently the only alternative which has a higher efficiency than HCFC-22. Propane and mixture of propane and butane are also used especially for refrigerators. Hydrocarbons reduce the refrigerant charge of a system by about 50 % and in contrast to HCFCs and HFCs they are hardly soluble in water.

Carbon dioxide is well known for low-temperature applications in refrigeration systems. The ideal application on the trans-critical CO_2 cycle is hot water production. The main problem with CO_2 is the high pressures up to 15 MPa, but due to the high density at high pressures, the size of the equipment decreases, the diameter of the piping can be reduced, heat exchanges become more compact. Due to low pressures in the range of 1.5 to 3, compressors become more efficient. CO_2 has a lower thermodynamic cycle efficiency. Its use may thus require a total redesign of low temperature units.

4. New refrigerants

Chlorofluorocarbons (CFCs) and hydrochlorofluorocarbons (HCFCs) have a high depletion potential (ODP) and hydrofluorocarbons (HFCs) are high global warming potential (GWP) working fluids and they are considered as greenhouse gases under Kyoto Protocol 1997 [7]. Natural refrigerants were expected to replace conventional refrigerants owing to its environmental benefits: zero ODP, low GWP. However, they have severe drawbacks including toxicity, flammability and lower thermodynamic cycle efficiency. As a result hydrofluoroolefins (HFOs), especially HFO-1234yf (2.3.3.3-tetrafluoroprop-1-ene), HFO-1234ze(E) (trans 1.3.3.3-tetrafluoroprop-1-ene) and hydrochlorofluoroolefins (HCFOs) that have zero ODP, low GWP and extremely short atmospheric lifetime, have been recently synthesized [8]. Hydrofluoroolefins (HFOs) and hydrochlorofluoroolefins (HCFOs) are unsaturated short-chain haloolefins. HCFOs contain one chlorine atom in their molecule. They have a slightly different molecular structure than HFCs, containing a carbon-carbon double bond. Four pure compounds for refrigeration and air-conditioning, chillers and heat pumps are commercialized (see Table 3); three HFO refrigerants HFO-1234yf, HFO-1234ze(E), HFO-1234ze(Z) and one HCFO refrigerant HCFO-1233zd(E). When a carbon-carbon double bond exists, stereoisomers may exist. For fluorinated propene isomers, stereoisomers exist if the second appended letter is «e». For example, HFO-1234ze has two stereoisomers referred to as «trans» (E) and «cis» (Z) isomers.

Table 3. Molecular mass (*M*), normal boiling temperature (T_b) and critical parameters of HFOs and HCFs haloisomers

HFO	*M*, kg·kmol^{-1}	T_b, K	T_{crit}, K	p_{crit}, kPa	GWP
R1234yf	114.04	243.70	367.85	3382.2	<1
R1234ze(E)	114.04	254.18	382.51	3634.9	6
R1234ze (Z)	114.041	282.90	423.27	3533.0	1.4
R1233zd(E)	130.50	291.11	439.60	3623.7	7

The first appended letter must be either «y» (refers to F-group substitution), or «z» (refers to H-group substitution). The second appended letter «d» refers to =CHCl-group substitution. HFOs do not include chlorine atoms and have an ODP value of zero. HCFOs might contribute in theory to local stratospheric ozone depletion, but the magnitude of this contribution will be extremely small due to their short atmospheric life-time. Chlorotrifluoropropene (HCFO-1233zd(E)) is oxidized rapidly in the lower atmosphere with atmospheric lifetimes of 26 days. There are different categories of flammability of HFOs/HCFOs. HFOs such as HFO-1234yf and HFO-12134ze(E) are generally classified as 2L (lower flammability) refrigerants. HCFOs-1233zd(E) are class 1 (non-flammable) refrigerants. One can reasonably conjecture that as the number of hydrogen atoms in the molecule relative to the total number of hydrogen and fluorine atoms increases the flammability of the molecule will increase. Although HFOs are not classified as toxic; they can decompose to form fluoride (HF) or carbon difluoride (COF_2) [9]. HFOs are also mixed with HCFs in a variety of blends proposed as alternatives to HCFC-22, HFC-134a, HFC-404A, and HFC-404A for use in refrigeration, air-conditioning and heat pumps applications. The most promising mixtures are composed of HFO-1234yf or HFO-1234ze(E) with HFC or natural refrigerants (CO_2) [10, 11].

5. Conclusions

This paper presents an overview of the current state of the art on the promising alternative refrigerants to substitute CFCs, HCFCs and HFCs, which is being applied in refrigeration, to conventional air-conditioning heat-pumping and organic Renkine cycle systems. Special emphasis is also given to the updated information about the fluorinated propene isomers that could potentially be used as low-GWP refrigerant candidates.

References

[1] UNEP 1987*Montreal Protocol on Substances that Deplete the Ozon Laye*r (Nairobi, Kenya, UNEP)
[2] UN 1997 *Kyoto Protocol to the UN Framework Convention on Climate Change* (NY, USA)
[3] 2017 International milestone agreement on the phase-down of HFC production and consumption in Kigali Editorial *Int. J. Refrigeration* **73** V-VI

[4] Kauffeld M 2012 Availability of low GWP alternatives to HFCs feasibility of a large phase-out of HFCs by 2020 *EIA report* (Environmental Investigation Agency)

[5] Ma Y, Liu Z and Tian H A 2013 Review of transcritical carbon dioxide heat pump and refrigeration cycle *Energy* **55** 156-172

[6] Person A 2008 Refrigeration with ammonia *Int. J. Refrigeration* **31** 545-551

[7] IPCC 2007 *Climate Change 2007* (UK, Cambridge, University Press)

[8] Brown J S 2009 HFOs-new, low global warming potentials refrigerants *ASHRAE-Journal* August **51(8)** 22-29

[9] Schuster P, Bertermann R, Rusch G M and Dekant W 2010 Biotransformation of 2,3,3,3-tetrafluoropropene (HFO-1234yf) in rabbits *Toxicology and Applied Pharmacology* **244 (3)** 247-253

[10] Kamiaka T, Dang C and Hihara E 2013 VLE measurements for binary mixtures of R 1234yf with R32, R125 and R134a *Int. J. Refrigeration* **36** 965-971

[11] Gremashi L, Wu X, Biswas A Decker P 2013 Experimental study of compressor operating characteristics and performance when using R32, R1234yf and two new low GWP refrigerants as drop-in replacements for R410A *8th Int. conference on Compressors* (London, UK)

The study of thermal processes of the skin of the object of the whole-body cryotherapy (WBC) using numerical analysis

D Yerezhep[1], A Minikaev[1], D Zhignovskaia[1], V Pronin[1] and A Yerezhep[2]

[1]ITMO University, 49, Kronversky Ave., St. Petersburg, 197101, Russia
[2]Ufa State Petroleum Technological University, 1, Kosmonavtov str., Ufa, 450062, Russian Federation

E-mail: darhan_13@itmo.ru

Abstract. The work investigated the thermophysical processes occurring in the cryosauna during cryotherapeutic effects on the human body. Experimental studies in the field of the whole-body cryotherapy (WBC) are complex, since studies are necessary in such a wide range of environmental parameters as phase state, temperature, heat flux density, gas flow velocity in the cryosauna. One of the solutions to this problem is the development of computer models of the processes of heat transfer of a biological object in a cryosauna, allowing a wide range of studies to be carried out without human experiments. In this paper, we analyze a modification of the design of a cryochamber with a biological object. In this work, several types of computer models will be created, the models of which vary depending on the geometry of the wall of the cryosauna chamber from 90 to 85 degrees. The paper will show how much it is necessary to pay attention to the geometry of the cryosauna and work on the modification of the geometry. This work shows that when the geometry of the wall of the cryosauna was changed, it turned out that it smoothed the temperature distribution along the length of the object, which, in turn, will positively affect further research in this area. The results can be applied to optimize the supply of cryogenic gas in saunas, further geometry modernization, as well as to increase the efficiency and safety of the WBC process.

1. Introduction

Whole body cryotherapy (WBC) is the process of cooling a biological object under investigation to low temperatures in a special unit (cryosauna). Under ideal conditions, in order to achieve the theoretical maximum cryotherapeutic effect, the temperature of the skin of the studied biological object should be cooled to a temperature of 270.5 K. With a local transition of the temperature barrier of 270.5 K, the probability of significant frostbite of the investigated biological object appears [1-3].

The application of this WBC procedure is diverse: from the treatment of a number of diseases associated with general effects on the body, the immune and endocrine systems, to the prevention of organism restoration in both athletes and ordinary people [4-6]. With inflammatory processes in the patient's body, the WBC procedure allows one to reduce the temperature of the inflamed tissue and limit inflammation, which favorably affects the overall treatment [7-9].

The experimental determination of the temperature field of the human body during WBC is extremely time-consuming and often fraught with the danger of local frostbite. In the literature, the most common and generally accepted approach for the numerical study of this class of problems is the use of thermal conductivity models of a system of multi-element multilayer cylindrical shells. Most of the work concerns the stationary formulation of problems, the use of fixed thermophysical parameters of the layers, as well as a limited range of environmental parameters. Moreover, such important features of thermophysical processes as the influence of internal heat sources and convective heat transfer between different layers and elements of the system, temperature, pressure and gas flow rate in the cryosauna are not considered in the literature. At the same time, taking these factors into account allows one to obtain more accurate and reliable results for further design and optimization of cryosauna elements.

In this paper, when modeling the cryotherapy process, additional equations were used that describe the internal heat sources, which made it possible to more accurately describe the cryotherapy process. The reliability of the results obtained is determined by a comparative analysis of the calculated data with the experimental and calculated data known in the literature, as well as testing of software models. Thus, the obtained estimate of the heat flux density is consistent with the data obtained in [1].

In this paper, we consider the influence of geometric factors, such as: the angle of the chamber wall (n = 3,4,5 °; m = 3,4,5 °) and the shape of the chamber (round and conical). These changes in geometry will affect the temperature of the body surface at a given point and the temperature heterogeneity along the length of the biological object. To collect information about the dynamics of the temperature field and control the heat flux in the volume of the cryosauna, as well as on the surface of the object, both at this point and along the entire length of the object before and after WBC, the Process, as during the process, can be obtained using computer simulation (finite element method), which is a safe method and a fairly reliable way to conduct an experiment.

2. Experimental setup WBC

Numerical analysis was carried out using the finite element method in a software package. Figure 1 schematically shows the different forms of the cryosauna. To simulate the work of WBC, conditional separation of the object into internal organs and muscles is used - “muscle layer - 0”, fat layer - “fat tissue - 1”, body integumentary tissues - “epithelium - 2”, cryogenic gas - 3, cryosauna - 4, cryosauna wall - 5, protective clothing for legs - 6.

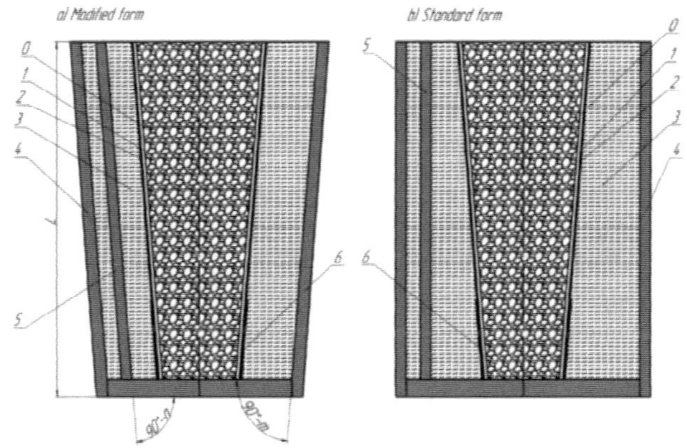

Figure 1. Schematic image: a – modified form; b – standard form.

In standard form, the geometric dimensions are equal (Figure 1b): gas supply hole (l_{out} = 75 mm, h_{out} = 160 mm), cryogenic gas outlet hole (l_{in} = 75 mm, h_{in} = 160 mm), free surface through which gas is discharged into the surrounding medium (d_{out} = 700 mm), camera body (l_{ch} = 1000 mm, h_{ch} = 1500 mm), wall (l_w = 400 mm, h_w = 500 mm), subcutaneous layer of the WBC object, internal fabric of the WBC object, the inner cavity of the chamber, filled gas mixture (d_g = 800 mm, h_g = 1400 mm) [2].

In a modified form, the angle of the wall of the cryosauna chamber changed from 90° to 90° - n (0 <n <5) and the camera angle from 90° to 90°-m (0 <m <5). The values of the dimensions of the cryosauna chamber body changed depending on the given angle.

The simulation was carried out in the time interval from 0 to 250 seconds, the intensive supply of cryogenic gas was from 0 to 180 seconds, the model was considered unsteady. The computer model used the parameters of the cryocamera "Kryon" (Russia). In this work, the circulatory system works in conjunction with a thermoregulation system and transfers heat from internal organs to the surface of the body.

3. Equations

3.1. *Heat Transfer in Solids*

For any selected elementary volume, it is assumed that the heat transfer process is stationary at each current point in time ($\partial\tau$). The heat transfer process is determined by the values of the effective coefficient of thermal conductivity (κ), specific heat (C) and density (ρ). The specific heat and density vary slightly inside the element; therefore, they are considered permanent.

General heat equation for solids is used in our calculations:

$$\rho C_p \frac{\partial T}{\partial t} + \rho C_p \mathbf{u}\nabla T + \nabla \cdot \mathbf{q} = Q \cdot \quad (1)$$

The heat flux density:

$$\mathbf{q} = -\kappa \cdot \nabla T , \quad (2)$$

where κ – a coefficient of thermal conductivity.

The Bioheat Equation:

$$\rho C_p \frac{\partial T}{\partial t} + \nabla \cdot \mathbf{q} = Q + Q_{bio} + Q_{ext} , \quad (3)$$

where Q_{bio} is the heat release of biological tissues:

$$Q_{bio} = \rho_b \cdot C_{p,b} \cdot \omega_b (\mathrm{T}_b - \mathrm{T}) + Q_{met} , \quad (4)$$

where ρ_b – blood density [10], $C_{p,b}$ – specific heat of blood at constant pressure, ω_b - blood perfusion rate [11], T_b - arterial blood temperature, Q_{met} - metabolic heat source.

Spatial heat source: Q_{ext} describes heat generation from an external source:

$$\begin{aligned} Q_{ext} &= H_{ab}(r,t)[T_{ab}(r,t) - T(r,t)] + \\ &+ H_{vb}(r,t)[T_{vb}(r,t) - T(r.t)], \end{aligned} , \quad (5)$$

where H_{ab} – coefficient for heat transfer between tissue and arterial blood per unit volume of tissue [W/(m^3·K)], H_{vb} - coefficient for heat transfer between tissue and venous blood per unit volume of tissue [W/(m^3·K)], subscript: b - blood, ab - arterial blood, and vb - venous blood.

We made assumptions, the heat dissipation Q_{ext} can be neglected, since we set H_{ab} and H_{vb} to be zero [12].

3.2. *Heat equation for a liquid*

The general heat equation for a liquid (in a gas N_2):

$$\rho \cdot C_p \cdot \frac{\partial T}{\partial t} + \rho \cdot C_p \cdot \mathbf{u} \cdot \nabla T + (\nabla \cdot \mathbf{q}) = Q, \tag{6}$$

where ρ - density, Cp - heat capacity, T - absolute temperature, t – time, **u** - velocity vector, **q** - heat flux, Q - heat sources.

The heat flux density:

$$\mathbf{q} = -\kappa \cdot \nabla T, \tag{7}$$

where κ – a coefficient of thermal conductivity.

4. Materials

All properties of materials were taken isotropic in the calculations of this model. The data on thermal conductivity and heat capacity of the epithelium, fat layer, muscle layer used in the model are presented as dependences on temperature and are shown in [13].

5. The study results obtained using numerical analysis

In this paper, we consider at what period of time a comparative analysis of future changes in the geometry of the cryosauna is necessary. Since, in this work, a biological object is placed in the cryosauna, and the biological object emits a significant amount of heat, which increases when the biological object is cooled, it is necessary to analyze and calculate the time interval, which has a significant effect during the WBC procedure. In this statement of the problem, it makes no sense to analyze all the modes, since we need to calculate the maximum cooled period of time, in all modes it will be one.

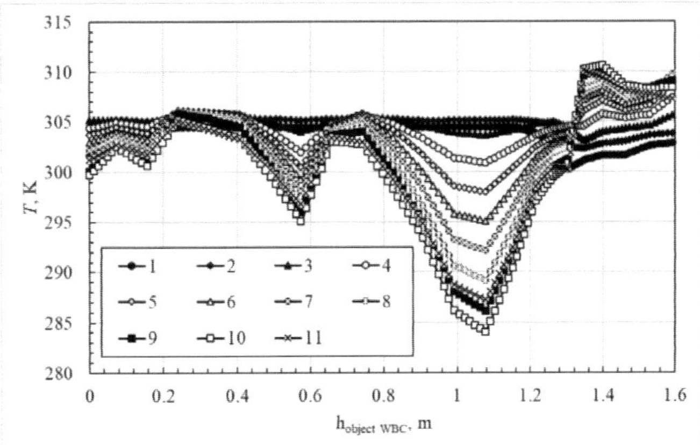

Figure 2. The distribution of skin surface temperature of the object length WBC. Time intervals: 1 – 0 s., 2 – 20 s., 3 – 40 s., 4 –60 s., 5 – 80 s., 6 – 100 s., 7 – 120 s., 8 – 140 s., 9 – 160 s., 10 – 180 s., 11 – 200 s.

The main parameter showing the effectiveness of the WBC procedure will be temperature. Also, during the analysis, the dependences of the heat flux and the temperature of the internal tissues of the biological object will be considered. It is necessary to create a uniform temperature field along the length of the object by changing the angle of inclination of the walls of the cryosauna (changing the geometry) and varying the modes of supply of cryogenic gas into the cryosauna chamber. The temperature distribution of the surface of the skin from the length of the biological object over time is shown in Figure 2. As can be seen from Figure 2, the biological object is cooled as much as possible in a time interval of "10" (180 seconds). In the subsequent analysis, we will use this period of time as the most dangerous and significant.

In Figure 3, we will arrange the temperature curves at different angles of the angle of inclination of the cryosauna wall, which is done to more clearly consider the temperature curves and conduct a comparative analysis. To facilitate the reading of the simulation results, we will consider the curves in one cryogenic gas supply mode. Following the safety conditions, we cannot use the 3 mode that was described earlier, since in this mode there is a local frostbite of the biological object. For comparison with standard geometry, we will use 2 mode, 1 mode is too weak and will also be excluded from the analysis. Also, on this graph we will place the results at the standard angle of the chamber wall (the right angle is taken as the standard angle, see Figure 1).

In Figure 3, it can be seen that curve 5 extends beyond the boundary of curve 6 in the region of 0.9-1.1 m, which corresponds to the critical temperature; therefore, it follows that for standard geometry it is necessary to reduce the supply of cryogenic gas, since there is a danger in the region of 0.9-1.1 m frostbite of integumentary tissues of a biological object.

As can be seen from Figure 3, the effect of a slight change in the angle of the wall of the cryosauna by several degrees leads to a significant change in the temperature curve, which characterizes the temperature of the integumentary tissues of the biological object.

In Figure 3 we see that, in comparison with other curves, curve 4 is acceptable and at a given angle, the supply of cryogenic gas can be increased for more intensive cooling, which in turn will increase the efficiency of this procedure.

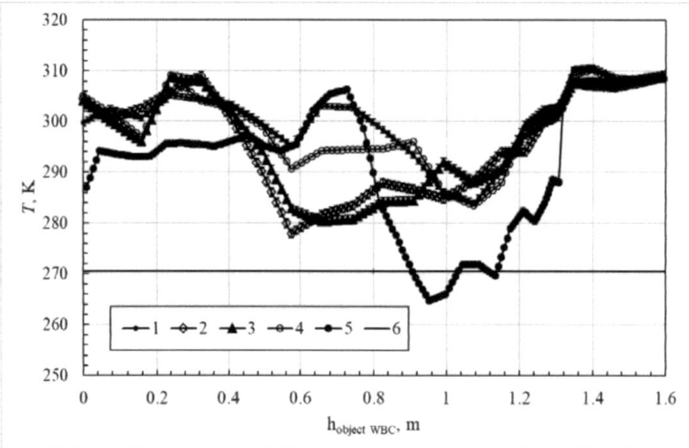

Figure 3. Temperature dependence on the length of the object under study. Geometry: 1 – 5° angle (n, m - 5°), 2 – 4° angle (n, m - 4°), 3 – 3° angle (n, m - 3°), 4 – 2° angle (n, m - 2°), 5 – 0° angle (n, m - 0°), 6 – critical temperature line (270.5 K).

6. Conclusion

In this article, several types of computer models were created, the model changed depending on the geometry, the geometry of the wall of the cryosauna chamber changed from 90 to 85 degrees. The created computer model allows one to estimate the thermal fields taking into account the internal heat source during and after the WBC process, as well as to study the effect of changes in the geometry of the cryosauna on such parameters as: velocity and temperature of the coolant, temperature of the integumentary tissues of the biological object, consider local frostbite and change the temperature regimes in depending on the results to increase the comfort and efficiency of the WBC process.

The work shows that it is necessary to pay attention to the geometry of the cryosauna and work on the modification of the geometry. It is shown that a change in the angle of inclination of the wall of the cryosauna chamber significantly affects the overall temperature field of a biological object. These experimental studies must be carried out with the participation of medical workers, since there is a high probability of obtaining local frostbite with an increase in the power of the cryosauna. This work shows that when the geometry of the cryosauna wall was changed, it turned out that it smoothest the temperature distribution along the length of the object, which in turn will positively affect further research in this area. The results can be applied to optimize the supply of cryogenic gas in saunas, further geometry modernization, as well as to increase the efficiency and safety of the WBC process.

7. Conflict of interest

The authors declare that they have no conflict of interest on the content of this paper.

8. Acknowledgments

The research was carried out with the support of the Ministry of Education and Science of the Russian Federation.

References

[1] Baranov A, Pakhomov O, Fedorov A, Ivanov V, Zaitsev A and Polyakov R 2019 Technique and Technology of Whole-Body Cryotherapy (WBC) *InTechOpen* **26** DOI: 10.5772/intechopen.83680
[2] Yerezhep D, Umutbaeva N R, Piterskov P, Baranov A Y and Yerezhep A 2019 *J. Phys.: Conf. Ser.* **1333** 032091 DOI:10.1088/1742-6596/1333/3/032091
[3] Yerezhep D, Yemelyanov A, Kuznetsov Y, Perminov A, Pronin V A, Minikaev A F and Yerezhep A 2019 *J. Phys.: Conf. Ser.* **1333** 032092 DOI:10.1088/1742-6596/1333/3/032092
[4] Banfi G, Melegati G and Barassi A 2009 *J Sci Med Sport.* **12(6)** 609-10
[5] Bouzigon R, Grappe F, Ravier G and Dugue B 2016 *J Therm Biol.* **61** 67-87
[6] Adam J 2014 *Kinesitherapie* **14(152-153)** 41-44
[7] Bertrand and Mesure S 2014 *Kinesitherapie* **14(152-153)** 45-55
[8] Mesure S, Catherin-Marcel B and Bertrand D 2014 *Kinesitherapie* **14(152-153)** 56
[9] Polidori G, Marreiro A, Pron H, Lestriez P, Boyer F C, Quinart H, Tourbah A and Taïar R 2016 *Med. Hypotheses* **96** 11-15
[10] Tungjitkusolmun S, Woo E, Cao H and Tsai J 2008 *Medical & Biolojical engineering & Computing* **38(5)** 562–568 DOI: 10.1007/BF02345754
[11] Chang I 2010 *The open biomedical engineering journal* **4(3)** 12 DOI: 10.2174/1874120701004020003
[12] Kakuta N, Yokoyama S and Mabuchi K 2002 *IEEE Engineering in Medicine and Biology Magazine* **21(6)** 65-72. DOI: 10.1109/MEMB.2002.1175140
[13] Tsyganov D I 2011 *Cryomedicine: Processes and Devices* (Moscow: SAINS-PRESS)

Numerical analysis of the heat flux density of the whole-body cryotherapy (WBC) object coming from the skin

D Yerezhep[1], A Minikaev[1], D Zhignovskaia[1], V Pronin[1] and A Yerezhep[2]

[1]ITMO University, 49, Kronversky Ave., St. Petersburg, 197101, Russia
[2]Ufa State Petroleum Technological University, 1, Kosmonavtov str., Ufa, 450062, Russia

E-mail: darhan_13@itmo.ru

Abstract. In technologies of cryotherapeutic effects, the main attention is paid to the choice of temperature of cryogenic gas. At the same time, along with the gas temperature, other process parameters are also essential: the gas velocity, the dimensions and spatial orientation of the biological cooling object, the humidity and gas velocity, etc. To develop the technique and technology of general cryotherapy, generalizing research is needed that can form the theoretical basis for the production and design of cryotherapy equipment. The most general results can be obtained by studying the effect that cooling has on a biological object with different intensities of heat removal along the length of the biological object. The work will conduct a chilled analysis of heat flux losses from the surface of the skin of a biological object with a slight change in the geometry of cryosauna during the whole-body cryotherapy process (WBC). To implement these tasks, a software package using the finite element method will be used. To describe the biological object with internal heat release, these biological heat equations were solved using the Penns approximation. The results of this work will help to give recommendations to cryosauna manufacturers to optimize the design of the cryotherapy unit, which in turn will help to increase the effectiveness of cryotherapy.

1. Introduction

Now, whole body cryotherapy (WBC) is used to treat and prevent a few diseases, such as dermatological, rheumatological, neurological, respiratory, cardiovascular diseases [1-5]. Unfortunately, at present, the WBC method has not gained wide popularity, although it is promising not only for medical needs, but also for services, beauty salons, etc. From a medical point of view, WBC is not yet included in generally accepted clinical methods and this is due to a small theoretical basis for the effect of the cooling effect on the body and the lack of control over the procedure. Only experimental data obtained on commercial equipment is not enough to introduce cryotherapy into wide medical practice, which is one of the main criteria of low popularity in the medical field. If we consider the use of WBC in the service sector, the main reasons for the lack of mass demand for cryotherapy are the high cost of the procedures, the bias of patients about the dangers of cryogenic temperatures, the unsatisfactory implementation of the procedure (lack of accurate temperature control on the skin surface, uneven distribution of heat flux over the height, etc. d.). The existing cryosauna is classified by the number of simultaneously accommodated patients (individual and group), as well as by the cooling mechanism (generating cold in a refrigeration machine with vapor compression or using cryogenic gas - liquid nitrogen). Group installations, such as ARCTIC (Poland), Crio Space Cabin (Germany), KGT LLC Krio Kholod (Ukraine), have shown their unpopularity and inefficiency due to the high cost of equipment and installation, high consumption of cooling medium or consumption of electric refrigeration power machine, significant heat gain. Thus, individual installations are more promising for modernization, due to their small size, efficiency, mobility, etc. A large number of such plants have been designed, for example, such as: KAEKT-01 "Kryon" (Russia), Icequeen, Cryomed-20/150-01, etc., but the cost and ease of operation directly depends on the supply of nitrogen and other thermal and economic factors. It should be borne in mind that it is possible to increase the efficiency of the procedure by lowering the temperature of the cooling gas around the patient, by increasing the gas supply rate and changing the geometry of the nozzle in the cryosauna. However, it must be understood that in the cryosauna the heat exchange conditions are formed between the cooling gas and the patient, which are close to natural convection and safety conditions cannot be neglected, and the fact that the gas flow distribution directly affects the uneven temperature distribution along the height of the cryosauna.

With a local transition of the temperature barrier of -2.5 °C, the probability of a significant frostbite of the object being studied appears [6-7]. There are various proposals for improving heat transfer in compliance with safe conditions, namely, maintaining the temperature at 12 °C> T (body surface temperature) > -2.5 °C, cooling rate $dT/d\tau >$ 0.01 °C/s [8] and limiting the duration of the procedure, the case of maximum risk in an experimental study on patients will only lead to the absence of a positive effect. The experimental determination of the temperature field distribution from the skin of the object during the WBC process is difficult, since in a single cabin the temperature decreases once to –130 °C at a rate of 8 °C/s [6]. During fluctuations, serial temperature sensors can give incorrect readings due to their own thermal inertia.

2. Experimental setup WBC

In numerical analysis, in this case, the most difficult moment is the description of the processes of heat transfer by the coolant (hemodynamic mechanism of the thermoregulation system). More often in such works, the thermoregulation system is not considered, to simplify the task and calculations. To solve this problem, in this work, additional terms, Q_{bio}, are added to the general heat equation, which characterizes the heat release of biological tissues through blood density, specific heat of blood at constant pressure, blood perfusion rate, arterial blood temperature, and metabolic heat source. Using computer simulation, we reproduce the WBC process with some assumptions; the results correlate with experimental data. The Bioheat module was used to mathematically describe biological tissues [9]. The module is used to simulate heat transfer in biological tissue and considers the following heat sources: blood perfusion and metabolism, which are included in the classical heat transfer equation in the form of Q_{bio} [10].

The model used the average density of these substances: nitrogen = 1.18 kg/m^3, epithelium = 1093 kg/m^3, fat layer = 916 kg/m^3 and muscle layer = 1041 kg/m^3. The model also considered heat sources from blood perfusion and metabolism: arterial blood temperature - 310.15 K, blood density - 1060 kg/m^3, blood heat capacity - 4200 J/(kg·K), blood perfusion - 0.003 1/s, metabolic heat source 7277 W/m^3 [8].

In this paper, assumptions will be introduced to describe biological processes, and the biological heat equation will be solved using the Penn approximation. The Penn approximation is described in more detail in [11]. This approximation is used to simulate heat transfer in biological tissue, considering heat sources from blood perfusion and metabolism in the classical heat transfer equation.

The simulation was carried out in the time interval from 0 to 250 seconds, the intensive supply of cryogenic gas was from 0 to 180 seconds, the model was considered unsteady. The computer model used the parameters of the cryocamera "Kryon" (Russia) [12]. In this work, the circulatory system works in conjunction with a thermoregulation system and transfers heat from internal organs to the surface of the body. This model is a multi-layer and multi-element model, where each part of the body is represented as a calculated element with the corresponding number and type of layers (Figure 1).

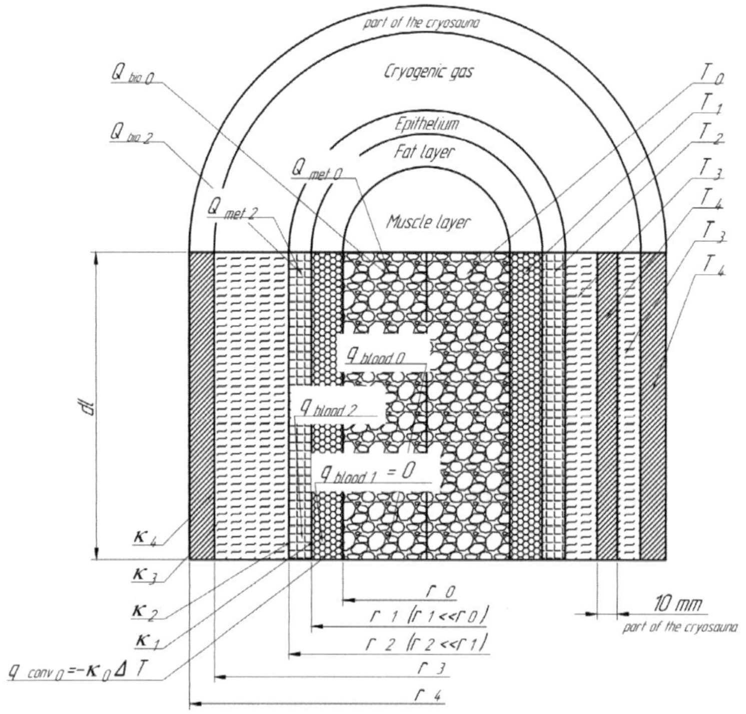

Figure 1. The simplified scheme used in the calculation (index number: 0 — internal organs and muscles, 1 — fat layer, 2 — epithelium, 3 — cryogenic gas (medium), 4 — cryosauna chamber).

3. Equations

3.1. Heat Transfer in Solids

General heat equation for solids used in our calculations:

$$\rho C_p \frac{\partial T}{\partial t} + \rho C_p \mathbf{u}\nabla T + \nabla \cdot \mathbf{q} = Q. \tag{1}$$

The heat flux density:

$$\mathbf{q} = -\kappa \cdot \nabla T, \tag{2}$$

where κ – a coefficient of thermal conductivity.

The Bioheat Equation:

$$\rho C_p \frac{\partial T}{\partial t} + \nabla \cdot \mathbf{q} = Q + Q_{bio} + Q_{ext}, \tag{3}$$

where Q_{bio} is the heat release of biological tissues:

$$Q_{bio} = \rho_b \cdot C_{p,b} \cdot \omega_b (T_b - T) + Q_{met}, \tag{4}$$

where ρ_b – blood density [13], $C_{p,b}$ – specific heat of blood at constant pressure, ω_b - blood perfusion rate [14], T_b - arterial blood temperature, Q_{met} - metabolic heat source.

3.2. Heat equation for a liquid.

The general heat equation for a liquid (in a gas N_2):

$$\rho \cdot C_p \cdot \frac{\partial T}{\partial t} + \rho \cdot C_p \cdot \mathbf{u} \cdot \nabla T + (\nabla \cdot \mathbf{q}) = Q, \quad (6)$$

where ρ - density, Cp - heat capacity, T - absolute temperature, t – time, **u** - velocity vector, **q** - heat flux, Q - heat sources.

(a) Turbulent flow of the k-ε model.

General equation:

$$\rho \frac{\partial \mathbf{u}}{\partial t} + \rho(\mathbf{u} \cdot \nabla)\mathbf{u} = \nabla \cdot \left[-\rho \mathbf{I} + (\mu + \mu_T)\left(\nabla \mathbf{u} + (\nabla \mathbf{u})^T\right)\right] + \mathbf{F}, \quad (7)$$

$$\rho \nabla \cdot (\mathbf{u}) = 0. \quad (8)$$

Turbulent viscosity is modeled as:

$$\mu_T = \rho C_\mu \frac{\kappa^2}{\varepsilon}, \quad (9)$$

where C_μ is a model constant.

The transport equation for *k* reads:

$$\rho \cdot \frac{\partial \kappa}{\partial t} + \rho \mathbf{u} \cdot \nabla \kappa = \nabla \left[(\mu + \frac{\mu_T}{\sigma_\kappa}) \nabla \kappa \right] + P_\kappa - \rho \varepsilon, \quad (10)$$

where the production term is:

$$P_\kappa = \mu_T \left[\nabla \mathbf{u} : \left(\nabla \mathbf{u} + (\nabla \mathbf{u})^T\right) - \frac{2}{3}(\nabla \cdot \mathbf{u})^2 \right] - \frac{2}{3} \rho \kappa \nabla \cdot \mathbf{u}. \quad (11)$$

The transport equation for ε reads:

$$\rho \cdot \frac{\partial \varepsilon}{\partial t} + \rho \mathbf{u} \cdot \nabla \varepsilon = \nabla \left[(\mu + \frac{\mu_T}{\sigma_\varepsilon}) \nabla \varepsilon \right] + C_{\varepsilon 1} \frac{\varepsilon}{\kappa} P_\kappa - C_{\varepsilon 2} \rho \frac{\varepsilon^2}{\kappa}, \quad (12)$$

where **u** is the velocity, μT is the eddy viscosity;

$$P_\kappa = \mu_T \left[\nabla \mathbf{u} : (\nabla \mathbf{u} + (\nabla \mathbf{u})^T) \right]. \quad (13)$$

The model constants in Equation (10), Equation (11), and Equation (13) are determined from experimental data [15].

4. The study results obtained using numerical analysis

Reflection of bioenergetic processes in the human body is the heat flux from the surface of the skin of a biological object [8]. To simulate thermal processes when changing the geometric parameters of the cryosauna wall, it is also necessary to analyze the change in the heat flux of the biological object WBC coming from the integumentary tissues. The heat flux will also be analyzed in height, since it is necessary to consider local heat losses. The heat flux emanating from the surface of a biological object during the WBC process, which varies with time, is shown in Figure 1 with a time step of 20 seconds, from 0 to 200 seconds.

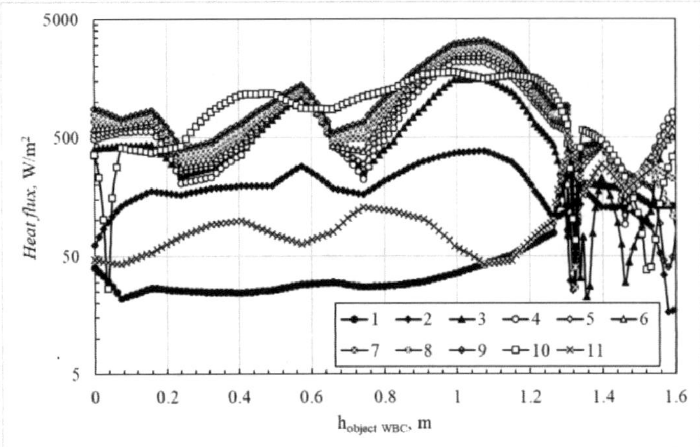

Figure 2. The dependence of the change in heat flux density on the length of the object WBC. Time intervals: 1 – 0 s., 2 – 20 s., 3 – 40 s., 4 –60 s., 5 – 80 s., 6 – 100 s., 7 – 120 s., 8 – 140 s., 9 – 160 s., 10 – 180 s., 11 – 200 s.

In Figure 2 it is seen that the largest heat loss occurs in a period of 160 seconds. In the future, we consider in more detail this period when changing the geometry of the chamber wall. In Figure 3 there is a graph of the heat flux of a biological object emanating from integumentary tissues.

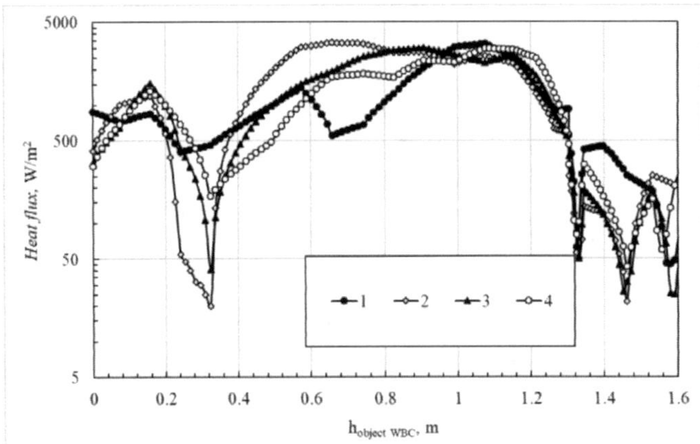

Figure 3. The dependence of the change in heat flux density on the length of the object WBC. Geometry: 1 – 5° angle (n, m - 5°), 2 – 4° angle (n, m - 4°), 3 – 3° angle (n, m - 3°), 4 – 2° angle (n, m - 2°)

5. Conclusion

In this paper, the reliability of the results obtained is determined using a comparative analysis of the calculated data with the experimental results, which are described in earlier works. Thus, the obtained estimate of the heat flux density is consistent with the data obtained in [6]. The value of the heat flux density allows us to estimate the duration of the effective exposure, that is, the time during which the analgesic effect of WBC is maintained. According to the method proposed in [7], the analgesic effect of WBC was evaluated for the case considered in this article. With an average calculated heat flux = 4 kW/m^2, the WBC analgesic effect will last 4 hours. Figure 3 shows a graph that has a maximum in the middle region, where the heat flux density locally increases to 4.85 kW/m^2. Such intense heat transfer can lead to local frostbite.

An increase in the heat flux density from the surface of a WBC biological object from 1.0 to 3.0 kW/m^2 is accompanied by an increase from 0.25 to 6, i.e., almost 25 times. Moreover, the main growth is observed in the range from 0.5 to 4.5 kW/m^2. The calculated values of the effective heat flux do not exceed 0.5 h., therefore it is necessary to modernize the design of the cryochamber to evenly remove the heat flux from the surface of the biological object at a level of from 3.5 to 4 kW/m^2. The different nature of the curve in different areas located to the right and left of the maximum indicates that an increase in the heat flux density leads to qualitative changes in the processes occurring in the layers of the WBC biological object. In order to obtain a high cryotherapeutic effect and compliance with patient safety requirements, it is necessary to ensure intensive and uniform heat removal from the body surface along the length of the WBC object. In this case, the only safe way to transfer heat from the surface of the housing to the heat receiver of the cooling system is gas convection. For this reason, it is necessary to use cryogenic temperatures to remove heat flux with a high density. Also, when designing real systems, it should be borne in mind that the average area of the entire human body is 1.5 m^2. The system of the cryotherapeutic complex should provide heat removal on average about 4 kW/m^2 per patient, and heat removal should be carried out at a temperature of no higher than 140 K. Determining the power requirements for cryotherapy complexes will protect the medical equipment market from low-quality WBC equipment.

6. Conflict of interest

The authors declare that they have no conflict of interest on the content of this paper.

7. Acknowledgments

The research was carried out with the support of the Ministry of Education and Science of the Russian Federation.

References

[1] Bouzigon R, Grappe F, Ravier G and Dugue B 2016 *J Therm Biol.* **61** 67-87
[2] Leonov V P, Kolishkin L M, Voronov V A and Shakurov A V 2018 *Refrigeration Science and Technology* **3rd IIR** 91-96 DOI: 10.18462/iir.CALS.2018.0022
[3] Yerezhep D, Yemelyanov A, Kuznetsov Y, Perminov A, Pronin V A, Minikaev A F and Yerezhep A 2019 *J. Phys.: Conf. Ser.* **1333** 032092 DOI:10.1088/1742-6596/1333/3/032092
[4] Polidori G, Marreiro A, Pron H, Lestriez P, Boyer F C, Quinart H, Tourbah A and Taïar R 2016 *Med. Hypotheses* **96** 11-15
[5] Carrard J, Lambert A and Genné D 2017 *BML Case Reports* DOI: 10.1136/bcr-2017-221431
[6] Baranov A, Pakhomov O, Fedorov A, Ivanov V, Zaitsev A and Polyakov R 2019 Technique and Technology of Whole-Body Cryotherapy (WBC*) InTechOpen* **26** DOI: 10.5772/intechopen.83680
[7] Lubkowska A 2012 Physical Therapy Perspectives in the 21st Century - Challenges and Possibilities *InTechOpen* 155-177 DOI: 10.5772/35055
[8] Tsyganov D I 2011 *Cryomedicine: Processes and Devices* (Moscow: SAINS-PRESS)
[9] Buzdov B 2014 *Applied Mathematical Sciences* **8 (57)** 2823-2831
[10] Haemmerich D, Chachati L, Wright A and Mahvi D 2003 *IEEE transactions on biomedical tngineering* **50 (4)** 493–500 DOI: 10.1109/TBME.2003.809488
[11] Lakhssassi A, Kengne E and Semmaoui H 2010 *Natural Science* **2 (12)** 1375-1385
[12] Yerezhep D, Umutbaeva N R, Piterskov P, Baranov A Y, and Yerezhep A 2019 *J. Phys.: Conf. Ser.* **1333** 032091 DOI:10.1088/1742-6596/1333/3/032091
[13] Tungjitkusolmun S, Woo E, Cao H and Tsai J 2008 *Medical & Biolojical engineering & Computing* **38(5)** 562–568. DOI: 10.1007/BF02345754
[14] Chang I 2010 *The open biomedical engineering journal* **4(3)** 12 DOI: 10.2174/1874120701004020003
[15] Patel K 2014 *IOSR Journal of Mechanical and Civil Engineering* **11(6)** 01-10.

Comparison of methods of evaluating materials' tendency to brittle fracture

S A Tcupka, S A Vologzhanina and A F Igolkin

ITMO University, 49, Kronverkskiy Ave., Saint Petersburg, 197101, Russia

E-mail: Sergey_spb92@mail.ru

Abstract. This article concerns different methods of evaluating material's tendency to brittle fracture. It includes results of impact bending tests of cold-resistant metastable austenitic steel 12Cr18Ni10Ti (AISI 321) and ferritic steel 0Ni9A in a broad temperature range. The article shows that evaluation of the effectiveness of the material should be carried out in consideration of numerous factors, especially the specific exploitation conditions of the materials.

1. Introduction

Equipment operating at low temperatures can be subject to severe damage or destruction when the additional load is applied. Whether the fracture would be brittle or ductile depends on two fundamentally different reasons. The first reason is related to purely physical factors, second - to mechanical. In certain cases, brittle or ductile fracture can occur in a state of the constant stress condition as the result of the change in material's properties. In other cases, fracture occurs due to change in the character of the stress condition while the material's properties remain unaltered. Understanding the nature of materials' embrittlement allows us to rationally choose methods of evaluating materials in terms of brittle fracture [1-9].

Even in the first studies concerning evaluation of fracture toughness of different metals, we already considered possibility of correlation between this parameter and other traditional mechanical tests. Attempts were made in the following areas: 1) replacement of K_{Ic} evaluation by measurement of more simple parameters of mechanical properties; 2) the possibility of interpretation of fracture toughness' essence by certain physical models of cracks propagation.

Studies [1, 10-13] proved the correlation between the size of the original austenitic grain in pearlitic steel and the frost fracture and fracture toughness. Authors [9] tried to find the correlation between the microstucture of the fracture and K_{Ic} of high-tensile steel 45Cr2Ni2Mo. The analysis of 223 combinations of K_{Ic} and $\sigma_{0,2}$ for steels, aluminium and titan alloys, conducted by N.V. Oleynik [13] indicates the impossibility of composing a universal correlation between different alloys.

Studies [1, 10] also included attempts of composing universal correlations between K_{Ic} and $\sigma_{в}$, which were unsuccesfull.

Correlation with toughness [11] has a limited, partial character. Impact strength results may be useful in predicting fracture toughness. Several authors [1, 11-14] analyzed the correlation between the dynamic fracture toughness and K_{Ic}. The correlation between G_{Id} and KC_p also occurs in the case of low-ductility high-carbon steels.

IOP Conf. Series: Materials Science and Engineering **826** (2020) 012020 doi:10.1088/1757-899X/826/1/012020

Studies conducted with ARMCO-iron (0,04% C) and 17ГС steel show that critical brittle temperature heavily depends on methods of evaluating steel's tendency to brittle fracture [1,11,12]. This is a well-known fact, as well as the fact that the critical brittle temperature even in the same tests depends on the sample and the notch geometry.

For materials used in low-temperature conditions, various stress concentrators should be considered, leading to difficult stress conditions, in addition to the possible dynamic load.

2. Materials and research methods

This study focuses on cold-resistant austenitic and ferrite steel of grades 12Cr18Ni10Ti (AISI 321) and 0Ni9A. See their chemical composition in table 1.

Table 1. Chemical composition of researched materials.

No.	Steel grade	Chemical composition, %										
		C	Mn	Si	P	S	Cr	Ni	Cu	Ti	N	B
1	12Cr18Ni10Ti (AISI 321)	0.11	1.3	0.65	0.035	0.020	18.9	10.2	0.21	0.30		
2	0Ni9A	0.07	0.67	0.42	0.015	0.021	-	9.3	-	-	-	-

Steels were smelted in the induction furnace with a main slag of a magnesite capacity of 150 kg. Then the steel is poured into 50-kg cast iron moulds and steep wafer-type rods. After grinding and stripping, ingots were forged into square billets with a cross section of 25x25 mm. Then, the preforms were subjected to heat treatment in order to obtain an austenitic structure in 12Cr18Ni10Ti steel and ferritic structure in 0H9A steel. Impact bending tests were carried out in accordance with GOST 9454-78 using regular samples with stress concentrators of types U (type 10), V (type 11) and T (type 15), irregular samples with lateral notches made to achieve complicated stress condition (type 25 – with a V-shaped notch, type 26 – with a T-shaped notch).

3. Results and their consideration

When evaluating the effectiveness of the materials in the most severe conditions, a natural overvaluation of the low boundary of the material is used. Analyzing the data from [1-6], the permissible temperature range, in which the material can be used, can be divided into several subareas.

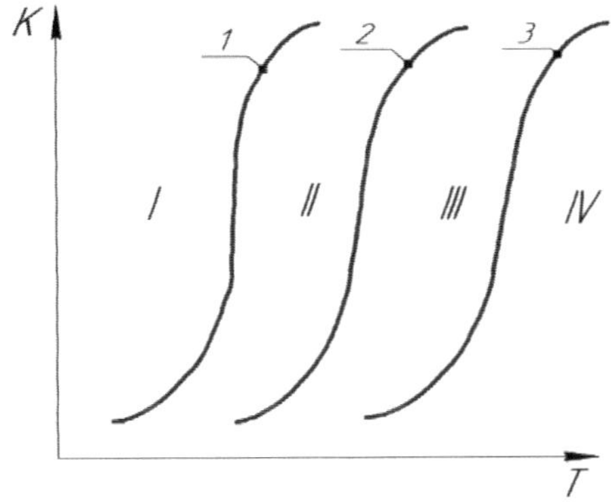

Figure 1. Threshold values of cold shortness (brittleness), determined by various methods: 1 - static bending tests according to Irwin; 2 - dynamic tests of standard samples with a notch Menaget; 3 - by measuring changes in the resistance to crack propagation under conditions of plane deformation under the action of impact forces.

In the area I, metals without stress concentrators and with large crack initiation can work under load. The same metals can be used in zone II with stress concentrators, but without impact forces.

The same materials can work in zone III even in the presence of impact forces if there are no critically sharp cracks. In area IV, the material can work even in presence of sharp cracks with the lowest probability of brittle fracture.

Typically, steels and alloys, used in low-temperature environment, do not have a distinct cold-shortness threshold. However, the presence of sharp stress concentrators (notches) causes local stresses to increase at the top of the notch and moves the serial position curve to the area of higher temperatures.

The research considered the effect of notch's top radius and test temperature on the impact strength of the studied steel grades. The impact test results of samples of types 1 and 11 (according to GOST 9454-78) are shown in table 2.

Table 2. Effect of the stress concentrator type and test temperature on the impact strength

No.	Steel grade	Impact strength, J/cm^2							
		20°C		-50°C		-100°C		-196°C	
		KCU	KCV	KCU	KCV	KCU	KCV	KCU	KCV
1	12Cr18Ni10Ti (AISI 321)	290	240	280	230	260	220	220	180
2	0Ni9A	290	180	200	180	180	140	160	100

As can be seen, the impact strength's value declines both for austenitic steel 12Cr18Ni10Ti and for ferritic 0Ni9A in the entire temperature range. Evaluation of notch's sharpness effect shows that the impact strength of the samples with sharp V-shaped notch is lower. This means that the temperature of brittle-plastic transition will move to the area of higher temperatures.

Decline in impact strength in correlation with decrease of the radius of the notch's top can be explained by the process that occurs during deformation and cooling of these steels [14-16]. The amount of martensite formed in the deformed part of the sample near the notch depends on the work absorbed during deformation before the crack initiation and the stress conditions' scheme in the notch area. The work of absorption during deformation before a crack emergence increases along with the notch's top radius, and the stress condition strengthens inversely to the notch's top radius increase.

Additional research was conducted to determine which factor exerts a greater impact on the reduction. We used non-standard samples with lateral notches, which lead to a more complicated stress condition in the notch area. The cross section of the specimens with side recesses was equal to the cross section of ordinary Charpy-type specimens.

The test results of standard and non-standard samples at a temperature of -196 ° C are shown in table 3.

Table 3. Impact values depending on the type of sample

No.	Steel grade	Impact strength KC, J/cm^2, for samples of types:			
		11	25	15	26
1	12Cr18Ni10Ti (AISI 321)	180	170	150	140
2	0Ni9A	100	57	75	40

An analysis of the results show that stressful conditions affect the martensitisation process occurring in the metastable austenitic steel 12Cr18Ni10Ti (AISI 321) under the influence of deformation and low temperatures much more than the volume of deformation. Thus, in the area of sharper stress, the concentrator amount of martensite is bigger ceteris paribus than it is in the soft notch area, which provides for the lower values of impact strength for both regular and irregular samples.

The decrease of the impact strength of samples made of metastable austenitic steels can be considered as a total decrease in the work, consisting of decrease in crack initiation work as a result of decrease in deformation volume, and the decrease in crack propagation work as a result of martensitisation. This can be explained by the fact that in martensite the crack propagates more slowly than it does in the austenitic matrix.

For ferritic steels, which after heat treatment have residual austenite in the structure, the decrease of test temperature and creation of the additional stress condition in the additional cracks' area results in the finishing of structural transformation $\gamma \rightarrow \alpha$. As a result, a decrease in impact strength occurs. For 09HA steel, a temperature of -100°C is obvious from the point of view of ductility. Therefore, we can predict a sharp decline in impact strength for the samples with both soft and harder notches at lower temperatures.

4. Conclusion

The analysis shows that evaluation of the tendency of a material to brittle fracture should be conducted differentially, considering the material's exploitation conditions.

Thus, the most stringent evaluation of materials should be applied in the case of material operating with impact forces and sharp stress concentrators (cracks). The method of impact strength assessment should be used to evaluate materials if in the future they will be used to produce components without sharp stress concentrators (without cracks) occurring during possible impact forces.

The static tension method with an extremely sharp notch should be used when working with the components working under stationary or cyclic loads, where cracks can occur.

References

[1] Soltsev Yu P and Titova T I 2002 *Steels for the North and Siberia* (Himizdat, St. Petersburg)
[2] Mockvichev V V, Mahutov N A and Chernyaev A P 2002 *Crack resistance and mechanical properties of structural materials of technical systems*
[3] Ermakov B S, Vologzhanina S A, Bobrovskij I N, Bobrovskij N M and Erisov Y 2018 Resistance to brittle fracture and availability of austenitic steels *IOP Conference Series: Materials Science and Engineering* 450
[4] Soltsev Yu P, Ermakov B S, Vologzhanina S A and Krutikov N V 2005 The problem of resource exhaustion of cryogenic objects *Journal of International Academy of Refrigeration* **2** 31-33
[5] Rybin V V, Malyshevskiy V A and Khlusova E I 2006 Structure and the properties of cold-resistant steels for the constructions of Northern design *Scientific and technical journal Materials Science Issues* [In Russian - Voprosy Materialovedeniya] 24-44
[6] Orlov V V, Malyshevskii V A, Khlusova E I and Golosienko S A 2014 Production technology for arctic pipeline and marine steel *Steel in Translation* **9** 696-705
[7] Bannykh O A, Bannykh I O and Lukin E I 2018 *Russian metallurgy (Metally)* 528-532
[8] Sagaradze V V and Uvarov A I 2013 Hardening and properties of austenitic steels *Editorial and publishing Department of the Ural branch of the Russian Academy of Sciences* 233–239
[9] Vologzhanina S, Ermakov B, Bobrovskiy S, Lukyanov A and Li R 2019 *Key Engineering Materials (Book Series)* 60-65
[10] Goritsky V M 2004 *Diagnostics of metals* (M.: Metallurgizdat)
[11] Panasyuk V V and Andreikiv A E 1988 *Fundamentals of fracture mechanics* (Kiev: Naukova Dumka)
[12] Solntsev Yu P, Ermakov B S and Sleptsov O I 2008 *Materials for low and cryogenic temperature* (Himizdat, St. Petersburg)
[13] Oleinik N V and Nu Van Kust 1976 *Problems of Strength* 72-78
[14] Wang Y L, Zhang H L, Li S X, Zheng K and Xue F Wang X T, 2013 *J Nucl. Mater.* 41
[15] Shin S Y, Hwang B, Lee S, Kim N J and Ahn S S 2007 *Mater. Sci. Eng.* 281.

[16] Tcupka S A and Vologzhanina S A 2017 The effect of low temperatures and deformation on the intensity of martensitic transformations in metastable austenitic steels *Kazakhstan-Kholod* 261-263

Study of steel hardness effect on the abrasive wear resistance

N M Berezin, A A Peregudov, A K Andreev, A F Igolkin and S A Vologjanina

ITMO University, 49, Kronverkskiy Pr., St. Petersburg, 197101, Russia

E-mail: akandreyev@yandex.ru

Abstract. In the course of developing the territories of the shelf of the Northern seas, the Far East and Siberia, the problem of cold-breaking materials was particularly acute. More than half of the territory of the Russian Federation is located North of the January isotherm, which has a temperature of - 20 °C. In such areas, the performance of equipment, especially in winter, falls sharply. Breakdowns of tractor and bulldozer parts in winter, compared to summer, occur more often up to 6 times, and excavator parts - up to 7 times. At the same time, it was revealed that the most dangerous period of starting machines after stops. To prevent major breakdowns at temperatures less than - 35 °C, excavators, drilling rigs, and construction vehicles are stopped. And this is despite the fact that in accordance with the operating regulations, they must be operated throughout the year. The performance of mining equipment is greatly affected by wear. Unsatisfactory resistance of cast parts at low temperatures leads to large downtime and large economic losses. The relationship between hardness and resistance to abrasive wear is considered. Abrasive with different particle sizes was used for the tests. It is shown that there is no linear relationship between hardness and resistance to abrasive wear, that in conditions of low abrasive rocks it is enough to have average hardness values of the order of 47-50 HRS, and when working in highly abrasive rocks it is necessary to strive to obtain high hardness values of about 50-55 HRS.

1. Introduction

Active development of the Northern territories requires the use of materials that provide a complex of physical and mechanical properties. Taking into account the wide application of mining equipment, it is necessary to use materials with high performance on different soils and in a wide temperature range of equipment operation.

During the analysis of failures of teeth of buckets of excavators it is established that the main reasons of it are abrasive wear and fragile destruction.

The mechanism and intensity of abrasive wear depends on the mineralogical composition of the natural mineral and its mechanical properties-primarily on its hardness. The main features of abrasive wear processes are considered in [1-7].

Physical and mechanical properties of abrasives depend on many factors of structure-homogeneity, presence of defects, fragmented or monolithic state.

Rocks have almost no plasticity and residual deformation. In practice, it is customary to consider the aggregate hardness and hardness of the minerals that make up these rocks. It should be noted that with the same aggregate hardness of different rocks, the hardness of the minerals in their composition can vary dramatically.

2. Materials and methods

The analysis of domestic and foreign practice of using steels for the teeth of excavator buckets allows us to establish the following limits of the chemical composition of steels, wt.%: carbon - 0.30 - 0.40; chromium - 0.9 - 3.5; nickel - 1.0 - 2.0; silicon 0.3 - 0.5; manganese - 0.8 - 1.2; molybdenum - 0.2 - 0.60; vanadium - 0.10 - 0.15. To search for the composition of wear-resistant steel, we used the method of mathematical planning, which allows us to conduct a relatively small number of experiments in order to find the optimal concentration of chemical elements in the alloy. The experimental melts used were melted at Izhora Plants OJSC in an open high-frequency induction furnace with a main lining. The temperature of the steel at the outlet from the furnace is 1600-1610 ° C, the casting temperature is about 1550 ° C. The casting was carried out by filling pre-dried sand-clay forms. The bars were homogenized at a temperature of 1100 - 1120 ° C for 0.5 hours, quenched from a temperature of 890-920 ° C with cooling of oil, followed by tempering at a temperature of 200 ° C and 600 ° C.

To study the mechanical properties, standard proportional samples of the first type were used according to GOST 22706-77. All tests were carried out according to the standards adopted for these types of tests. Based on the analysis of existing methods for the study of wear resistance, the installation shown in Figure 1 and the test scheme of Figure 2 are adopted.

Figure 1. Appearance of the installation

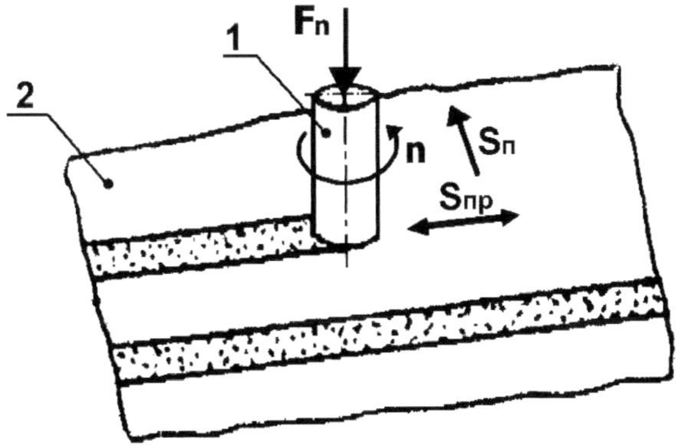

Figure 2. Durability Test Scheme: 1 - sample; 2 - abrasive paper; Fn is the load on the sample; n is the rotation of the sample around the axis; Spr - longitudinal movement of the table with the abrasive paper attached to it; Sp - lateral movement of the sample

3. Research results and discussion

The conducted abrasive tests of steels allowed to establish the degree of its wear resistance from the surface hardness of the samples. Figure 1 shows the dependence of abrasive wear (Z) of steel of different hardness on the size and size of abrasive particles.

The tests showed that at the initial stage, an increase in the hardness of the metal leads to a sharp increase in wear resistance (figure 3).

Further increase in hardness leads to a slowdown in the growth of wear resistance, and even some decline, which is associated with embrittlement of the material and its dyeing.

Figure 4 shows photographs of the surface of the samples after abrasive wear tests. As shown by fractographic studies, the decrease in the growth of wear resistance with increasing hardness values may be associated with a decrease in the depth of introduction of abrasive into the metal surface. This leads to a transition from micro-cutting (figure 4, a) to plastic repulsion (figure 4, b) and subsequent dyeing (figure 4, c).

IOP Conf. Series: Materials Science and Engineering **826** (2020) 012021 doi:10.1088/1757-899X/826/1/012021

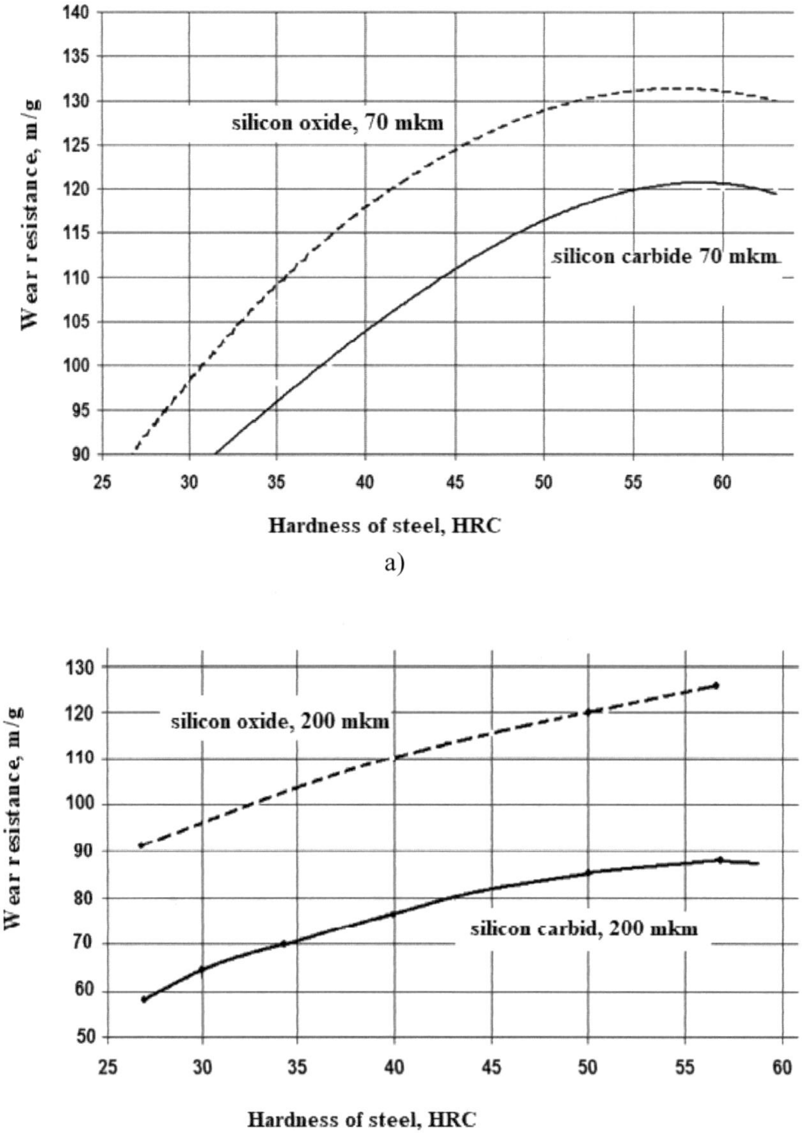

b)

Figure 3. Effect of steel hardness on abrasive wear resistance depending on SiO_2 particle size: a - 70 microns, b - 200 microns

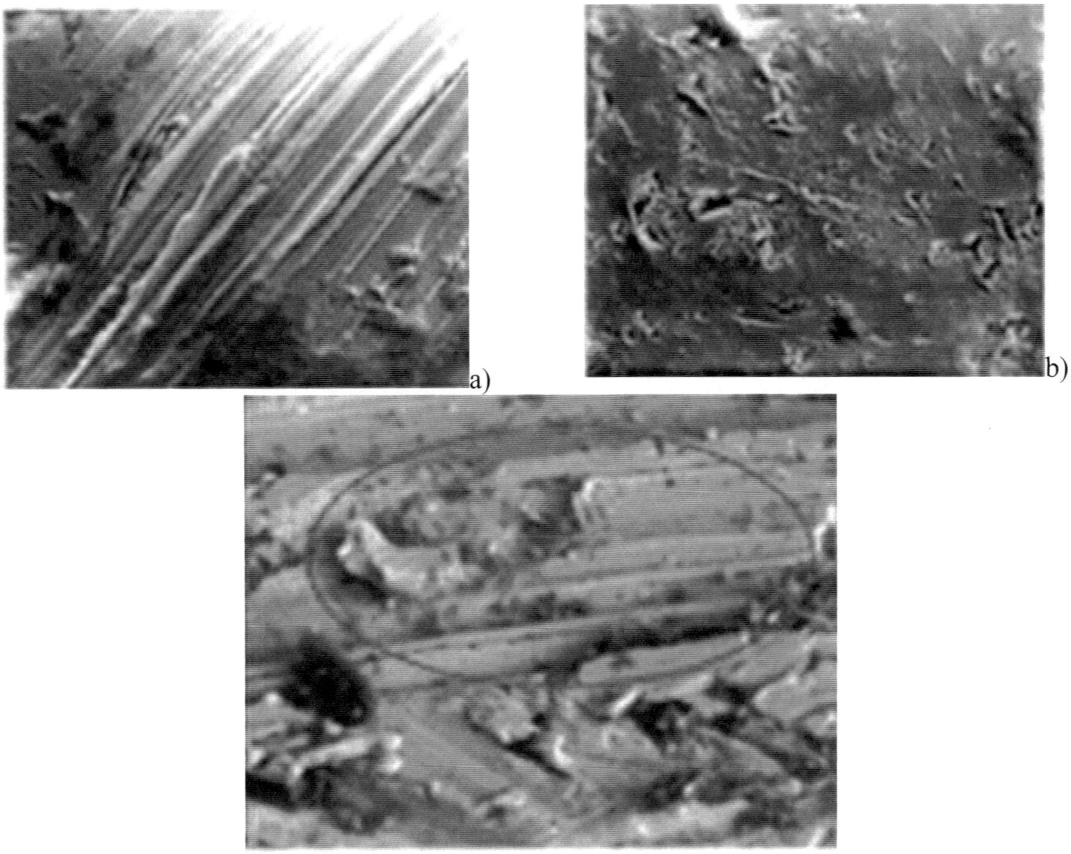

c)

Figure 4. Fractograms of wear surface of samples of different hardness, x200: a – micro-cutting and plastic extrusion, hardness 35 HRC; b – plastic extrusion, hardness 48 HRC; c – surface painting, hardness 57 HRC

4. Conclusion

Thus, the analysis of tests showed that in the conditions of low-abrasive rocks it is enough to have average values of hardness of the order of 47-50 HRC, and when working in highly abrasive rocks it is necessary to strive to obtain high values of hardness of about 50-55 HRC.

References

[1] Kragelsky I V 1998 *Friction and wear. Ed. 2* (Mockow: Engineering)
[2] Wellinger K and Uetz H 1995 Verschliss durch körnige und mineralische Stoffe unter Berücksichtigung des Mahlverschlcißcs in Kugelmühlen *Zement -Kalk -Gips* **18** (**2**) 3-14
[3] Safronov B P 1991 *Scientific and methodical bases of synthesis of tribosystems in relation to wear of steels by abrasive* (MPT)
[4] Vinogradov V N, Sorokin D A and Dotsenko V A 1995 *Abrasive wear of drill tools* (Moscow: Science)
[5] Kashcheev V N 1999 *Abrasive destruction of solids* (Moscow: Science)

[6] Serditov A E, Lebedev V V *2004 Materials X Intern. science.-technical Conf. Institute of refrigeration and food engineering* 97-102
[7] Wellinger K, Uetz H 2007 Verschliss durch Wirkung von körnigen mineralischen Stoffen *Sonderdruck aus "Materialprüfung"* **9 (5)** 25-34

Power interactions of scroll compressor elements

A Minikaev[1], D Yerezhep[1,2], D Zhignovskaia[1], V Pronin[1] and A Kovanov[1]

[1]ITMO University, 49, Kronversky Ave., St. Petersburg, 197101, Russia
[2]Al-Farabi Kazakh National University, Scientific Research Institute of Experimental and Theoretical Physics, 71, al-Farabi Ave., Almaty, 050040, Republic of Kazakhstan

E-mail: artyr_minikaev@mail.ru

Abstract. The subject of our research is a scroll compressor. With its large number of advantages over reciprocating compressors in the form of overall, mass and production characteristics, they have their own difficulties associated with the theory of stability of a movable scroll during the operation process. As you know, a complex system of forces acts on the spiral, is very variable in magnitude and direction, for the period of one orbital revolution of the spiral. It should be noted that in modern designs of scroll compressors in the axial direction, a movable scroll has most often a one-sided rigid support - a thrust bearing. Moving the movable spiral in the opposite direction is limited only by the force of the gas axial pressure, which does not guarantee it saving its original position, at least within the gaps, touching the movable and stationary spirals, which is unacceptable. Next, we consider the basic calculation equations associated with the stability of a moving spiral.

1. Introduction

The emergence in 1985 of Scroll Compressors (SCC) in the Japanese market aroused great interest in commercial and industrial circles [1]. The Japanese company Hitachi managed to realize the idea of spiral machines, patented by the Frenchman Leon Kreuks back in 1905, although before that it had taken more than ten years of searching for structural forms and testing the technology for making the machine [2]. During this time, serial production of three standard sizes of SEC for domestic and industrial air conditioners with a cooling capacity of up to 12 kW, operating on R22 refrigerant was established [3-4].

Currently, a number of companies in the USA and Western Europe conduct intensive research and development. The cooling capacity of the SEC was significantly increased to 35-50 kW and higher [5-6].

The main advantage of SECs is their high energy, production and operational efficiency. Compared with low-performance piston refrigeration compressors, SECs have higher (by 10-15%) indicator efficiency and feed rate (20-30%); smaller dimensions (30-49%) and compressor mass (10-15%); 5 dB lower sound pressure level (A); fewer details and a number of other advantages [7-8].

One of the important questions of the theory of SEC is the stability of a mobile spiral (PSP) during compressor operation, since it is affected by a complex system of forces, variable in magnitude and direction, for a period of one orbital revolution of the spiral, namely [9-10]:

- gas tangential forces directed perpendicular to the axis of rotation of the PSP (and the axis of the eccentric shaft) and do not intersect the latter. They determine the effect of torque on the compressor shaft. The vector of these forces, variable in magnitude and direction;

- centrifugal forces $\sum R_C$ directed parallel to the eccentricity line and also perpendicular to the axis of the shaft. They are constant (at ω = const) in magnitude and also variable in direction;
- gas axial forces $\sum R_A$ directed parallel to the shaft axis, variable in magnitude.

2. Materials and methods

Gas tangential and centrifugal forces create a tipping moment M_{OPR} acting on the PSP. At the same time, the gas axial forces create a moment opposite to the overturning moment of the sign. We call it the moment of stability M_y. Obviously, it is necessary (Figure 1):

$$K_y \cdot M_{OPR} \le M_y, \tag{1}$$

where the stability coefficient $K_U = 1.1 \div 1.2$.

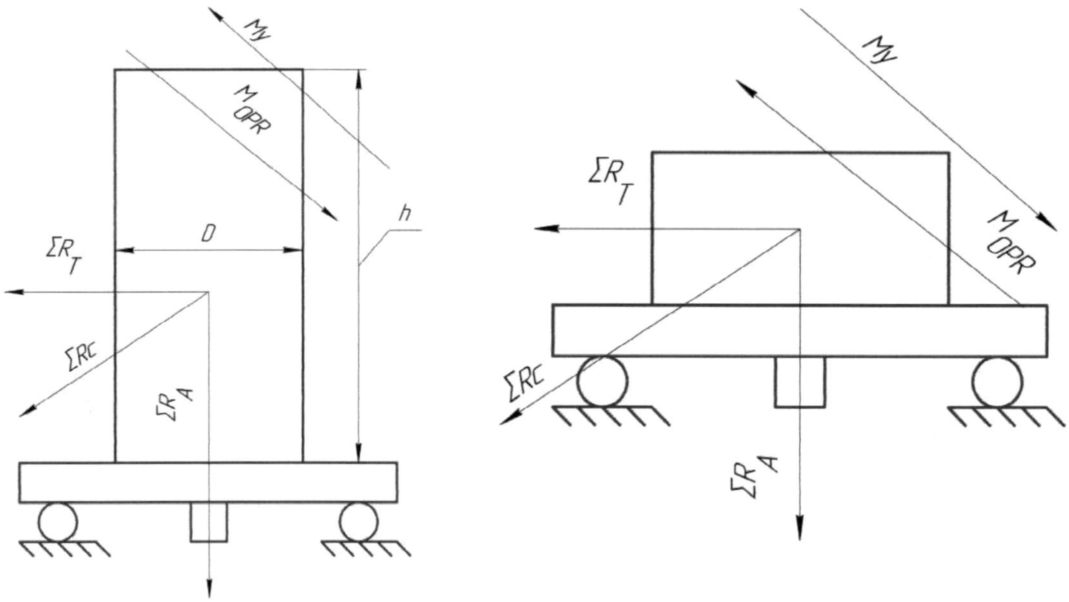

Figure 1. The total forces and moments acting on the moving spiral at different D/h ratios with the same volume of W_B cells.

In Figure 1(a-b), two possible variants of movable spirals are presented, which provide the same given SPK productivity.

Let us compare these options at the same gas pressures, performance, and frequency of orbital motion of the PSP. It can be argued that (with the same scale of drawings):

$$M^a_{OPR} > M^b_{OPR}, \tag{2}$$

$$M^a_y < M^b_y. \tag{3}$$

Deflection of the tops of the feathers of spirals:

$$\Delta^a_{PSP} > \Delta^b_{PSP}. \tag{4}$$

On the other hand:

$$\sum R^a_A > \sum R^b_A; \tag{5}$$

$$m^a > m^b \text{ (due to the size of the platform).} \quad (6)$$

The dimensions of the thrust bearings, the load on them and the moment of friction in the bearings and the anti-rotation device (PPU) are higher in the second version (Figure 1, b). At the same time, the technological difficulties of ensuring the accuracy of manufacturing spirals (taking into account thermal and force deformation) are more significant in the first version (Figure 1, a).

From this list of pros and cons of the two options for spirals, one can see how complicated the issue of optimizing the size of the PSP SPK is.

To simplify the problem, we restrict ourselves to considering the stability of the PSP position under the influence of the forces acting on it.

We recall that in modern designs of the SPK in the axial direction, the PSP has most often one-sided rigid support - a thrust bearing. The movement of the PSP in the opposite direction is limited only by the force of the gas axial pressure, which does not guarantee it from losing its original position, at least within the gaps, to touching the movable and fixed spirals, which is unacceptable.

3. The discussion of the results

Assuming the same loss as a first approximation, we assume that the theoretical volumetric performance of both options is the same:

$$V_T = W_B n_C, \quad (7)$$

where, W_B – the volume of two suction cells SPK; n_C rotation frequency $W_B = f(\mathrm{D}, \mathrm{h}, \mathrm{t}, \delta)$.

In its turn:

$$D = f(\theta_o, t, \delta), \quad (8)$$

where t - spiral pitch, h and δ the height and thickness of the spiral, respectively.

We can accept $\delta = const;$ θ_o – twist angle of spirals.

We accept $\theta_o = 4\pi$ for both options.

Thus, the volume of the suction cells, which determines the theoretical volumetric productivity under the indicated conditions.

To ensure the remaining parameters of the SPK, to determine the forces and moments acting on the PSP and other elements of the compressor, it is necessary to reveal this dependence. According to the dependences obtained in the theory of SPK:

$$W_B f = 2h\left[(f_{\varphi 2} - f_{\varphi 1}) - f_{CN}\right], \quad (9)$$

where $(f_{\varphi 2} - f_{\varphi 1})$ – helix area between twist angles φ_2 and φ_1; f_{CN} – the end area of the feather of the spirals within the specified angles.

Spiral area:

$$f_\varphi = 0.5 r_o^2 \int_{\varphi_1}^{\varphi_2} \varphi^2 dy = \frac{(\rho_2^3 - \rho_1^3)}{6 r_o}, \quad (10)$$

where r_o - radius of the main circle of the spiral; ρ_1 and ρ_2 the radius vectors of the spiral, relevant angles φ_2 and φ_1.

At $\varphi_2 = 4\pi$ area bounded by external spiral arcs $\varphi_2 = 4\pi$ and $\varphi_1 = 3\pi$, according to the formula (10), will be:

$$f_{4\pi} - f_{3\pi} = 6.17\pi^3 r_o^2 = 4.84 t^2 = C_o t^2, \quad (11)$$

as $r_o = \dfrac{t}{2\pi}$.

Area:

$$f_{CN} = \delta S_{\varphi 1}^{\varphi 2} f(\varphi) \tag{12}$$

where $S_{\varphi 1}^{\varphi 2} f(\varphi)$ – spiral length between corners $\varphi_2 = 4\pi$ and $\varphi_1 = 3\pi$.

$$S_{3\pi}^{4\pi} f(\varphi) = r_o \int_{3\pi}^{4\pi} \sqrt{1+\varphi^2}\, d\varphi = 0.5 r_o \left[\varphi\sqrt{1+\varphi^2} + \ln(\varphi + \sqrt{1+\varphi^2}) \right]_{3\pi}^{4\pi} = c_1 r_o \tag{13}$$

where $c_1 = const$,

$$f_{CN} = \delta c_1 r_p = \frac{\delta c_1 t}{2\pi} = c_2 t \tag{14}$$

where $c_2 = const$, (c_2 is a named number).
So:

$$W_B = 2h(c_o t^2 - c_2 t^2) \tag{15}$$

According to equation (2), the volumetric theoretical productivity of the SPK:

$$V_t = 2 n_c h (c_o t^2 - c_2 t^2). \tag{16}$$

This dependency allows you to build graphs $t=f(h)$ and $\varepsilon_o = f(h)$ at $V_t = const$ and $n_c = const$ (Figure 2, ε_o - radius of the orbital circle).

By choosing t and h you can calculate the dimensions of the spirals, find the required cell areas and their volumes, then the location of the cells relative to the tipping axes, determine the forces acting on the spiral, and the $\sum \vec{R}_c$ forces are reduced to the plane of the force $\sum \vec{R}_T$.

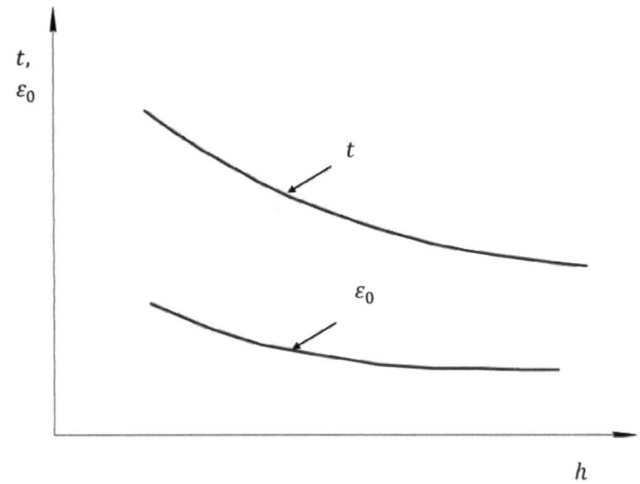

Figure 2. The relationship between the main parameters of the spiral

After that, we should calculate M_{OPR} and M_y, build graphs in the function h, find K_y (Figure 3-4).

When calculating M_{OPR} you must first determine the resultant force vector $\sum \vec{R}_T$ and $\sum \vec{R}_c$, i.e. tipping force:

$$\sum R = \sqrt{(\sum R_T)^2 + (\sum R_2)^2 + 2\sum R_T \sum R_C \cos(\sum R_T ; \sum R_c)} \quad (17)$$

and its direction.

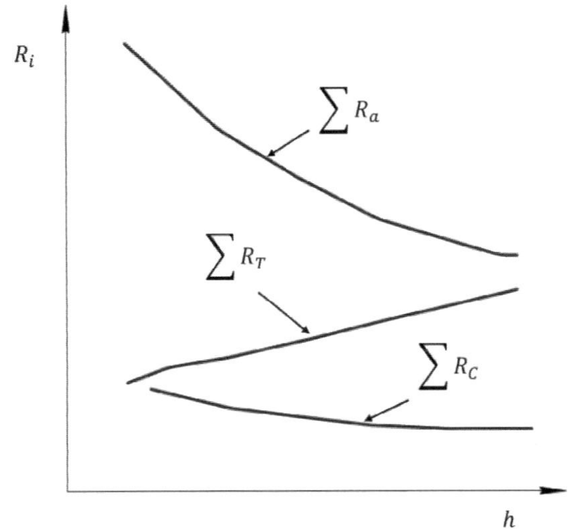

Figure 3. The nature of the change in the forces acting on the moving spiral when changing ***h***

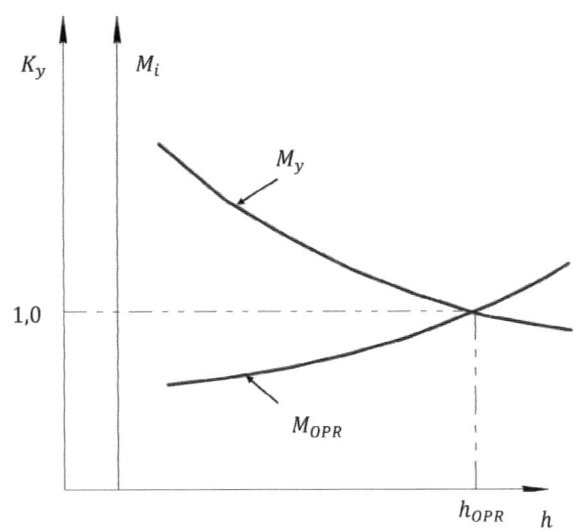

Figure 4. Graph for determining the optimal value of the height of the ribs of the spirals

Next, several checks of the SRP for stability should be performed. To do this, without changing the thermal regime (t_0 and t_K), it is necessary to check the stability of the SRP at other angles of rotation of the SRP, i.e. at other positions of the SRP in the limit of one orbital revolution. So, if the previous calculation was made, for example *φ=90°*, it is necessary to do the same calculations with other values *φ*, for example, through *45°* or *90°* from *0°* to 2π. We find $(M_{OPR})_{max}$ and $(M_{OPR})_{min}$, corresponding values M'_y ,M''_y *and* $K_y^{(i)}$.

4. Conclusion

The determination of the optimal value of h does not end there, since under other thermal conditions the ratio between M_y and M_{OPR} (i.e., the K_y value) will be different. Since with a decrease in the pressure difference $\Delta p = p_K - p_o$, the stability coefficient K_y will decrease.

In fact, with a decrease in Δp, M_y also decreases in proportion to the change in Δp (for each position of the SRP on the orbital path).

M_{OPR} will also decrease, but in a smaller proportion, because of the two components of the overturning force, the tangential force $\sum \vec{R_T}$ decreases in proportion to change Δp, and the total centrifugal force $\sum \vec{R_C}$ does not change (at ω = *const*). Therefore, when decreasing Δp will decrease and K_y.

From the foregoing, it follows that after a preliminary determination of the optimal value ***h*** it is necessary to check this parameter in other thermal conditions with lower values Δp and upon receipt $K_y \geq \left[K_y\right]$ finally take the value ***h***.

5. Conflict of interest

The authors declare that they have no conflict of interest on the content of this paper.

References

[1] Mozgina V 1986 *Scroll compressor* (Moscow.: TSINTIkhimneftemash) **2** 1-4
[2] Morichita E and Sigihara M 1986 *Mech. Eng.* **32** 2121-2130
[3] Yerezhep D, Umutbaeva N R, Piterskov P, Baranov A Y and Yerezhep A 2019 *J. Phys.: Conf. Ser.* **1333** 032091 doi: 10.1088/1742-6596/1333/3/032091
[4] Mateu-Royo C, Navarro-Esbría J, Mota-Babilonia A, Molés F and Amat-Albuixech M 2019 *Applied Energy* **253(1)** 113504. doi: 10.1016/j.apenergy.2019.113504
[5] Yerezhep D, Yemelyanov A, Kuznetsov Y, Perminov A, Pronin V A, Minikaev A F and Yerezhep A 2019 *J. Phys.: Conf. Ser.* **1333** 032093 doi: 10.1088/1742-6596/1333/3/032093
[6] Shunan Hu et al 2019 *IOP Conf. Ser.: Mater. Sci. Eng.* **612** 032126 doi: 10.1088/1757-899X/612/3/032126
[7] Ahn S, Choi S and Rhim Y 2018 *International Journal of Refrigeration* **93** 205-212 doi: 10.1016/j.ijrefrig.2018.07.006
[8] Zhao B, Chen S, Xu L, Shi C, Gao D and Zhang Y 2017 *Numerical Heat Transfer, Part B: Fundamentals* **72(4)** 311-342 doi: 10.1080/10407790.2017.1394127
[9] Aksenov A et al 2017 *IOP Conf. Ser.: Mater. Sci. Eng.* **232** 012042 doi:10.1088/1757-899X/232/1/012042
[10] Tsukamoto K, Hiradate K, Sakamoto K, Yasushi Shinkawa 2017 *Proceedings of the ASME Turbo Expo* **2C** 130041 doi: 10.1115/GT2017-64210

Study of the rheological characteristics of the potato-based fish and vegetable mixture melt in the pre-die zone of a single-screw extruder

O I Aksenova, G V Alexeev and D D Temershin

ITMO University, 9, Lomonosova st., St-Petersburg, 191002, Russia

E-mail: oksi280491@yandex.ru

Abstract. This paper presents study of the rheological characteristics of the melt of the fish and vegetable mixture biopolymer in the pre-die zone of a single-screw extruder. The analysis of the flow index, being less than one, and that of the flow curves of the fish and vegetable mixture biopolymer melt, intended for production of expanded potato snacks, showed that the melt in a certain range of deformation rates can be attributed to pseudoplastic fluids. The results of the study represent important data in the design of the extruders molding units, on which the stability of the finished products quality depends. The results of the study can be used for mathematical modelling of the flow of the biopolymer melt in the pre-die zone of the extruder and can serve as justification of the screw rotational speed range in order to obtain extrusion foodstuffs of stably high quality.

1. Introduction

Extruded foodstuffs represented all over the world with a wide range of products, from breakfast cereals to protein textures, have recently been in great demand. The extrusion allows to process a variety of raw materials of plant and animal origin, mixtures and by-products of the food industry, which characterizes extrusion as a universal process.

The variety of processed raw materials, which qualitatively changes its properties during thermoplastic extrusion, causes problems of stabilization of the product yield and of obtaining a product of a predicted quality level, as well as problems associated with the creation of new designs of extruder molding units that would increase production efficiency, problems associated with the development and implementation of methods for assessing the quality of finished extrudates in the production. Rheological research methods can play a significant role in solution of the above-mentionned problems.

Among the complex of physical properties, rheological ones are main, since they determine the behavior of the biopolymer melt in the technological process. They are an external expression of the internal essence of objects, that is, they characterize the aggregate state, dispersion, composition, structure and type of interactions within the product [1]. The rheological properties of the biopolymer melt from fish-vegetable raw materials depend on the moisture content of the feedstock, on the temperature mode of the extrusion process and on the pressure in the pre-die zone of a single-screw extruder [2].

The aim of the paper is to research the rheological properties of the melt of the fish and vegetable mixture in order to predict the behavior of the material in the forming channel of a single-screw extruder.

An analysis of rheological studies [3-9] shows that, as for the nature of the flow, melts into which mixtures based on starch raw materials or mixtures with a predominant content of starch raw materials when processed in the field of high temperatures and pressures are transferred are abnormally viscous fluids obeying the power law of Oswald-de-Ville:

$$\tau = k \cdot \dot{\gamma}^n, \tag{1}$$

τ is the shear stress at the capillary wall, Pa; $\dot{\gamma}$ is the shear rate near the capillary wall, s^{-1}, k is the consistency coefficient, Pa·s^n; n is the flow index. At the same time, the consistency coefficient is a measure of the liquid viscosity, and the flow index characterizes non-Newtonian behavior. Based on the value of the flow coefficient, power-law fluids are divided into three groups: into pseudoplastic and thixotropic ones for $n < 1$, into Newtonian one for $n = 1$, into dilatant one for $n > 1$.

In extrusion processing, the coefficients k and n included in equation (1) can be determined experimentally based on capillary viscometry. Traditionally, capillary viscometry is used in order to measure the viscosity of melts at medium and high shear rates and at high temperatures up to 773 K [10]. The rheometer in this measurement is the rheometer-channel system with means for measuring the pressure difference, which is proportional to the shear stress and to the volumetric flow rate. At the same time, the extruders used to conduct the study should provide a sufficiently high pressure for forcing the biopolymer melt through long capillaries of small radius.

2. Materials and methods

The methodology for determination of the flow index n and of the consistency coefficient k is based on the direct use of the power law of the flow which is realized through the equation for calculating the pressure loss arising from the viscous flow of the polymer melt in any given simple channel (a capillary), and on the application of the two capillary and the least square methods.

The method of two capillaries allows to eliminate the determination of the pressure loss at the input to the forming channel, because capillaries are geometrically similar systems with different lengths of the forming channel, to neglect the turbulence arising in the initial section of the measuring capillary, as well as the final effects [10].

For the study, a laboratory unit based on a SFE-2 extruder was used. A pair of matrices was made, shown in Figure 1. The forming channels of the matrices had a circular cross section of equal diameter of 0.003 m and different lengths of 0.027 m for the short capillary and of 0.078 m for the long one. A ratio of capillary lengths to diameter equal to 9 and 26, respectively, was chosen so that in one of the cases a high ratio of capillary lengths to diameter was observed. The use of a high ratio completely eliminates the influence of the most significant input effects due to the stored elastic energy in the biopolymer melt subjected to shear at a high speed, the non-laminar flow in the capillary entrance area and the unsteady flow conditions when the flow is accelerated at the entrance into a small diameter capillary [10].

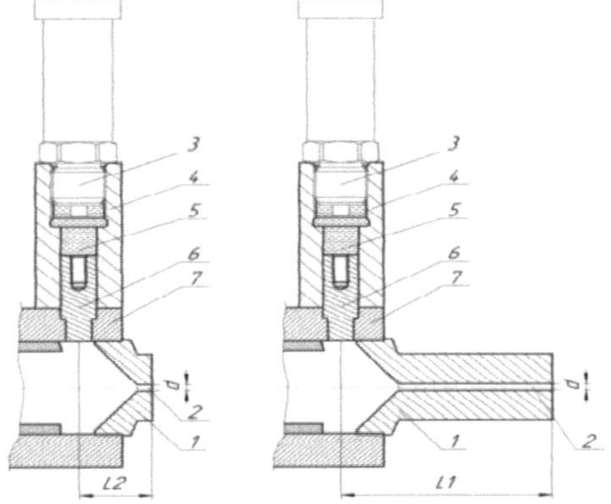

Figure 1. Dies with short and long round capillaries: 1 – die, 2 – capillary, 3 – pressure sensor PD100-DI12,0-111-1,0, 4 – boss, 5 – incompressible fluid, 6 – pressure transmission piston, 7 – barrel.

In order to determine the pressure at the input to the capillary, a pressure measuring system was installed, based on a membrane-pressure transducer. The pressure measurement system is a multilayer construction in which pressure is transferred to the sensor membrane from the biopolymer melt in the pre-die zone by a piston through a layer of incompressible fluid. Direct connection of the pressure sensor to the extruder pre-die zone is complicated by the constant sticking of the biopolymer melt flow and by the limited diameter of the extruder pressing case. Also, during rheometry in systems with a round capillary, we estimated the pressure difference in the extruder body and at the exit of the die (the atmospheric pressure), since the pressure difference inside the capillary cannot be determined, due to the difficulty to install pressure sensors in a round capillary of such a small diameter.

The medium used was a fish and vegetable mixture intended for the production of potato snacks with an increased content of protein and of dietary fiber. The recipe mixture had the following percentage of components: 79% of dehydrated ground potatoes, 13% of dry malt grains, 8% of fish powder from by-products of salmon processing (growths, flanks, napes, fish bellies, cuts from ridges, slices and non-standard pieces of salmon).

In the experiment, pressure losses Δp_1 and Δp_2 were determined for each capillary at a fixed volume flow and at variable temperature values in the pre-die zone of a single-screw extruder in the range from 373 to 433 K and at the initial humidity of the mixture from 12 to 18%. The selected range of temperature and humidity is typical of the hot extrusion mode.

The constancy of the flow rate was set by the rotational speed of the screw shaft of the extruder. The volumetric flow rate was determined based on the mass of the extrudate flowing out per unit time, at a known melt density in a given temperature range.

Below is the calculation procedure.

The difference between the obtained pressure loss values is calculated as follows:

$$\Delta p_{1,2} = \Delta p_1 - \Delta p_2 \tag{2}$$

where Δp_1, Δp_2 are the experimental values of pressure loss at the exit of the extruder at an equal volume flow, Pa.

It is known that the equation for a power-law fluid using capillaries of circular cross section can be written as follows:

$$\Delta p_{1,2} = 2k \cdot (L_1 - L_2) \cdot Q^n \cdot (\tfrac{A}{4} \cdot r^A)^{-n} \qquad A = \frac{3 \cdot n+1}{n}, \tag{3}$$

L_1, L_2 are the lengths of the long and short capillaries, respectively, m; r is the radius of the capillary, m, Q is the volumetric flow rate, m^3/s. We denote:

$$N = 2k \cdot (L_1 - L_2) \cdot (\tfrac{A}{4} \cdot r^A)^{-n} \qquad \Delta p_{1,2} = N \cdot Q_n. \tag{4}$$

Let us suppose that in processing the experimental data, a system of points Q_k, $\Delta p_{1,2}$, $k = 1 - m$, was compiled, which theoretically correlates with the equation (4). Then, using the least squares method, we can find estimates of the pair (N, n):

$$\ln N = [\textstyle\sum_{k=1}^m \ln^2 Q_k \cdot \sum_{k=1}^m \ln\Delta p_{1,2} - \sum_{k=1}^m \ln Q_k \cdot \sum_{k=1}^m \ln Q_k \cdot \ln\Delta p_{1,2}] \cdot [m \cdot \sum_{k=1}^m \ln^2 Q_k - (\sum_{k=1}^m \ln Q_k)^2]^{-1} \tag{5}$$

$$n = [m \cdot \textstyle\sum_{k=1}^m \ln Q_k \cdot \ln\Delta p_{1,2} - \sum_{k=1}^m \ln Q_k \cdot \ln\Delta p_{1,2}] \cdot [m \sum_{k=1}^m \ln^2 Q_k - (\sum_{k=1}^m \ln Q_k)^2]^{-1} \tag{6}$$

It is known that for an annular forming channel, the consistency coefficient is defined as follows:

$$k = \frac{N}{2(L_1 - L_2)} \cdot (\tfrac{A}{4} \cdot r^A)^{-n}. \tag{7}$$

3. Results and discussion

The results of calculating the coefficients k and n, according to the procedure described above, based on the experimental data are shown in Table 1.

Table 1. Results of calculating the coefficient of consistency and of flow index.

Temperature in the pre-die zone T (K)	Initial humidity of the mixture H (%)	Flow index n	Consistency coefficient k (Pa·s^n)
373	12	0.522	492507
373	14	0.556	398121
373	16	0.589	327810
373	18	0.621	275341
383	12	0.532	420725
383	14	0.566	341960
383	16	0.599	283231
383	18	0.631	239417
393	12	0.542	359989
393	14	0.576	294234
393	16	0.608	245176
393	18	0.640	208604
403	12	0.552	308532
403	14	0.586	253621
403	16	0.618	212643
403	18	0.649	182138
413	12	0.562	264878
413	14	0.595	219015
413	16	0.627	184791
413	18	0.658	159368
423	12	0.572	227795
423	14	0.605	189484
423	16	0.637	160914
423	18	0.667	139754
433	12	0.582	196252
433	14	0.615	164249
433	16	0.646	140412
433	18	0.676	122830

An analysis of the results of calculating the flow index n in Table 1 shows that for all humidity and temperatures in the considered range, the biopolymer melt can be assigned to abnormally viscous fluids (pseudoplastic or thixotropic), since $n < 1$.

Data from Table 1 were processed using mathematical statistics methods, on the basis of which the following regression equations were obtained:

$$k = 105.34 \cdot 10^4 - 22.39 \cdot 10^3 \cdot H + 3.59 \cdot 10^3 \cdot T - 2.84 \cdot H^2 - 0.409 \cdot T^2 \quad (8)$$

$$n = 0.253 + 1.61 \cdot 10^2 \cdot H + 0.95 \cdot 10^3 \cdot T. \quad (9)$$

T is the melt temperature of the biopolymer in the pre-die zone of a single-screw extruder, K; H is the initial moisture content of the fish mixture, %.

An analysis of the regression equations shows that humidity has a greater effect on the rheological properties of the melt than temperature, which is consistent with the results of studies [3, 5, 6, 9]. The direct dependence of viscosity on humidity in the raw material and the inverse one on the temperature of the barrel are explained by the plasticizing effect of moisture during the extrusion of starch-containing raw materials, which reduces the degree of degradation during shear.

The obtained equations describe the change in rheological constants with a 95% probability level and make it possible to identify patterns of change in the rheological characteristics of the biopolymer melt in a given range of the studied parameters.

An analysis of the dependences of the rheological constants on the mixture humidity and temperature in the pre-die zone shown in Figure 2 showed that an increase in temperature from 373 K to 433 K leads to an exponential decrease in the consistency coefficient, which may be due to the overcoming of the viscosity by the kinetic energy of the molecules.

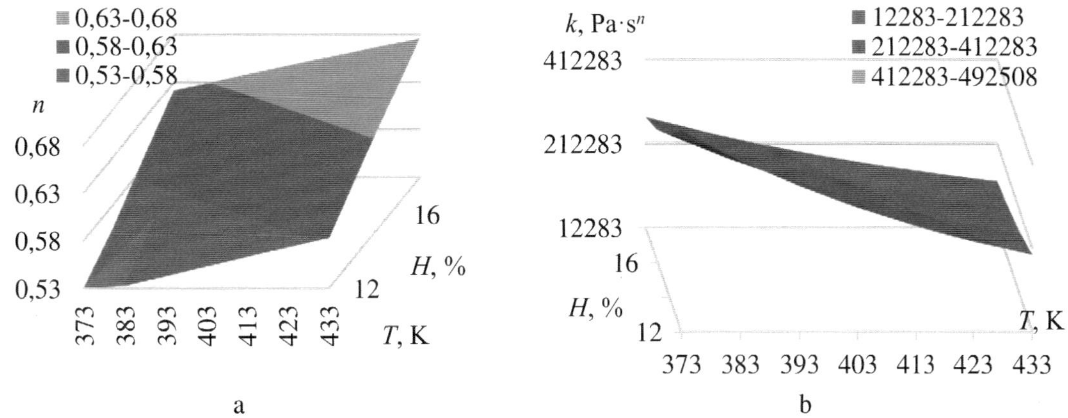

Figure 2. Dependence of the flow index n (a) and of the consistency coefficient k (b) on the humidity H of the fish and vegetable mixture and the temperature T in the pre-die area.

Based on the obtained values of the flow index n and of the consistency coefficient k, the change in shear stress near the capillary wall at various humidity and temperature values in the pre-die zone of a single-screw extruder was determined by the power-liquid equation (1). The flow curves for the calculated parameters at a constant temperature and at a constant humidity are shown in Figure 3. An analysis of the flow curves in Figure 3 shows that the structure of the biopolymer melt breaks down in the range of shear rates of 10–20 s^{-1}, which indicates the rearrangement occurring in the melt of the product under study. Figure 3b shows that an increase in temperature causes a decrease in the shear stress τ of the mixture melt, which is also consistent with studies [6, 9, 10].

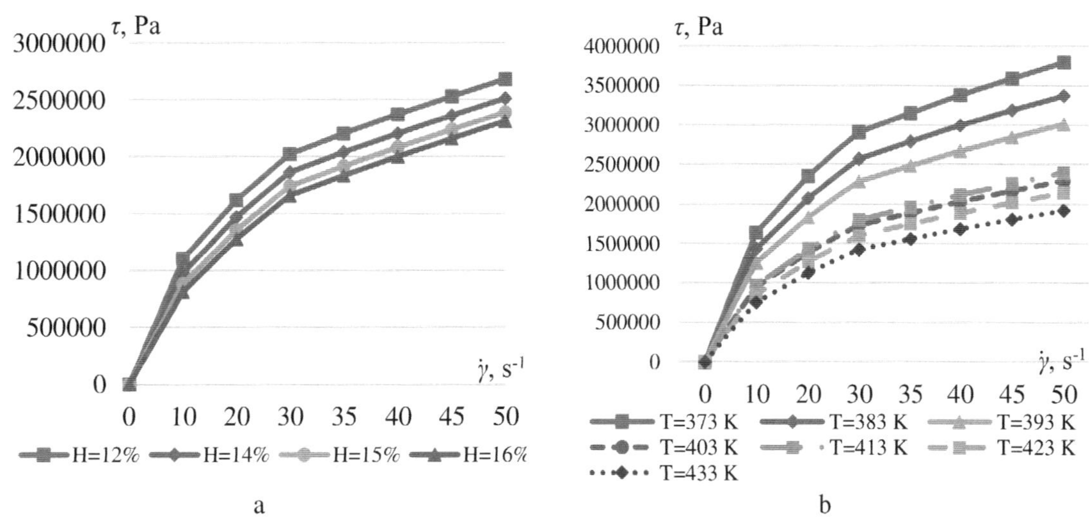

Figure 3. Curves characterizing the currents: a - at a constant temperature of 403 K and at various humidity H values; b - at a constant humidity of 12% and at different temperatures T.

4. Conclusion

In a series of experiments, it was confirmed that all components of the recipe mixture, for the production of potato snacks, when processed in the field of high temperatures under high pressure, pass into the melt, which refers to abnormally viscous liquids obeying the Oswald-de-Ville power law. As a result of studying the rheological properties of the melt of the fish and vegetable mixture located in the pre-die zone of the extruder, based on the calculation results at n <1 and the nature of the flow curves, it was determined that the melt in a certain range of deformation rates can be attributed to pseudoplastic fluids. It was found that the maximum deformation rates necessary to obtain extrudable products of the required quality correspond to the smooth transition to the region of the destroyed structure and are of 10-20 s^{-1}.

The obtained results of the study of the rheological properties of the biopolymer melt helped to better understand the ongoing changes in the raw material during its transformation into a finished product and the influence of extrusion process factors on the formation of the finished product quality.

The authors will direct further research to study the relationship between the rheological properties of the melt and the microstructure of the finished product and the design of a unified forming unit for single-screw extruders.

References

[1] Vlachopoulos J and Strutt D 2003 *Extrusion Minitec and Conference: From Basics to Recent Developments* 1-25
[2] Chen Y, Struble L J and Paulino G H 2008 *Int. J. Appl. Ceram. Tech. Applied.* **5** 513-21
[3] Chen H, Xie F, Chen L and Zheng B 2019 *J. Food Eng.* **244** 150-58
[4] Alam M S, Kaur J, Khaira H and Cupta K 2016 *Crit. Rev. Food Sci.* **56** 445-73
[5] Thandavathi Y L N, Wassen S and Kadar R 2019 *J. Food Eng.* **245** 112-123
[6] Xie F, Yu L, Su B, Lui P, Wang J, Liu H and Chen L 2009 *J. Food Eng.* **49 (3)** 371-77
[7] Wani I A, Farooq G, Qadir and Wani T A 2019 *Int. J. Biol. Macromol.* **131 (15)** 850-57
[8] Martinez-Bustos F, Agullar-Palazuelos E and Galicia-Garcia T 2012 Thermoplastic Behavior of Biopolymers during Extrusiom *Biopolymer Engineering in Food Processing* (CRC Press, Taylor & Francts Group) chapter 7 pp 245-78
[9] Cheyne A, Barnes J and Wilson D I 2005 *J. Food Eng.* **66** 1–11
[10] Schramm G 1994 *A Practical Approach to Rheology and Rheometry* (Karlsruhe: Gebueder HAAKE GmbH)

IOP Conf. Series: Materials Science and Engineering **826** (2020) 012024 doi:10.1088/1757-899X/826/1/012024

A secondary heat effect on the properties of K70 strength class steel for trunk pipelines

D M Sharapova[1], O Y Ganzulenko[1] and M G Sharapov[2]

[1] Saint Petersburg Mining University, 2, 21st line of Vasilyevsky Island, St. Petersburg, 199106, Russia
[2] NRC «Kurchatov Institute» - CRISM «Prometey», 49 Shpalernaya str., St. Petersburg, 191015, Russia,

E-mail: mthi@spmi.ru

Abstract. The article presents the results of the technological weldability studies for K70 strength class steel. A gradation of the studied semi-finished products by their repairability (by the number of the allowed local reheat of the material during the welding and repair processes) is proposed. The viscosity reducing of the metal in the overheating zone, determined by the magnitude of the impact test with Charpy samples at the temperature – 40° C was a basis for the gradation condition. The influence of the welding heat input and the number of reheats on the HAZ metal viscosity is established. Thus, the obtained results allow us to state the low technological weldability of this steel, taking into the account such an important indicator as the repairability of the material.

1. Introduction

The intensive exploration of the northern seas shelf and regions with a cold climate creates an interest in the elaboration of the K70 strength class steels for the trunk pipelines and determines not only the development and success of the materials science, but the industrial technologies improvement for the semi-finished products manufacturing (smelting and rolling) as well [1 - 4].

When designing new domestic pipelines, there were well-known research results and the developed requirements for the welded pipe joints taken into account [5]. Increased requirements were set for the welding procedure by the impact bending tests at low temperatures. The test temperatures of – 40° C normalize impact energy on Charpy specimens with a sharp notch for base metal and welded joint.

The authors noticed the degradation problem of the properties in the heat-affected zone (HAZ) when the results of the dynamic tests at low temperatures on Charpy specimens with an incision in the place of overlapping heat-affected zones, became known. The specimens were cut from welded joints of the standard parts from strength class K70 steel, had been produced by various manufacturers [6]. Among the results, there were drops up to 37 J, which was lower than the regulatory requirements (KV_{min} = 47 J at the test temperature - 40° C). It should mentioned that on the studied details the obtained drops on impact energy below the permissible level in the places of a single heat exposure were not noted.

The fact that various researchers, both in Russia and in other countries, are actively involved in issues of the properties degradation of the metall materials in the heat-affected zone confirms the urgency of this problem [7 - 11].

2. Materials and requirements for the pipes mechanical properties

We used semi-finished steel of strength class K70 with a thickness of 20 mm as the base metal.

The developed in Russia low-alloyed high-strength steel K70 belongs to the manganese alloying system, it contains relatively small additives of Ni, Cu, Mo and microalloying elements Ti, Nb, V with the limitation of the harmful impurities, such as S$\leq$ 0.001% and P $\leq$ 0.010%. The studied steel meets the applicable requirements to the steels of this strength class being used abroad [4].

According to the requirements, K70 steel should provide tensile strength at the level of σ_B = 690-790 MPa, 0,2 %-proof strength $\sigma_{0,2}$ = 590-690 MPa with a ratio of $\sigma_{0,2}$ / $\sigma_B \leq 0.9$ and an elongation after fracture $\geq$ 16%. The impact energy of the base metal at - 40° C should be not lower than 80 J (the average of 3 tests with a minimum value of $KV_{min} \geq 60$ J). The impact energy of the weld metal should be $\geq$ 65 J (with a minimum value of $KV_{min} \geq 47$ J).

3. Research methodology

There were plates of 1200x500x20 mm in size cut out of the steel for the research, after that the plates were welded with a symmetrical V-shaped edge preparation without a gap into the samples of 1200x1000 mm in size. A groove angle was 60°, a root face was 5 mm.

The welding mode was selected in a such way as to avoid the complete penetration of the root face. After the welding, the weld convexity was mechanically removed and the original edge preparation was restored. After that, the welding was repeated in the selected mode, which ensured the imposition of the HAZ from the second pass on the HAZ of the first one.

Thus, the sample metal was heated in the heat-affected zone up to 10 times (see the scheme in Figure 1). In the experiments, not only the number of heatings (welding passages) was changed, but also the heat input. The studies were carried out with a heat input of 1.5 kJ / mm, 3.0 kJ / mm and 5.0 kJ / mm. The welding was realized by an electric arc with a consumable electrode under a flux on a 2-arc machine ESAB Master A6 (2 arcs in one bath).

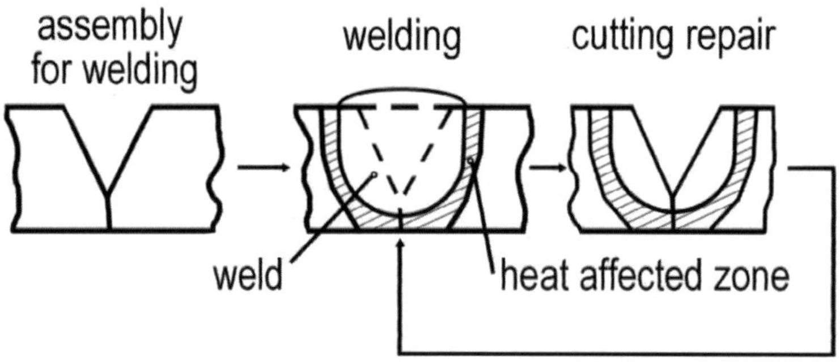

Figure 1. An experimental scheme

Impact bending tests were performed on a Walter + Bai PH-450J pendulum impact testing machine with an impact energy of 450 J. The metal viscosity in the HAZ was determined according to DNV OS-F-101 on transverse samples with a V-shaped notch. The incision was made in the samples in the perpendicular direction to the surface of the plates.

4. Results and discussion

HAZ preliminary measurements on the study samples showed the dependence of its size on the heat input according to the linear law $\delta_{haz} = 1 + 0.8$ Qn, which was confirmed by a calculation [12-13], based on the unsteady heat equation (calculation was performed jointly with V.N. Startsev) according to the proposed tested method in [14-15].

Taking into account the rectilinear uniform motion of the heat source (arc) along the weld seam (in the direction of the coordinate OX), a mathematical model of the form was used for calculations:

$$\rho\ C_p\ V\ dT/dx = div\ (\lambda\ grad\ T{+}S),$$

where ρ is the density, C_p is the heat capacity, λ is the thermal conductivity, S are the volume heat sources, and T is the temperature.

The numerical solution was performed using the Flex-EDF program.

For the modes under consideration the thickness of the HAZ was 2.2 mm $\leq \delta_{haz} \leq$ 5 mm, which allows us to confidently assert that the zones of the thermal influence from the subsequent passes to the previous ones are completely superimposed.

When testing the HAZ metal on transverse Charpy specimens, the incision was positioned so that it fell on a section of large grain (near the fusion line) on the studied semi-finished products of K70 steel, the impact work obeys the law:

$$KV \geq 75 - 5n,$$

where n is the number of reheats.

As it was established in previous reseach [6], the decrease in the properties of HAZ was associated with the formation of a fragile high-strength martensitic or martensitic-bainitic structure, which contained a high level of carbon and alloying elements, having appeared during the repeated heating along the grain boundaries at the overheating site of the HAZ (Figure 2).

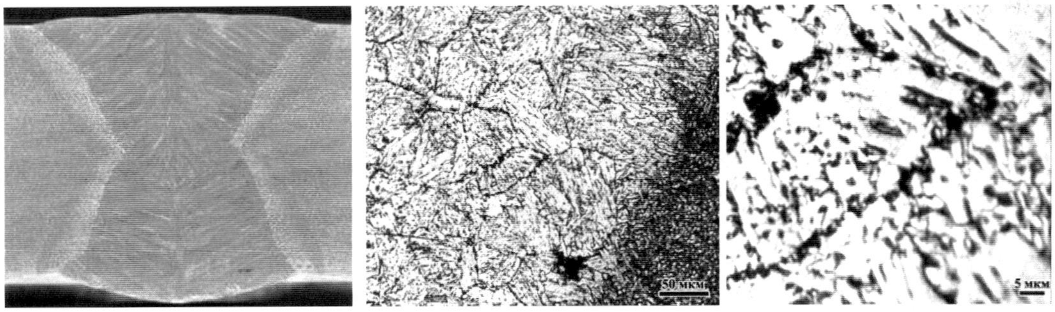

Figure 2. Weld joint structure of the longitudinal seam pipe from steel K70

Based on the obtained results, the parts, made of steel K70 strength class, were graded according to their repair ability, namely, according to the number of permissible repeated local heating of the product material, for example, carried out during repair work (see Table 1).

Table 1. Permissible amount of the local reheats of the steel with K70 strength class, depending on a heat input and a degree of the parts reliability

Degree of the parts reliability	Heat input, kJ / mm		
	1,5	3,0	5,0
critical parts	3	2	-
parts of increased reliability	5	3	1
1.5	unlimited	4	2

The gradation was made taking into account the fact that the viscosity of the metal in the heat-affected zone, determined by the average value of the impact energy, must satisfy the presented requirements depending on the minimum value level KV $\geq$ 65 J, which is pointed in the regulatory documentation:

- for critical parts $\geq$ 0.9 [KV];
- for parts of increased liability $\geq$ 0.8 [KV];

- for non-critical parts ≥ 0.7 [KV];
- for non-essential parts, KV is not regulated.

In an analysis of the obtained results, there is a direct dependence of the HAZ metal impact testing energy value on the heat input during under the welding and the number of reheats. In particular:

- at Qn = 5 kJ / mm, the average value of the impact energy varies from 49 J (with 1 reheat) to 37 J (with 2 reheats);
- at Qn = 3 kJ / mm, the average value of the impact energy varies from 59 J (with 2 reheats) to 47 J (with 4 reheats);
- at Qn = 1.5 kJ / mm, the average value of the impact work varies from 61 J (for 3 reheats) to 53 J (for 5 reheats).

After 10 reheats the average value of the impact energy is fixed at 45.5 J.

The average decrease in the impact energy with an increasing in the number of reheats practically fits into 8 ± 2% for one repeated welding pass.

Although, in order to improve the technological weldability during the development of the K70 strength class steel, it was envisaged to increase the purity of harmful impurities S and P, as well as to reduce the content of C <0.10% (at Mn ~ 1.7-1.9%, Cu ≤ 0, 5%, Mo ≤ 0.5%, Ni ≤ 0.5%, microalloying elements (V, Nb, Ti not more than 0.12%) in steel, so the obtained results allow us to insist on the low technological weldability for this steel, taking into account such a most important indicator as the repairability of the material.

5. Conclusion

The metal reheats are typical for a number of many industrial technologies, primarily such as welding and repair work, using electric arc gouging, especially applied on high-strength steels of K70 strength class, which leads to a decrease in the properties of the material in the heat-affected zone.

The described degradation of properties is observed in the form of decreasing in an impact energy on Charpy specimens with an incision, located in the coarse grain area. Each subsequent heating for the welded joints made from steels of K70 strength class leads to a decrease in an impact energy by 8 ± 2% and it appears to be lower than the regulatory level requirements.

Thus, the obtained results allow us to state the low technological weldability for this steel, taking into account such an important indicator as the repairability of the material.

References

[1] Golovanenko S A 2002 New steels and technological schemes for the production of a thick plate for gas pipes of large diameter *Metals* **5** 36–46

[2] Hillenbrand H-G, Gras M and Kalwa C 2001 Development and production of high strength pipeline steels *Proc. Int. Symposium on Niobium* **Dec.** 2–5

[3] Meimeth S, Grimpe F and Meuser H 2006 Development, state of the art and future trends in design and production of heavy plates in X80 steel grades *9th intermationsl & 4th European Conferences* Steel Rollin **June** 19-21

[4] Nemtinov A A, Korchagin A M and Popkov A G 2008 et al. Mastering the production of K70 strip for large-diameter pipes at mill *Metallurg* 5000 **11** 61–67

[5] Kiryan V I and Mikhoduy L I 2002 The problems of using new steels of increased and high strength in welded structures *Automatic welding* **3** 10–17

[6] Kruglova A A , Orlov V V and Sharapova D M 2014 Modeling heating effects in the heat-affected zone of high-strength pipe steel K70 with double-pass submerged arc welding *Metallurgist* **9** 98-104

[7] Velichko A A, Bortsov A N , Shabalov I P et al. 2014 The relationship of thermal processes with the morphology of welded joints and promising types of welding in relation to thick-walled electric-welded pipes *Metallurg* **3** 72–77

[8] Pryakhin E I and Sharonov N I 2018 The main provisions and problems of ELS technology in relation to the manufacture of structures from aluminum-magnesium alloys *Proceedings of the*

Mining Institute **229** 84-91
[9] Golosienko S A, Il'in A V, Lavrentiev A A, Mikhailov M S, Motovilina G D, Petrov S N and Sadkin K E 2019 Resistance of high-strength medium-alloy steel to brittle fracture and its connection with structural state parameters *Materials Science Issues* **3 (99)** 128-147
[10] Hamada M 2003 Control of strength and toughness at the heat affected zone *Welding international* **17 (4)** 265-270
[11] Shi Y and Han Z 2008 Effect of weld thermal cycle on microstructure and fracture toughness of simulated heat-affected zone for a 800 MPa grade high strength low alloy steel *J. Mater. Proc. Technol.* **207** 30-39
[12] Erofeev V A and Maslennikov A V 2005 Physico-mathematical model of the multi-pass arc welding process *Izvestiya TulGU, ser. Computer technology in the combination of materials* **3** 246-255
[13] Ueda J, Muracawa H and Luo Ju A 1995 Computational model of phase transformation for welding processes *Transation of JWRI* **1** 95-100
[14] Makarchuk A V, Makarchuk N V and Startsev V N The development of a mathematical model of multi-pass arc welding 2017 *Bulletin of the State University of the Sea and River Fleet. Admiral S.O. Makarov* **9 (1)** 121-131
[15] Pryahin E U, Sharapova D M and Sharapov M G 2017 Weldability of low-alloy steels in submersion-arc welding at high linear energies *Welding International* **31(4)** 291-302

IOP Conf. Series: Materials Science and Engineering **826** (2020) 012025 doi:10.1088/1757-899X/826/1/012025

A study of recovery rheological properties of egg yolk after freezing in making table salt

A S Bobko[1], A L Bobko[2] and O I Sergienko[1]

[1] ITMO University, 49б Kronverksky Pr., Sankt-Petersburg, 197101, Russia
[2] Prima Melange LLC, 45A, Centralnaya str., Leningrad Region, Vsevolozhsk District, pos. Lepsari, Sputnik industrial zone, 188671, Russia

E-mail: a.s.bobko@itmo.ru

Abstract. We studied the effect of salt on the quality indicators of egg yolk at various storage temperatures (–6; –12 and –18 ° C) in order to determine the optimal storage parameters with minimal violation of its rheological properties. The object of study was an egg yolk made from standard five-day eggs of the first category, produced at the poultry farm in the Leningrad Region, and intended for the meat processing industry. The used salt met the requirements in accordance with GOST R 51574-2000. The yolk was frozen using shock freezing at a temperature of -35 ° C and an air velocity of 15 m /sec. It was stored in refrigerators at the temperatures of –6, –12 ° C for 10 months, and at –18 ° C for 12 months. After thawing, the viscosity of the yolk was determined on the Engler viscometer; using a laboratory pH meter, organoleptic indicators for compliance with the requirements specified in GOST 30363-2013 "Liquid and dry egg products. Technical conditions". Tests were conducted in accordance with TR TS 021/2011 "On food safety".

1. Introduction

Currently, chicken eggs are being used less and less in food production, semi-finished products are being replaced - egg melange and egg yolk, in liquid or dry form, with or without additives. This significantly reduces the energy consumption of food production, due to the lack of additional technological operations - sorting, washing, and breaking eggs, separation into protein and yolk, if necessary. It is much more convenient to use liquid egg products, packed in aseptic bags with lids, which allow you to dose the product and take up significantly less storage space [1,2].

For producers of egg products, the question arises of preserving the value of products throughout the year, avoiding the mismatch of the seasons for the supply of raw materials and demand for the product. One way is to freeze egg melange in spring and summer during the season of laying chickens and use this semi-finished product in autumn and winter, during the period of demand for products and high prices for an egg. But when using this technology, it should be taken into account that the egg yolk, when processed at a temperature lower than -6 ° C, is irreversibly gelled and delaminated, which complicates its further use after defrosting [3,4].

Studies on the freezing of egg products were intensively carried out in the 1970s and 1980s, which was associated with the plan for the development of the national economy, namely, with an increase in egg production. The data published in domestic and foreign literature on storage conditions, thawing and the study of the physicochemical and organoleptic properties of frozen egg products are very limited and contradictory. Irreversible changes in the rheological properties of egg products during freezing and cold storage were revealed, the restoration of which at that time was not possible [5].

The purpose of this work is to investigate the restoration of the rheological properties of the yolk after freezing by adding table salt.

The yolk considered in this paper is used for the production of mayonnaise. The use of salt in this case is due to the fact that it plays the role of a natural preservative, and also positively affects the rheological properties of the yolk. The salt concentration was selected empirically, the effect of various concentrations on the rheological properties of the yolk at various negative storage temperatures was studied. The concentration used was discussed with customers to adjust their formulation of the finished product.

As you know, the quality of the yolk depends not only on the conditions of its freezing, but also on storage conditions [5]. In accordance with the GOST 30363-2013 "Liquid and dry egg products. Technical conditions "[6], egg products should be stored at a temperature not exceeding -18 ° C for not more than 15 months, at a temperature not exceeding -12 ° C - not more than 10 months.

The research objectives were to develop a new product - salted egg yolk, which has the ability to restore rheological properties after thawing. In particular, we studied the effect of different storage temperatures (-6, -12, -18 ° C) of yolk with and without added salt on quality indicators in order to determine the optimal salt concentration in the product and the optimal storage parameters of the frozen yolk with minimal violation of its rheological properties. Temperatures –12 and –18 ° C were selected based on the data of GOST 30363-2013 "Liquid and dry egg products. Technical conditions". At a temperature of –6 ° C, according to foreign studies [7-9], irreversible changes occur in the yolk.

2. Materials and methods

For experiments, we took a yolk prepared from standard five-day eggs of the first category. The selection of raw materials and technological operations for the production of yolk was carried out in accordance with the current technological instructions for the production of frozen egg products.

Freezing was carried out in the Criodor heat chamber using shock freezing at a temperature of –35 ° C and an air velocity of 15 m/ sec.

The yolk with an initial temperature of 4 °C was packed in packages of a polymer barrier film with a thickness of 60 μm, with dimensions of 320x150 mm and a capacity of 5 kg. The thickness of the product was 80 mm. They were stored in refrigerators at a temperature of –6, –12 °C for 10 months, at –18 °C for 12 months. They were thawed without opening the package [10] in air at + 18–20 °C and air velocity of 7 m/ sec.

The quality and property of the yolk after thawing and storage was evaluated by indicators characterizing the changes in its main components: by viscosity, pH, organoleptic, and microbiological indicators [11].

Viscosity was determined on an Engler viscometer in degrees Engler (°E) of conditional viscosity.

pH was measured using a laboratory pH meter, pH-150MI. Test procedure was conducting according to the GOST 31469-2012 "Food products of processing of eggs of poultry. Methods of physico-chemical analysis".

The requirements for organoleptic indicators are given in the GOST 30363-2013 "Liquid and dry egg products. Technical conditions". The tests were carried out in a production laboratory.

Microbiological analyses were carried out by a third-party accredited laboratory - Branch of the Center for Hygiene and Epidemiology in the Leningrad Region in the Vsevolozhsk Region. Tests were conducted in accordance with the TR TS 021/2011 "On food safety".

Salt was added to the yolk before pasteurization and thoroughly mixed with a stirrer for 30 minutes. The salt concentration was selected as 6, 9 and 12%. Dosage was carried out using attorney scales.

To assess the quality of the yolk during storage, the egg yolk parameters were taken as a base for comparison before freezing.

3. Results and discussion

The results of studies indicate that the process of freezing and storing the yolk without adding salt is accompanied by irreversible changes in the initial properties of the product. Table 1 shows the change in viscosity of the yolk during storage at different temperatures. As one can see, the viscosity increases from the very first days of freezing and then changes slightly. So, at –6 ° C the viscosity of the frozen yolk in the first month increased 11 times.

Table 1. Change in viscosity of the yolk during storage at different temperatures

Temperature, °C	Before freezing	Viscosity, m^2/sec, $x10^{-4}$							
		The duration of the storage process, months							
		1	6	7	8	9	10	11	12
-6	0.22	2.39	2.43	2.43	2.43	2.43	2.43	Samples removed from storage	
-12	0.22	2.67	2.67	2.68	2.68	2.68	2.68	Samples removed from storage	
-18	0.24	2.89	2.89	2.89	2.89	2.89	2.89	2.90	2.90

Table 2 shows the change in viscosity of the yolk after adding salt at various concentrations during storage at different temperatures. The table shows that the viscosity of the yolk with the addition of salt at a concentration of 9% and 12% changes slightly at all studied temperatures.

During storage of egg yolk at a temperature of –6; –12 and –18 ° C there is a slight pH shift to the alkaline region. As indicated in Table 3, at –18 ° C, the changes are less pronounced, which is obviously associated with a slowdown in microbial and enzymatic processes that affect the pH of the product [12]. The addition of salt to the egg yolk shifts the pH to an acidic environment.

During storage, there is a decrease in carotenoids [13] which act as antioxidants in the oxidation of fats and substances that affect the color of the yolk. During storage, the color of the frozen yolk changes from bright orange to orange, especially at –6 ° C. At this temperature, the color after defrosting the product practically does not recover. At –18 ° C, the color of the frozen product changes with a lower intensity and is practically restored after defrosting. The addition of salt helps to fix the color, and the color of the yolk during storage does not change.

Table 2. Changing the viscosity of the yolk with the addition of salt in various concentrations during storage at different temperatures

Temperature, °C	Concentration,% to mass	Viscosity, m^2/sec, $x10^{-4}$								
		Before freezing	The duration of the storage process, months							
			1	6	7	8	9	10	11	12
–6	6	0.23	0.48	0.51	0.59	0.68	0.79	0.85	Samples removed from storage	
	9	0.23	0.25	0.25	0.26	0.26	0,27	0,27		
	12	0.23	0.24	0.25	0.25	0.26	0.26	0.26		
–12	6	0.24	0.46	0.51	0.57	0.61	0.68	0.79	Samples removed from storage	
	9	0.24	0.24	0.25	0.25	0.25	0.25	0.26		
	12	0.24	0.24	0.24	0.24	0.25	0.25	0.25		
–18	6	0.23	0.41	0.44	0.47	0.50	0.52	0.55	0.58	0.60
	9	0.23	0.23	0.23	0.24	0.24	0.24	0.24	0.24	0.25
	12	0.23	0.23	0.23	0.23	0.23	0.23	0.24	0.24	0.24

Table 3. Changing the pH of the yolk during storage at different temperatures

Temperature, °C	pH								
	Before freezing	The duration of the storage process, months							
		1	6	7	8	9	10	11	12
–6	5.92	5.95	5.98	5.99	6.04	6.12	6.25	Samples removed from storage	
–12	5.92	5.94	5.97	5.98	5.99	6.02	6.19	Samples removed from storage	
–18	5.92	5.93	5.94	5.95	5.96	5.97	5.97	5.98	5.99

In terms of organoleptic characteristics, consistency and taste, omelettes baked from a frozen yolk stored at various temperatures were no different. All samples showed a rubbery consistency, an unpleasant appearance and a specific taste.

Assessing the quality of the yolk during storage would be incomplete without taking into account the results of bacteriological studies, which led to the following conclusions. The limiting factor in assessing the quality of the yolk was the results of bacteriological laboratory tests. Coliforms remained normal - not detected in 0.1 g.

However, during long-term storage at –6 and –12 ° C, an increase in total viable count is observed, and the earlier it is, the higher the storage temperature is. So, in a product stored at –6 ° C, total viable count growth began after 8 months of storage, at –12 ° C - after 9 months of storage, at –18 ° C for 12 months no growth was detected.

According to laboratory studies, the addition of salt helped to slow down the growth of total viable count during storage at various temperatures.

4. Conclusion

The study revealed a positive effect of salt on the restoration of the rheological properties of the yolk after thawing it.

A salt concentration of 6% does not have a sufficient effect on the change in the properties of the yolk. Concentrations of 9% and 12% lead to a positive effect, however, from an economic point of view, the optimal salt concentration of 9% was chosen at which the rheological properties of the yolk are restored after it is thawed and there is no increased salt consumption.

The optimum storage temperature of the frozen yolk –18 ° C was determined, at which the change in viscosity is less pronounced, the color of the product is less intense, and the growth of KMAFAnM is not detected.

In the course of the research, a technological scheme was developed for the manufacture of salted egg yolk. As a result of the introduction of this product in the range of production of liquid egg products located in the Leningrad region, it was possible to increase sales of egg yolk by 13%.

References

[1] Agafonychev V P 2011 Egg production is a promising direction *Ptitseprom* **3** 54-58

[2] Agafonychev V P and Karenin A I 2006 Improving the functional properties of egg products by fermentation *Materials of the international scientific-practical conference New world trends in the production of products from poultry meat and eggs GU VNIIPP* 125–127

[3] Agafonychev V P, Krugalev S S, Petrova T I 2006 Egg production: key trends *GU VNIIPP* 121–125

[4] 2016 Vencomatic Group. Egg Focus: Careful collection *International Poultry Production* **24 (3)** 35

[5] Frolov V L 1983 The effect of refrigeration processing and storage on the properties of egg melange *Refrigeration* **2** 51-54
[6] GOST 30363-2013 "Liquid and dry egg products", Russia
[7] O'Keefe T 2014 The Americas dominate rankings of the world's largest egg companies *Egg Industry* **119 (11)** 10-12
[8] Seuss-Baum I 2010 Variability in estimation of egg nutritional value throughout Europe: how to control it? *Proceedings of XIII European Poultry Conference* (Tours, France, August 23–27, 2010)
[9] McNamara D J 2010 Eggs: A world of possibilities *World Poultry* **26 (7)** 36–37
[10] Kavtarashvili A Sh, Karapetyan R V, Golubov I I 2013 New methods for determining production efficiency *Animal husbandry in Russia* **4** 1–12
[11] Krasyukov Yu N 2013 Egg products: quality indicators and control methods *Poultry and Poultry products* **4** 34–34
[12] Kiokias S, Gordon M H, Oreopoulou V 2017 Effects of composition and processing variables on the oxidative stability of protein-based and oil-in-water food emulsions *Critical Reviews in Food Science and Nutrition* **57 (3)** 549–558
[13] McKenn B M 2008 *The structure and texture of food. Products of emulsion nature* (St. Petersburg: Profession)
[14] Agafonychev V P 2011 Egg production is a promising direction *Ptitseprom* **3** 54-58

IOP Conf. Series: Materials Science and Engineering **826** (2020) 012026 doi:10.1088/1757-899X/826/1/012026

Innovative online technologies as a tool of qualified specialists training in the field of waste management

A A Kakabayev[1], I B Fakhrudenova[1], A S Kurmanbayeva[1], A S Yevloyeva[1] and A A Saparbekova[2]

[1]Kokshetau State University named after Sh. Ualikhanov, 76, Abay st., Kokshetau, 020000, Republic of Kazakhstan
[2]South Kazakhstan State University named after M. Auezov, 5, Tauke-khan ave., Shymkent, 160018, Republic of Kazakhstan

E-mail: koksuprojects@mail.ru

Abstract. The article discusses new technologies that are used in creating content for online training modules, as well as specific examples of using certain tools in the formation of courses for online modules in the field of waste management in the framework of the EduEnvi project. The article describes online tools that can be used for visualizing, structuring and systematizing of various types of information, such as describing the course structure, showing a list of topics, possible tasks, etc. The need for visualization and systematization of information becomes very important when it comes to online training, which requires developers to be an extremely clear and detailed presenting their educational material, teaches them to be more open and communicative. The article also contains information about the formats of work in the project which helps partners from five different countries to exchange experience, learn something new and develop online courses. These online courses will be the main result of the EduEnvi project.

1. Introduction

The transition of the Republic of Kazakhstan to sustainable development is an urgent need because the country needs development that meets the requirements of the present generation and does not constrain the ability of future generations to meet their needs. Economic growth due to the exploitation of natural resources can occur only at a certain stage. In modern conditions, more progressive mechanisms are required for growth and development.

In all Kazakhstani settlements, especially in large cities, the problem of collection, storage and processing of increasing volumes of municipal waste is acute. The main method of waste management today is the placement of waste in landfills. At the same time, the operation of most landfills and municipal waste dumps in Kazakhstan does not meet regulatory requirements. About 90 % of the existing landfills in the country do not meet the requirements of environmental and sanitary legislation [1]. Only a part of settlements of the Republic is covered by services of the specialized enterprises on collecting and export of waste, and the others are left without service.

2. Materials and methods

One of the main factors influencing the solution to the problem of effective waste management in the

Republic is the training of qualified specialists. Modern trends in education are closely related to the online environment, for example, the use of information technologies in the educational process, the introduction of distance learning, etc. At the moment, universities are reviewing existing teaching methods, are in search of the benefits of contact with students to further redistribute resources in favour of more effective teaching methods in combination with the use of online courses in the educational program [2]. Innovative technologies are important in the process of online learning, due to the rapid obsolescence of existing professional knowledge and the need for continuous improvement. One of the most optimal forms of modern education, equally effective in obtaining basic and additional education, is distance learning or online learning, called by some researchers the educational system of the 21st century [3]. Thus, in terms of improving the competencies of future specialists in the field of waste management in 2018 the international project EduEnvi has been launched, the main idea of which is to train competitive specialists in the field of waste management through the development of online educational modules, placing them on national online platforms, as well as introducing them into educational programs at the master's level [4].

3. The study of the innovative technologies usage opportunities in the online educational process

The EduEnvi project announces specialists from five countries and aims to increase competence in the field of sustainable waste management in the training of specialists in universities of Russia and Kazakhstan and internationalization of education. The project coordinator is Finland (Tampere University of Applied Sciences), methodological assistance is provided by specialists from Denmark and Spain.

In addition to the "live" meetings, during which the specialists of 6 universities of Russia and Kazakhstan get acquainted with the latest European pedagogical technologies, as well as improve their skills in the formation of curricula, selection of teaching tools and evaluation criteria, the work within the project is carried out using online technologies, for example, programs that allow virtual meetings, so-called webinars. By the way, a large-scale webinar was organized with the participation of three Kazakh universities: Kokshetau State University named after sh. Ualikhanov, Kazakh National University named after al-Farabi, South Kazakhstan State University named after M. Auezov. The participants of the meeting – teachers, students and undergraduates, representatives of the industry, as well as stakeholders – discussed topical issues related to waste management, exchanged views and came to generalized conclusions.

As mentioned earlier, the work in the project is based on the development of eight modules that cover a wide range of competencies necessary for successful waste management, from introduction to environmental risks and their assessment to business process modelling and environmental management in General. Currently, content developers are engaged in applying the knowledge and skills to form a thematic plan of disciplines within each of the modules, as well as the selection and development of suitable tasks and types of training activities that would be most effective in the online environment, tools and evaluation criteria [5].

The courses will draw on the latest research, best practices and the latest technologies in sustainable waste management [6]. Increasing the competence of the teaching staff in the field of sustainable waste management will modernize existing curricula and create new educational programs in the field of sustainable waste management. The online platform will be international, multilingual, and can be used as continuous training for individuals employed in the sustainable waste management industry. The project involves closer cooperation between universities and industrial enterprises and integration between universities.

Focusing on the tools that can be used in the implementation of online learning, it is necessary to note their great variety, different functionality and breadth of application: writing texts and blogging, searching for the necessary content, editing and creating images, tools for questionnaires and newsletters, creating lists with the help of task managers and planners, the use of interactive maps, etc [7,8].

So, as part of the work on the content of future modules, to familiarize students with the discipline,

IOP Conf. Series: Materials Science and Engineering **826** (2020) 012026 doi:10.1088/1757-899X/826/1/012026

its structure, competencies, a special resource was selected that allows you to organize all the necessary information for the student – the platform canva.com, which can also be used for summarizing information by students, presenting the results of activities in the framework of group work, project activities, etc. (Figure 1).

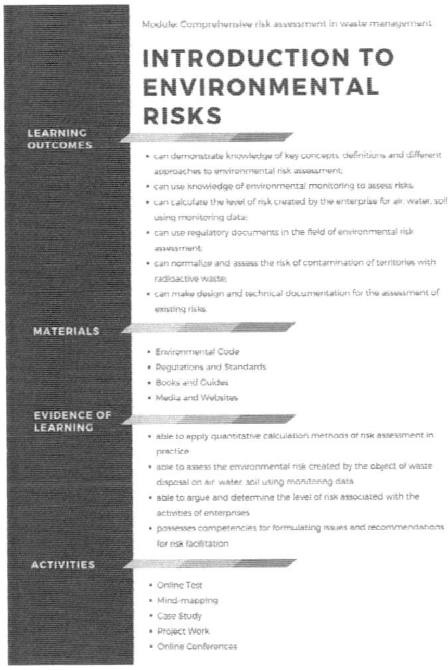

Figure 1. Canva.com platform usage example. Brief information about the discipline "Introduction to environmental risks"

For a structured visualization of the process or a certain route that the student will have to go for the development of a particular course, as well as for the creation of mental maps, fixing the results of brainstorming, the most appropriate service is coggle.it (Figure 2).

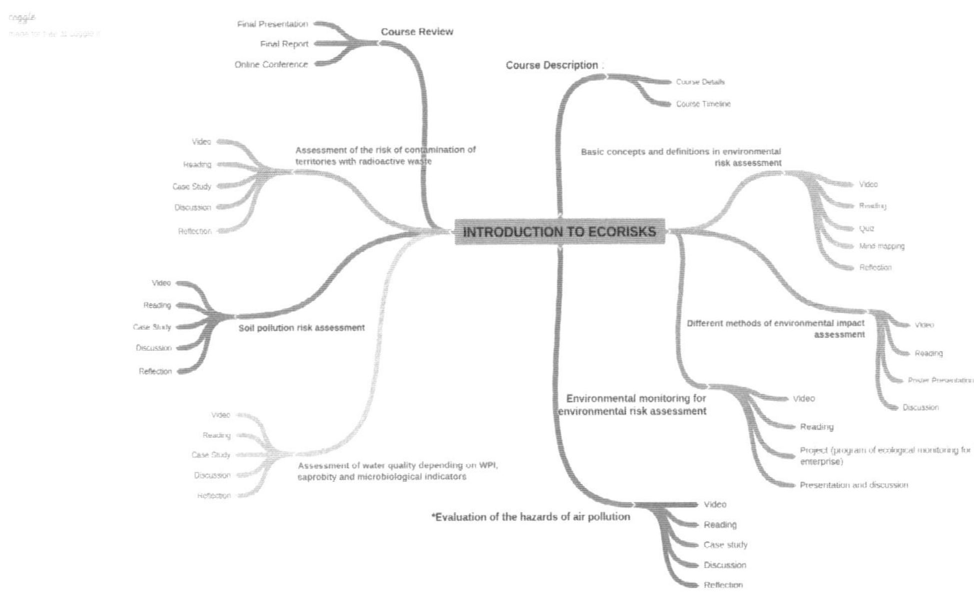

Figure 2. Description of the "Introduction to environmental risks" course structure using the service coggle.it

4. Conclusion

As a result, participation in the work of such projects has many positive aspects, such as improving the skills of the teaching staff, exchange of experience, a new look at existing methods of creating educational content, dissemination of knowledge, as well as improving the quality of teaching using interactive teaching methods in the work. The experience gained will be useful in the further implementation of distance learning and the development of increasingly popular open online courses. The undoubted advantage of the project is also the intensification of joint activities of higher education institutions and industrial enterprises to find the best algorithms for rational use of resources, the search for opportunities for the use of secondary raw materials in production and reduce the formation of new waste.

The training of highly qualified specialists in sustainable waste management will give impetus to the development of research in cooperation with local enterprises and the creation of innovative technologies to reuse valuable components from waste and thus reduce the use of natural resources and reduce the waste generation in production.

5. Acknowledgments

The article was prepared in the framework of the project: "Enhancing competences of sustainable waste management in Russian and Kazakh HEIs /EduEnvi" with the financial support of the program "Erasmus+".

References

[1] *Ministry of National Economy of the Republic of Kazakhstan. Statistics committee*, available at: http://stat.gov.kz/

[2] Vyushkina E G 2015 *Saratov University News. Series: Philosophy. Psychology. Pedagogy* **15 (2)** 78-83

[3] Chekalina T, Tumandeeva T and Maksimenko N 2018 *Contemporary tendencies in professional education development* **3 (31)** 44-52

[4] Zakharova U and Tanasenko K 2019 *Education Studies* **3** 176-202

[5] Kakabayev A A, Kurmanbayeva A S and Fakhrudenova I B 2018 *Proc. Int. Conf. "Internationalization of higher education: experience in implementing joint European projects"* (Kokshetau) p 24-29

[6] Morrisseya A J and Browneb J 2004 *Waste Management* **24** 297-308

[7] Martinez-Cerda J F, Torrent-Sellens J and Gonzalez-Gonzalez I 2018 *Behaviour & information technology. SI.* **10-11** 1055-1071

[8] Akcaoglu M and Kale U 2016 *Contemporary Issues in Technology and Teacher Education* **1** 60-81

Energy efficient technology of waste management by micro-cogeneration

Y A Rakhmanov, O I Sergienko and G N Gorbunov

ITMO University, 9, Lomonosova st., St. Petersburg, 191002, Russia

E-mail: rahmanovua2010@gmail.com

Abstract: The paper considers possible options for energy-efficient waste management technologies using micro-cogeneration. A typology of energy technologies with micro-cogeneration is proposed depending on the main method of disposal, type of reactor, products of the implementation of the main recycling process, as well as processes and equipment of micro cogeneration technology. A brief overview of the current market situation and trends in the development of energy technologies for waste management is presented. The proposed typology and block diagram can be used in choosing rational energy-efficient technologies for waste management.

1. Introduction

In the process of human life, a large amount of production and consumption waste is generated. After preliminary sorting, non-recyclable waste is mainly disposed of at special landfills, which are complex environmental protection structures that protect the atmosphere, soil, surface and ground waters. Landfills and supporting infrastructure require large capital investment at the construction stage, occupy territories that are being withdrawn from land use and subsequently require large restoration costs.

In this regard, nowadays various methods of local waste treatment are used: thermal oxidation or neutralization, gasification, pyrolysis, anaerobic fermentation of organic waste [1], the products of which have a certain energy resource and can be used by means of cogeneration technologies to produce electrical and heat energy, and, if necessary, cold within three-generation technologies.

Due to this, the consumption of fossil fuel resources, whose reserves are not infinite and located in hard-to-reach areas, is reduced, environmental pollution by products of their use, in particular, CO2 emissions, is reduced also. In the "Energy Strategy of Russia for the Period until 2035", solid municipal waste is assigned to important regional types of fuel resources.

The purpose of the paper is to consider possible options for energy-efficient waste management technologies using micro cogeneration and propose their typology and block diagram, which can be useful for choosing rational waste management schemes.

2. Materials and methods

As the main method for finding the rational options for waste management by micro turbine technologies the authors used creation of its typology, based on the literature review. The typology was proposed depending on the main method of treatment, type of reactor, products of the implementation of the main treatment process, as well as processes and equipment of micro cogeneration technology (table 1).

Table 1. Energy efficient waste management technologies based on cogeneration

Treatment method	Type of reactor	Process Products	Cogeneration technologies for heat recovery
Thermal oxidation (fire neutralization, burning)	Layered, chamber, rotary drum furnaces, furnace devices, cyclone reactors	High temperature flue gases (800–1200°C), ash	Steam boiler, steam engine (steam piston engine, steam screw machine, steam turbine); ORC technology with micro turbines; Hot-water, thermo-oil boiler, thermoelectric generator; Air turbine and gas turbine with reverse sequence technology processes.
Gasification	Gasgenerators	Generator gas with a calorific value of 3.5-6.5-16MJ/m^3, ash (slag)	Gas piston, gas turbine engines with heat recovery of flue gases
Pyrolysis	Pyrolysis Reactors	Pyrolysis gas with a calorific value of 15-22 MJ/m3, solid coke residue with a calorific value of up to 30 MJ /kg	Gas piston, gas turbine engines with heat recovery of flue gases
Anaerobic OrganicWaste Fermentation	Bioreactor	Biogas with a calorific value of about 22MJ/m^3, fertilizers	Gas piston, gas turbine engines with heat recovery of flue gases

Considering the suggested typology resulted in creating the block diagram, which can be useful for choosing rational waste management schemes based on cogeneration.

3. Results and discussion

During thermo-oxidative neutralization of waste to ensure their high-quality combustion, the process temperature is maintained at the level of 900 - 1200 ° C [1]. In this case, a high temperature flue gas is formed containing solid fly ash particles. Heat generated by cooling to a temperature acceptable for gas purification and removal systems (depending on the purification technology from 300 to 120 ° C) can be useful in a steam boiler (PC) for producing saturated or superheated steam. The resulting water vapor enters the steam engine (PD), where, expanding, it performs the work used to drive the electric generator. Wastewater vapors enter the heat consumer. The flow rate and parameters of water vapor

IOP Conf. Series: Materials Science and Engineering **826** (2020) 012027 doi:10.1088/1757-899X/826/1/012027

depend on the thermal power of the thermal oxidative neutralization system, power, heat consumption regimes and type of steam engine (Figure 1).

Wet - steam microturbines with a power of 5 - 30 kW and energy complexes developed on their basis by the Don Production Technologies Enterprise are quite interesting for use [7].

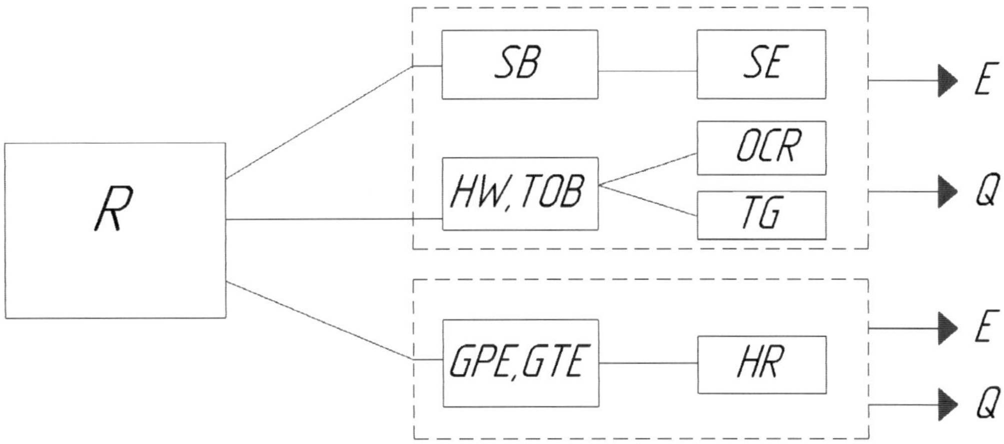

Figure 1. Structural diagram of possible energy-efficient waste disposal technologies
R - reactor, where the main methods of waste disposal are carried out (burning, gasification, pyrolysis, anaerobic fermentation of organic waste); SB - steam boiler; HW, TOB - hot-water, thermo-oil boiler; SE - steam engine (steam piston engine, steam screw machine, steam turbine); OCR - Renken organic cycle technologies; TG - thermoelectric generator; GPE, GTE - gas piston and gas turbine engines; HR- heat recovery system; E - electric energy; Q - thermal energy

Research in the field of their application in the conditions of a mini-thermal power plant and the development of a more compact and advanced design are carried out by the joint research group Promteploenergetika of the Moscow Aviation Institute, the All-Russian Research Institute of Electrification Agriculture, Royal College of Space Engineering and Technology [3]. Operational experience of steam piston engines [5] indicates the reliability of their operation on wet saturated steam at moderate speeds of rotation (up to 3000 rpm), stability of operation with fluctuations in steam parameters and the appropriateness of their use at low powers.

When working on saturated water vapor at a steam flow rate of 3.6 tons of steam or more, it is also advisable to use steam screw machines [6]. Research and development conducted by Eco-Energy CJSC and Peter the Great University show that when steam pressure is triggered 0.3 - 0.6 MPa and steam consumption is 6-50 t/ h using a steam screw machine, one can get an electric power of 200 - 1500 kW [2,6].

If there is no need for water vapor, the heat of cooling of the flue gases can be used advantageously in a hot water boiler (WB) to produce hot water, the required parameters, or in a thermo-oil boiler (TOB) for heating thermal oil. The useful heat obtained in this case can be used in cogeneration using ORC - technologies that implement the Renken thermodynamic cycle on organic liquids boiling at lower temperatures compared to water [8,9,10]. Infinity Turbine developed microturbines operating on organic working fluids with an electric power of 2 kW or more [9]. CALNETIX [10] developed the TGX100 steam turbine power plant with an electric power of 125 kW, operating on the R245 FA working fluid. At the same time, to obtain an electric power of 125 kW, 750 kW of heat in the form of hot water at a temperature of 110 ° C are required.

When using a thermal oil boiler, hot oil with a temperature of 280–300 ° C can be supplied to the thermoelectric generator modules to generate electrical energy [11]. Such a system was introduced at the Bio Kibor power plant with an electric power of 5 kW to 1 MW produced by the KIBOR plant.

Also, air turbines and gas turbines with the reverse order of processes, technological schemes, the analysis of energy and environmental efficiency of which we described earlier, can be used to utilize the heat of cooling of flue gases [12].

In the process of gasification of organic waste in the reactor (P), generating combustible gas [13] and a solid residue are obtained. The composition of the generator gas depends on the type of gasified waste and the gasification method, the heat of combustion can reach 3.5-6.5-16 MJ /m3. generator gas can be burned in boiler furnaces to produce water vapor or hot water, or in gas reciprocating, gas turbine engines to produce electrical energy. [13,14,15,16,17,18]. The gas energy complexes include a gas generator, a gas treatment system, and a gas reciprocating or gas turbine power plant with heat recovery devices. The gas-generating energy complexes GEC - 001 - GEC -1 are proposed, operating on wood, briquetted organic waste with a moisture content of 6 to 20%, based on gas piston engines with an electric power of 3 to 350 kW and a thermal power of 100 to 1000 kV, a waste flow rate of 4 up to 400 kg / h, generator gas from 13 to 1300 m3 / h [16]. CATERPILLAR manufactures gas-piston power plants with electric power from 70 to 6520 kW [14]. GEIbenbacher gas piston power plants can run on heavy types of gaseous fuels (associated petroleum gas, biogas, pyrolysis gas, etc.). Their electric power ranges from 330 to 4034 kW, respectively, thermal power from 363 to 3634 kW [15]. In MSTUunder the name of N.E. Bauman a gas-turbine energy converter with the production of electric energy was developed for an installation for utilization of solid household and industrial wastes of an unconventional scheme that does not require the purification of gases from solid particles in front of a turbine that runs on clean high-temperature air heated in a gas-air heater by the products of combustion of generator gas [17].

Using gas microturbines Capstone electric power from 30 to 200 kW [18]. In cogeneration mode, in addition to electricity, heat can be obtained with a capacity of 84.7 to 394.5 kW. To do this, they must be equipped with devices that utilize the heat of flue gases. Capstone microturbines can run on gaseous fuels with a calorific value of 10,475 kJ / m3 (biogas, low-calorie gases).

When disposing of waste by the pyrolysis method, it is generally possible to obtain heating oil, pyrocarbon, and pyrolysis gas [19, 20]. Moreover, pyrocarbon with a calorific value of 25 -31 MJ / kg can be used directly as solid fuel or in the form of coal water. Pyrolysis plants of KB Klimova LLC allow processing from 3 to 25 tons of waste per day. When the reactor is operating in gasification mode with generating times from the processing of solid waste and garbage in the amount of 5000 kg / day, an electric 150 kW and a thermal 200 kW power can be obtained using a gas piston engine [19]. The Safe Technologies industrial group (St. Petersburg) developed pyrolysis plants for the thermal destruction of continuous heat treatment units with a capacity of up to 1,500 kg / h, which allow processing various types of waste [20,21]. The tests of the UTD-2-200 installation showed that the pyrolysis gas generated in this process can be used in cogeneration technologies based on Capstone microturbines and the FAS gas piston engine (Fasenergomash LLC, Russia) [22].

4. Conclusion

Obtained from anaerobic fermentation of organic waste biogas can be rational to use in cogeneration after additional preparation on the basis of the previously listed gas reciprocating and gas turbine engines with a waste heat recovery system for flue gases. The feasibility study of a cogenerator based on the small-sized gas turbine unit GTES-200 manufactured by Kaluga Engine using the technology of microbiological utilization of municipal solid waste [22] confirms a payback period of no more than three years, which allows us to consider this technology as cost-effective.

The results of studies of trends in the development of the energy technology market with microcogeneration, the proposed typology and block diagram can be used to select rational energy-efficient technologies for waste management.

References

[1] Bernadiner M N and Shurygin A P 1990 *Fire processing and neutralization of industrial waste* (Moscow: Chemistry)
[2] Trokhin I S 2012 Mini-combined heat and power plants with steam engines for uninterrupted power supply of responsible consumers *Industrial Energy* **9** 15-20
[3] Dubinin V S 2009 *Ensuring the independence of electric and heat supply of Russia from electric networks based on piston technologies*: Monograph (Moscow: MPEI)
[4] Zhigalov V A 2017 *Thermal power plant with a reciprocating steam engine,* available at: https: //www.rodteplo24.ru
[5] *Spilling steam engines for steam boilers*, available at: https: //www.hansaenergo.ru/parovye-dvigately.ru
[6] Berezins S R, Borovkov V M, Vedayko V I and Bogacheva A I 2009 *Steam screw machine as a means of energy conservation,* available at: http: //www.energosovet.ru
[7] Efimov N N 2013 A microenergocomplex based on a wet steam turbine *Energy Saving* **6** 54-56
[8] Mokhov O V 2016 *Application of OCR - technology in waste heat recovery*, available at: http://aqua-therm.ru/articles/articles
[9] 2015 *Microturbines on the Renken organic cycle,* available at: www.joule– watt/mikroturbiny-na-organicheskom-tsicle–renkina
[10] *Microturbines from CalnetixPowerSolutions* , available at: www.electrosystems.ru
[11] KIBOR *Autonomous power plants on wood waste based on thermoelectric converters. Factory - manufacturer KIBOR* (Moscow), available at: BioPower.kibor.ru/biokibor
[12] RakhmanovYu A, Sergienko O I, Vasilenok V L and Gorbunov G N 2019 On the economic efficiency of the use of gas turbine waste disposal technologies *Scientific journal NRU ITMO Series "Economics and Environmental Management"*, available at: http://economics.ihbt.ifmo.ru/file/article/18523.pdf **1 (36)** 93-101
[13] KIBORBioPower *Power plant KIBORBioPOWER on wood waste. Factory manufacturer Kibor* (Moscow), available at: kibor.ru/biokibor.
[14] *Gas engine power plants CATERPILLAR,* available at: http://ehc-group.ru/catalog/gpu/caterpillar/
[15] *Gas engine power plants GEIenbacher*, available at: https//www.roltpower.ru/ equipment/ge-ienbacher
[16] *Gas reciprocating power plants with generator gas,* available at: https://gisee.ru/articles/low-energy/831/
[17] Ivanov V L 2012 Gas turbine power converter for the installation of solid domestic and industrial waste utilization by gasification *Bulletin of MSTU. N.E. Bumana Ser. "Engineering"* 384-399
[18] 2010 *Advanced Capstone technology*, available at: http://iescorporation.org/about /novosti.htmi?6%
[19] The development catalog of KB Klimova LLC 2018, available at: https://www.potram.ru/img/catalog.
[20] Installation of pyrolysis of UTD, available at: https://i-pec.ru/about
[21] Ladygin K V, Stompels S I and SpektorYu A 2016 Alternative energy from waste: only facts *Solid municipal waste* **12** 22-23
[22] Morozenko M I, Kusacheva S A and Chernyaev S I 2016, Feasibility study of a cogenerator based on a small gas turbine unit using microbiological waste disposal technologies *Fundamental Research*, available at: *http://www.fundamental-resarch.ru/ru/article/view?Id 40693* **9-1** 48 - 52

The ecotoxicological aspects of using the marine benthic invertebrates skeleton in the production of active calcium substances

A A Oripova, E A Ovsuk, O I Sergienko, N B Uliyanov and U Oripov

[1] ITMO University, 9, Lomonosova st., St. Petersburg, 191002, Russia

E-mail: a.a.oripova@niuitmo.ru

Abstract. Marine benthic invertebrates are susceptible to the accumulation of heavy metals and other pollutants from the environment, leading to the impossibility of their use as raw materials in the food production and pharmaceutical industry. In this paper, the accumulation of heavy metals in the shells and needles of *Strongylocentrotus* sea urchins was studied to assess the safety of the use of this raw material in the production of active calcium substances for food and medical purposes with the use of voltammetry analysis and capillary electrophoresis. High levels of strontium accumulation in the studied samples were revealed, indicating the unfavorable situation of commercial waters with respect to the presence of mobile forms of strontium.

1. Introduction

Marine invertebrates represent approximately 60% of the total diversity of marine animals and are rich sources of biologically active compounds. The biotechnological potential of marine benthic invertebrates is increasingly attracting scientific and economic interest worldwide. Biological features, chemical composition, mechanical characteristics of benthic invertebrates determine a great interest in them in different areas of industry: they are the sources of raw materials for food production, the sources of biologically active substances and medicines, animal feed, fertilizers, etc. Great interest is shown in them in chemistry, microbiology and pharmacy.

The Arctic region of Russia, including the Barents Sea, is one of the main sources of fish wealth (including marine benthic invertebrates), surpassing other marine regions in terms of fish stocks.

The most common genus of sea urchins *Strongylocentrotus* in the Arctic region, belonging to the class of echinoderms (*Echinodermata*), is an object of active fishing. Sea urchins are mainly caught in order to extract gonads from them. The gonads (caviar) of a sea urchin are the sources of raw materials for food production, and extraction of valuable biologically active substances for pharmaceuticals, chemistry and microbiology.

Due to the fact that sea urchins are the sources of biologically active substances, it is necessary to have information about the pollution in their populations habitats. This article is devoted to the study of heavy metals accumulation in the shells and needles of the *Strongylocentrotus* sea urchins to assess the safety of their use as raw materials in the production of active calcium substances.

2. Materials and methods

2.1. *Study area and research materials*

The Arctic ecosystems with all their richness and diversity are the most fragile and vulnerable due to climatic and geographical features of the region. The problem of rational use of marine resources has arisen for a long time and includes the great economic pressure on the main commercial species of hydrobionts. Rational nature management implies complexity in the processing of natural resources, which is not sufficiently applied in the field of marine benthic invertebrates.

In the extraction process of gonads from the sea urchins about 70% of their body weight goes to waste consisting mainly of the shells and needles, which are known for their high content of unique biologically active substances of various classes, such as proteins, quinoid pigments, and calcium [1, 2].

The main component that forms the skeleton of a sea urchin is calcium carbonate: in the *S. franciscanus* species, 91.08% of their skeleton consists of calcium carbonate [3], *S. droebachiensis* – 90.77%, *S. intermedius* and *S. nudus* – 73.4% [4], *Strongylocentrotus* genus – 83-90% [5]. The shells of sea urchins also contain magnesium carbonate - in sea urchins of the *Strongylocentrotus* genus it varies between 2-14% [6].

The carbonate skeleton of sea urchins is easily neutralized with decomposition by solutions of organic acids. The calcium salts of organic acids obtained by neutralizing the carbonate skeleton can be widely used in the food and pharmaceutical industries. Calcium isolated from the considered raw materials can serve as an additive to feed in animal husbandry, as well as the basis for biologically active substances (BAS) and functional food products.

Marine benthic invertebrates are susceptible to the accumulation of heavy metals and other pollutants from the environment, which can lead to the impossibility of their use as raw materials for the above described purposes and to a decrease in their commodity characteristics. Therefore, the measurement of heavy metals in the shells and needles of sea urchins is an important component in assessing the safety of their use as raw materials for the isolation of biologically active calcium.

This article is devoted to the study of heavy metals accumulation in the shells and needles of the *Strongylocentrotus* sea urchins to assess the safety of their use of as raw materials in the production of active calcium substances.

The conducted research involved the samples of shells and needles from two species of the sea urchins genus *Strongylocentrotus* with 50-57 mm in diameter caught in the waters of the Barents Sea in winter and summer 2017: a green sea urchin *S. Droebachiensis* (10 specimens) and a lilac sea urchin *S. Purpuratus* (10 specimens).

2.2. *Methods and equipment*

The study obtained data about heavy metals accumulation in research materials (samples of shells and needles) by carrying out two different methods of analysis: voltammetry [7], and capillary electrophoresis [8].

Studies to quantitatively determine the copper, lead, and cadmium content in the samples by a voltammetry method of analysis were carried out on an ABC-1.1 polarograph. This analyzer is designed to measure the mass concentration of various elements such as copper, lead, cadmium, zinc, mercury, nickel, bismuth, arsenic, iodine, etc., in drinking, natural, and waste waters, food, metals, and other materials according to certified or standardized measurement procedures.

Sample preparation for the quantitative analysis of the copper, lead, and cadmium content on the ABC-1.1 analyzer was carried out using the MC-6 microwave sample preparation system, in which the samples under study were subjected to wet mineralization. The MC-6 microwave sample preparation system is used to decompose samples of food products, food raw materials, soils, environmental materials, biological fluids, etc., when analyzing the chemical composition of samples by various

IOP Conf. Series: Materials Science and Engineering **826** (2020) 012028 doi:10.1088/1757-899X/826/1/012028

methods. The principle of the system is to use microwave energy for rapid heating of samples with the addition of oxidizing agents. The prepared sample was then analyzed by the standard addition method and processed according to the standardized measurement procedure (08-01-MVI).

In the conducted studies to determine the content of strontium, calcium and magnesium in the samples of shells and needles using capillary electrophoresis methods, the Kapel-103RT capillary electrophoresis system was used. This system is intended for the qualitative and quantitative determination of the composition of samples in aqueous and aqueous–organic solutions. The capillary electrophoresis method is implemented in capillaries and is based on differences in the electrophoretic mobilities of charged particles in both aqueous and non-aqueous buffer electrolytes.

Sample preparation for the quantitative analysis of the content of strontium, calcium and magnesium was carried out using a microwave sample preparation system MC-6 (subjected to wet mineralization). Then, prepared samples were analyzed by recording electropherograms according to a standardized, approved methodology (PND F 14.1:2:4.167-2000).

3. Results and Discussion

This research covers the sample analysis of shells and needles taken from a green sea urchin *S. Droebachiensis* and a lilac sea urchin *S. Purpuratus*. The obtained data on heavy metals accumulation in the stadied samples of shells and needles is shown in Table 1.

Table 1. Heavy metal accumulation in the shells and needles of studied species, mg/kg.

Species	Sample	Zn	Cd	Pb	Cu	Sr
S. Purpuratus	needles	13.26±1.46	0.03±0.002	0.23±0.01	1.54±0.14	54000.00±4850.00
	shells	14.17±1.59	0.07±0.005	1.83±0.12	1.03±0.10	26153.85±1875.54
S. Droebachiensis	needles	17.39±1.87	0.03±0.001	0.25±0.01	0.30±0.02	69565.22±5467.65
	shells	18.68±1.95	0.10±0.009	3.17±0.22	0.33±0.02	30825.22±2650.23

±SD, standard deviation.

According to the data obtained, the quantitative content of cadmium and lead in the studied samples corresponds to the standards of the permissible level of toxic elements, presented by Technical Regulation of the Customs Union (TR CU 021/2011).

Analysis of shells and needles samples of the studied sea urchin species revealed the presence of high concentrations of strontium, magnesium and calcium (Table 2).'

Table 2. Accumulation of strontium, calcium and magnesium in the shells and needles of the studied sea urchin species, mg/kg.

Species	Sample	Sr	Ca	Mg
S. Purpuratus	needles	54000.00±4850.00	760.00±68.45	7700.00±742.45
	shells	26153.85±1875.54	923.08±87.59	9230.77±894.73
S. Droebachiensis	needles	69565.22±5467.65	857.25±74.14	7826.09±612.56
	shells	30825.22±2650.23	5095.65±479.83	10260.87±985.47

±SD, standard deviation.

The quantitative content of magnesium in the studied samples is confirmed the levels of magnesium carbonate presence in the shell and needles of the studied species [6]. At the same time, the levels of strontium accumulation in the studied samples significantly exceed the content of calcium and magnesium, while the quantitative content of calcium is the lowest one (Figure 1).

The potential reason for high levels of strontium in the samples of shells and needles of the studied sea urchin species is the increased ability of strontium to replace calcium cations in the living organisms: strontium accumulates in the body after ingestion and cannot be excreted. According to available studies, the accumulation of strontium in the aquatic organisms is closely related to the concentration of calcium, and the strontium accumulation coefficient is inversely proportional to the concentration of calcium in the water bodies [9, 10].

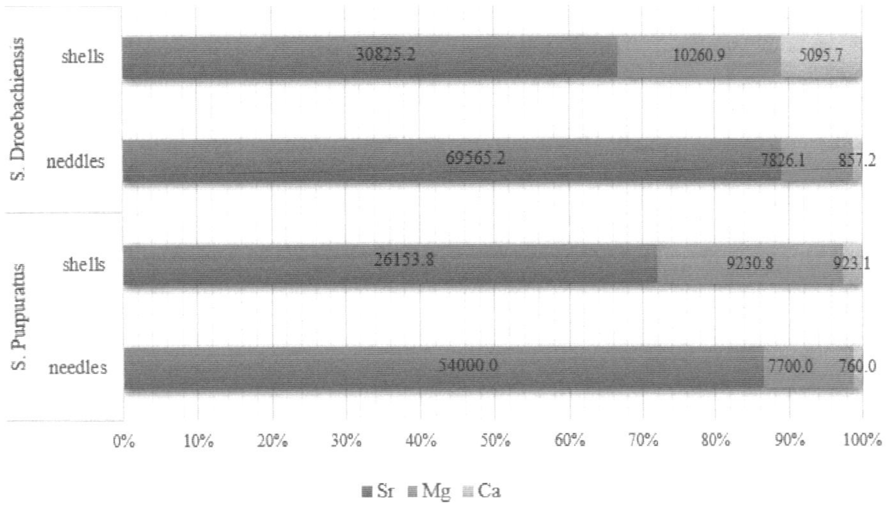

Figure 1. Normalized distribution diagram of strontium, magnesium and calcium in the studied species of the sea urchins.

4. Conclusion

The revealed low levels of calcium compared with the concentration of strontium in the studied samples of shells and needles of *S. Droebachiensis* and *S. Purpuratus* sea urchins indicate an unfavorable situation in the habitat of these species in relation to the presence of mobile forms of strontium.

The results of the study indicate the need for strict regular monitoring of strontium content in the Arctic waters, particularly in the Barents Sea. The obtained dependences in the accumulation and distribution of heavy metals within the studied objects (*S. Droebachiensis* and *S. Purpuratus*) can serve as a scientific basis for more effective safety control and ensuring high quality calcium salts production in the food and pharmaceutical industries when using marine benthic invertebrates, especially sea urchins, as raw materials.

References

[1] Amarowicz R, Synowiecki J, and Shahidi F 1994 Sephadex LH-20 separation of pigments from shells of red sea urchin (Strongylocentrotus franciscanus) *Food Chem.* **51** 227–229

[2] Hou Y et al. 2016 Marine shells: Potential opportunities for extraction of functional and health-promoting materials *Crit. Rev. Environ. Sci. Technol.* **4(11–12)**1047–1116

[3] Amarowicz R, Synowiecki J, and Shahidi F 2012 Chemical composition of shells from red (Strongylocentrotus franciscanus) and green (Strongylocentrotus droebachiensis) sea urchin *Food Chem.* **3** 82–86

[4] Goodwin T W and Srisukh S 1950 A study of the pigments of the sea-urchins, Echinus esculentus L. and Paracentrotus Lividus Lamarck *Biochemistry Journal* **47** 69–76

[5] Dincer T and Cakli S 2007 Chemical composition and biometrical measurements of the Turkish

sea urchin (Paracentrotus lividus, Lamarck 1816) *Critical Review in Food Science and Nutrition* **47** 21–26

[6] Magdans U, Hermann G 2004 Single crystal structure analysis of sea urchin spine calcites *European Journal of Mineralogy* **16(2)** 261–268

[7] Yuanyuan Lu et al. 2018 A review of the identification and detection of heavy metal ions in the environment by voltammetry *Talanta* **178** 324–338

[8] Zemann A, Rohregger I and Zitturi R 2018 Determination of small ions with capillary electrophoresis and contactless conductivity detection *Capillary Electrophoresis. Methods In Molecular Biology* 384 ed Schmitt-Kopplin Ph (Humana Press) 3–19

[9] Selda O T and Nursah A 2012 Relationship of heavy metals in water, sediment and tissue with total length, weight and seasons of Cyprinus carpio L., 1758 from Isikli Lake (Turkey) *Pakistan J. Zool.* **44(5)** 1405–1416

[10] Zimmerman C E 2005 Relationship of otolith strontium-to-calcium ratios and salinity: Experimental validation for juvenile salmonids. *Canadian Journal of Fisheries and Aquatic Sciences* **62** 88–97

IOP Conf. Series: Materials Science and Engineering **826** (2020) 012029 doi:10.1088/1757-899X/826/1/012029

Study of submicrocapsules structure stabilized by modified silica dioxide nanoparticles

B Zh Mutaliyeva[1], A B Tleuova[2], G M Madybekova[1,3] and A S Kurmanbayeva[4]

[1] M. Auezov South-Kazakhstan State university, 5, Tauke khan av., Shymkent, 160012, Republic of Kazakhstan
[2] University of Chemistry and Technology Prague, 5, Technická, Prague 6, 166 28 Czech Republic
[3] South-Kazakhstan State Pedagogical University, 13, Baitursynov str., Shymkent, 160012, Republic of Kazakhstan
[4] Sh. Ualikhanov Kokshetau State university, 76, Abay str., Kokshetau, 020000, Republic of Kazakhstan

E-mail: mbota@list.ru

Abstract. In this research the structures of Miglyol 812 submicrocapsules, stabilized by silica dioxide nanoparticles, modified by oleic acid, were studied. It was found that the size distribution of silica particles modified with oleic acid depends on the amount of oleic acid. There is a limit to the ratio of oleic acid to silica at modification of silica particles with oleic acid. Experimental results have shown that the optimal ratio of oleic acid to silica is in the range from $1 \cdot 10^{-2}$ to $5 \cdot 10^{-1}$. The use of silica particles modified with oleic acid allows to obtain an emulsion with a 30 % of oil phase, with an average size from 80-300 nm. Thus, modification of hydrophilic silica particles with surfactants such as oleic acid in order to make them more hydrophilic and able to be adsorbed at the oil / water interface can be used to prepare a stable Pickering emulsion. As a result the stable Pickering emulsions which can be used for encapsulation of various substances were obtained, and the use of microcapsules allows one to decrease their impact on environment and increase efficiency of usage.

1. Introduction

Microencapsulation and controlled release of various active substances, for example agricultural chemicals, can help to reduce encapsulated agent toxicity, to reduce the impact of substances on the environment, and to improve agent efficiency and efficacy [1].

That is why microencapsulation has found a great application in agriculture as an active substances delivery system. The different plant protection products were microencapsulated, for instance, insecticides [2-4], herbicides [5-8], fungicides [9], essential oils pesticides [10, 11]. Also, microencapsulation of microbial cells [12] can help to prevent stress caused by environment and to increase their stability [13, 14]. Technologies of microencapsulation can improve the stability and residual action of bioactive substances, and this could increase their field use.

Thus, encapsulation is being attracting method because it improves cell's stability and tolerance to inhibitors, increases the amount of biomass inside of reactor, and reduces the cost of cell restoration, utilization and further processing. There are many studies have reported on microencapsulation of active ingredients by use of several methods including spray drying [15], simple coacervation [16], complex coacervation [17], emulsion extrusion [18] and supercritical fluid deposition [19]. However, these methods do not yet solve the problems of the duration of use due to unreliability of used capsules. Therefore further researches have to be carried out to find methods that allow to ensure mechanical strength of capsules and reliable release. For this, it is necessary to find methods that will ensure the mechanical strength of the capsules and reliable release.

This research aims to study the potential of emulsions, including Pickering emulsions, to maintain the effectiveness of various biocides in relation to pathogenic microflora for use in various industries.

2. Methods of research

In case of using oleic acid to modify the surface of silica nanoparticles the following procedure was used [20]. The Ludox TM-40 silica particles were ultrasonicated (Bandelin, Sonopuls HD 3100) for 10 min at 30 W power prior the use to avoid aggregation. The particle dispersion had a transparent and slightly bluish appearance. The oleic acid-water mixture was also ultrasonicated for 10 min at 30 W power to effectively disperse oleic acid in water as micrometer-sized droplets. Afterward, the proper amounts of this mixture were added to the silica particles suspension. The resulting sample was ultrasonicated at 30 W power in pulse mode (3 s pulses with 1 s pauses) for 10 min to provide further energy for mixing and adsorption.

To prepare the emulsions, the modified silica particle dispersions were transferred to a vial, to which water, and then the lipophilic phase, were added. For pure oil dispersions, 30% of the total volume was oil (Miglyol 812). The resulting mixture was ultrasonicated at 60 W to disperse the oil phase for 10 min, which led to a milky suspension. All measurements were performed at 25 °C

3. Results and discussion

Figure 1 shows the mean diameter of silica nanoparticles by intensity. The results correspond with specification sheet of the product. Only one pick with maximum value at of 20-30 nm can be seen.

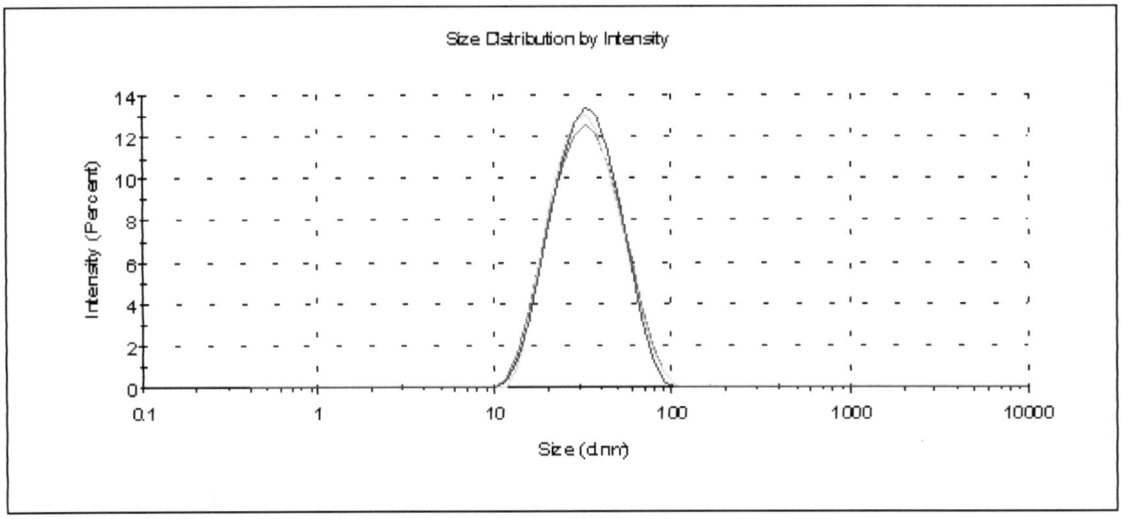

Figure 1. Size distribution by intensity of 5 wt. % dispersion of silica particles Ludox TM-40, nm.

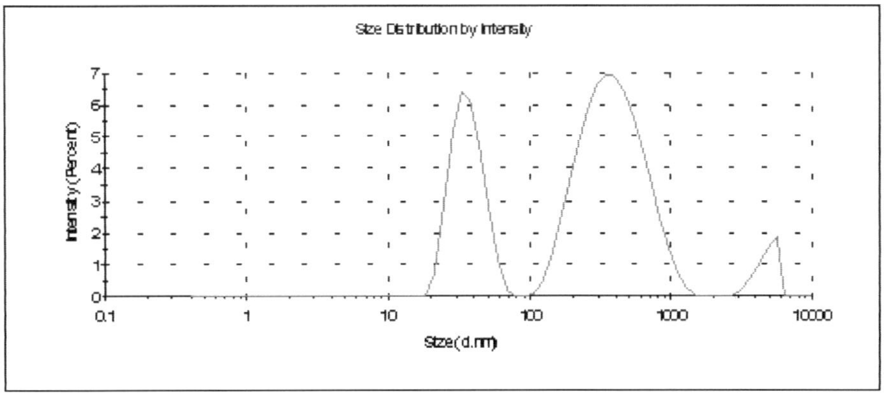

Figure 2. Size distribution by intensity of 5 wt. % dispersion of silica particles modified by 12 mM oleic acid, nm.

These particles were modified by oleic acid. In Figure 2 the results of size distribution by intensity of 5 wt. % dispersion of silica particles modified by 12 mM oleic acid are shown. Measurements show that two more picks appeared.

According to DLS results the first pick corresponds to size of 36 nm, the second – 418 nm, and the third – 4.7 µm. The value of first pick can be explained by modification of separate silica nanoparticles by oleic acid, however the second pick might be aggregates of modified nanoparticles. There is the third pick with a 4.7 µm diameter that is most probably oleic acid droplets that were coagulated before modification of silica nanoparticles. It can also indicate that all silica particles were modified and there is a certain extra amount of oleic acid and it is possible to decrease amount of oleic acid.

The second pick can be emulsion droplets stabilized by silica nanoparticles that can be seen from SEM photograph (Figure 3).

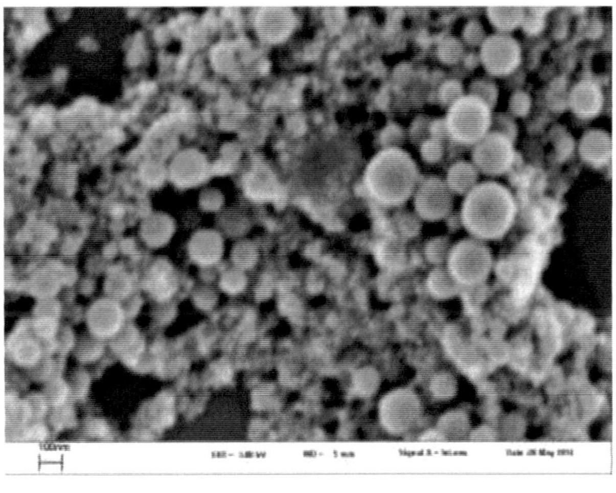

Figure 3. SEM photograph of oleic acid-modified silica nanoparticles, 100 nm.

However, the results of size distribution by volume show that in case of modified particles the number of particles with size of 418 nm and 4.7 µm can be neglected because of its extremely small amount (Figure 4).

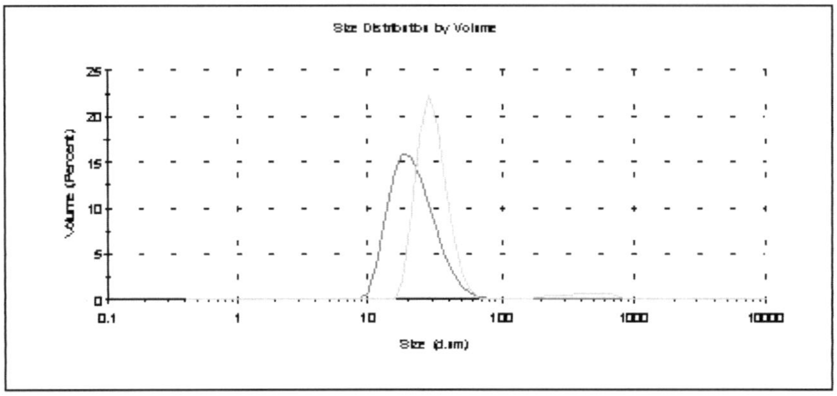

Figure 4. Size distribution by volume of non-modified and modified silica particles, nm. Red - 5 wt. % dispersion of silica particles, green - 5 wt. % dispersion of silica particles modified by 12 mM oleic acid

Considering red line (Ludox TM-40 nanoparticles) and the highest pick of green line (silica particles modified by 12 mM oleic acid) it can be seen that the diameter of separate silica nanoparticles increased from 22 to 36 nm that indicated success of the modification process. The experimental results showed the formation of uniform emulsion with white colour and average size of droplets of 350 nm. Comparing (Figure 5) modified silica particles size (green line) with emulsion droplets size (red line) by intensity it can be clearly seen that the number of particles with the diameter of 36 nm reduced significantly, meanwhile the pick corresponding to particles with the diameter of 300-400 nm increased.

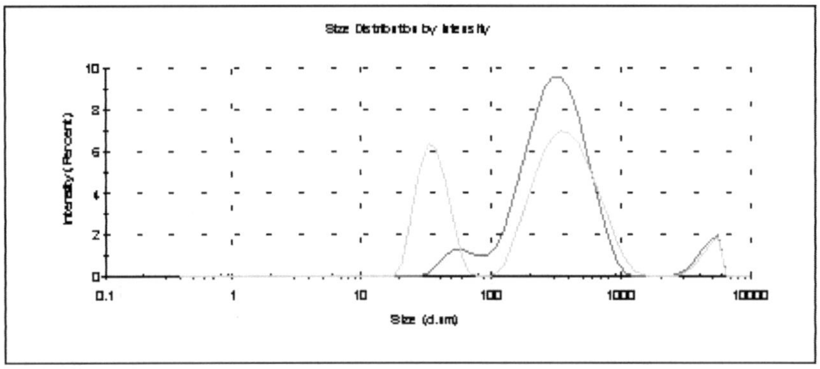

Figure 5. Comparison of silica particles modified by 12mM oleic acid and emulsion obtained using these particles. Green – modified silica particles 5% with 12mM oleic acid; Red – Miglyol 30% submicrocapsules stabilized with 5 wt. % dispersion of silica particles modified by 12 mM oleic acid

4. Conclusion

Thus, emulsions were prepared using silica nanoparticles modified with oleic acid. Experimental results showed that use of modifier leads to formation of stable emulsions with a diameter of 80 – 300 nm. The presence of more than one picks says that it is necessary to adjust the optimal ratio, and modification and emulsification modes.

On the basis of experimental results Middle chain triglyceride Miglyol 812 allowed one to obtain stable Pickering emulsion that can be used for encapsulation of various substances, and use of microcapsules allows one to reduce their impact on environment and increase efficiency of using.

5. Acknowledgments
Authors would like to express their sincere gratitude to the Ministry of Education and Sciences of the Republic of Kazakhstan for the support of project AP 05132810 "Scientific-practical bases of microencapsulation of bioactive substances and principally new stimulators of plant development with the purpose of agricultural production intensification", as well as to PROJECT 585761-EPP-1-2017-1FI-EPPKA2-CBHE-JP EduEnvi 'Enhancing Competences of Sustainable Waste Management in Russian and Kazakh HEIs" of the Erasmus + program.

Disclaimer: "This project has been funded with the support of the European Commission. This publication reflects the views only of the author, and the Commission cannot be held responsible for any use which may be made of the information contained therein".

References

[1] Shirley I M, Scher H B, Perrin R M, Wege P J, Rodson M, Chen J L and Rehmke A W 2001 Delivery of biological performance via micro-encapsulation formulation chemistry *Pest Management Science* **57** 129-132
[2] Sims S R, Appel A G and Eva M J 2010 Comparative toxicity and repellency of microencapsulated and other liquid insecticide formulations to the german cockroach (Dictyoptera: Blattellidae) *Journal Of Economic Entomology* **103** **(6)** 2118-25
[3] Bagle A V, Jadhav R S, Gite V V, Hundiwale D G and Mahulikar P P 2013 Controlled release study of phenol formaldehyde microcapsules containing neem oil as an insecticide *International Journal of Polymeric Materials and Polymeric Biomaterials* **62** **(8)** 421-25
[4] De Oliveira J L, Campos E V R, Bakshi M, Abhilash P C and Fraceto L F 2014 Application of nanotechnology for the encapsulation of botanical insecticides for sustainable agriculture: Prospects and promises *Biotechnology Advances* **32** 1550–1561
[5] Faria D M, Dourado Júnior S M, do Nascimento J P L, da Silva Nunes E, Marques R P, Rossino L S and Moreto J A 2017 Development and evaluation of a controlled release system of TBH herbicide using alginate microparticles *Materials Research* **20** **(1)** 225-235
[6] Garrido J J, Cagide F, Melle-Franco M, Borges F and Garrido E M 2014 Microencapsulation of herbicide MCPA with native b-cyclodextrin and its methyl and hydroxypropyl derivatives: An experimental and theoretical investigation *Journal of Molecular Structure* **1061** 76–81
[7] Sopeña-Vázquez F, Morillo E, Villaverde J and Maqueda C 2008 Reduction of pesticide pollution in agricutural soils by using controlled release formulations: Efficacy and persistence studies *Fresenius Environmental Bulletin* **17** **(10B)** 1705-1710
[8] Undabeytia T, Sopeña F, Sánchez-Verdejo T, Villaverde J, Nir S, Morillo E and Maqueda C 2010 Performance of slow-release formulations of alachlor *Soil Science Society of America Journal* **74(3)** 898-905
[9] Fernandes L P, Turatti I C C, Lopes N P, Ferreira J C, Candido R C and Oliveira W P 2008 Volatile retention and antifungal properties of spray-dried microparticles of Lippia sidoides essential oil *Drying Technology* **26** **(12)** 1534-1542
[10] Chung S K, Seo J Y, Lim J H, Park H H, Yea M J and Park H J 2013 Microencapsulation of essential oil for insect repellent in food packaging system *Journal of Food Science* **78** E709–E714
[11] Devi N, Maji T K 2011Neem seed oil: encapsulation and controlled release – search for a greener alternative for pest control *Pesticides Use and Management. – InTechOpen* pp 191–232
[12] Schoebitz M, López M D and Roldán A 2013 Bioencapsulation of microbial inoculants for better soil–plant fertilization A review *Agronomy for Sustainable Development* **33** 751–765
[13] He X, Sun Z, He K and Guo S 2017 Biopolymer microencapsulations of Bacillus thuringiensis crystal preparations for increased stability and resistence to environmental stress *Applied Microbiology and Biotechnology* **101** 2779–2789
[14] John R P, Tyagi R D, Brar S K, Surampalli R Y and Prévost D 2011 Bio-encapsulation of microbial cells for targeted agricultural delivery *Critical Reviews in Biotechnology* **31** **(3)** 211-226

[15] Shaikh J, Bhosale R and Singhal R 2006 Microencapsulation of Black Pepper Oleoresin *Food Chemistry* **94** (**1**) 105-110
[16] Lazko J, Popineau Y and Legrand J 2004 Soy Glycinin Microcapsules by Simple Coacervation Method *Colloids and Surfaces B:Biointerfaces* **37** (**1**) 1-8
[17] Chang C P, Leung T K, Linc S M and Hsu C C 2006 Release Properties on Gelatin-Gum Arabic Microcapsules Containing Camphor Oil with Added Polystyrene *Colloids and Surfaces B:Biointerfaces* **50** (**2**) 5-20
[18] Yuliani S, Torley P J, D'Arcy B, Nicholson T and Bhandari B 2006 "Extrusion of Mixtures of Starch and D- Limonene Encapsulated with β-Cyclodextrin: FlavourRetention and Physical Properties *Food Research International* **39** (**3**) 318-331
[19] Martín Á, Varona S, Navarrete A and Cocero M J 2010 Encapsulation and Co-Precipitation Processes with Supercritical Fluids: Applications with Essential Oils *Open Chemical Engineering* **4** 31-41
[20] Sadeghpour A, Pirolt F and Glatter O 2013 Submicrometer-Sized Pickering Emulsions Stabilized by Silica Nanoparticles with Adsorbed Oleic Acid *Langmuir* **29** 6004–6012

IOP Conf. Series: Materials Science and Engineering **826** (2020) 012030 doi:10.1088/1757-899X/826/1/012030

Environmental assessment of the Akmola region territories relative to the exposure to diseases of the population by the sigmal deviation method

A S Kurmanbayeva, A A Kakabayev, Z E Bayazitova, G E Baikenova, and B U Sharipova

Kokshetau State University named after Sh. Ualikhanov, 76, Abay St., Kokshetau, 020000, Republic of Kazakhstan

E-mail: aygul6868@mail.ru

Abstract. This paper presents the statistical indicators of the population' sickness in various districts of Akmola region. The object of the work is to conduct an environmental assessment of Akmola region's districts on the relative risk of morbidity. The following morbidity indicators were studied: blood, blood-forming organs and individual disorders involving the immune mechanism; respiratory organs; oncopathology; digestive organs and general morbidity. The sigmal deviation method was used to find out environmental assessment and ranking of territories. The conclusion is made about the importance of risk assessment for environmentally related diseases for ranking territories, identifying crisis zones and ensuring the well-being of the population. It was revealed that the territory of Akmola region has an uneven anthropogenic impact. It is shown that in areas with a high level of atmospheric pollution and adverse water supply conditions, the medical and environmental situation is extreme or critical. The necessity of constant monitoring of the incidence rate for environmentally caused diseases and the adoption of environmental measures in the identified crisis areas is substantiated. The methods for assessing the risks of morbidity considered in the article can be used to improve the activities of the executive branch of Kazakhstan.

1. Introduction

The transition of the Republic of Kazakhstan to sustainable development is an urgent need because the country needs development that meets the requirements of the present generation and does not constrain the ability of future generations to meet their needs. Economic growth due to the exploitation of natural resources can occur only at a certain stage. In modern conditions, more progressive mechanisms are required for growth and development [1 - 2].

In all Kazakhstani settlements, especially in large cities, the problem of collection, storage and processing of increasing volumes of municipal waste is acute. The main method of waste management today is the placement of waste in landfills. At the same time, the operation of most landfills and municipal waste dumps in Kazakhstan does not meet regulatory requirements. About 90 % of the existing landfills in the country do not meet the requirements of environmental and sanitary legislation

[3 - 4]. Only a part of settlements of the Republic is covered by services of the specialized enterprises on collecting and export of waste, and others are left without service.

2. Materials and methods

The starting materials to study the health indicators were the reporting documents of the Department of Health of Akmola region for ten years. The medical and environmental situation in various territories of Akmola region was evaluated by the method of sigmal deviations based on methodological recommendations for a comprehensive hygienic assessment of the degree of tension in different territories [5].

The data was subjected to statistical processing using the Excel software package. Districts and cities of Akmola region have different anthropogenic impact. The basic sources of air pollution are located in the cities of Kokshetau and Stepnogorsk. There are very high levels of air pollution on their territory [6]. A 10-year massive analysis of annual data of statistical reports on the diseases extent according to primary circulation, taking into account the influence of local living conditions were studied by determining the relative risk (the ratio of the frequency of diseases in the study area to the frequency of the same diseases in the background (or control) territory). The indicators of health disorders in the region having the lowest indicators of health disorders were in the quality of background indicators.

The medical and environmental situation was assessed according to one of 5 categories: 1 - satisfactory environmental situation (SES 1 -1.5); 2 - relatively tense environmental situation (RTES 1.5 -2); 3 - significantly tense environmental situation (STES 2.1 - 2.5); 4 - critical or emergency environmental situation (CEES 2.6-3.5); 5 - catastrophic or environmental disaster situation (CEDS more than 3.5) [7-8].

3. Findings

Assessing the state of the territories according to changes in the incidence of blood, blood-forming organs and individual disorders involving the immune mechanism, we revealed: in five districts of the region, the ecological state is satisfactory; relatively tense situation in four areas. A significantly tense environmental situation was observed in 5 districts. In the city of Kokshetau and 4 districts (Korgalzhinsky, Sandyktau, Tselinogradsky and Burabaysky) the ecological situation is extreme or critical (Table 1) [9].

The medical and environmental situation regarding respiratory diseases is satisfactory only in two districts of the region; the situation is relatively strengthening in Korgaldzhinsky district. A critical situation has developed in eight districts of the region, and in many districts of the region, the situation is at the level of environmental disaster.

There are three districts at a satisfactory level in the oncology. Relatively tense situation in seven districts of the region, significantly tense situation in other 6 districts of the region. A critical environmental situation has developed in the Tselinograd region, and in the cities of Stepnogorsk and Kokshetau.

The morbidity of the digestive organs, satisfactory revealed the situation in the Sandyktausky district, in Tselinograd - relatively tense. The situation in the Bulandinsky and Enbekshildersky districts are substantially tense, and in 11 districts and 2 cities of the region, the situation is critical. The situation in the Akkol and Yereimentau districts is catastrophic for this type of incidence.

An analysis of the overall incidence in three districts of the region revealed a satisfactory situation, in nine regions it was relatively tense. In three districts, it is significantly tense, in four districts it is critical and catastrophic in the city of Kokshetau, where the highest level of general incidence was revealed.

A general analysis of the medical and ecological situation in the territories of Akmola region using the sigmal deviation method showed that the most disadvantaged areas are Arshali, Atbasarsky, Egindykolsky, Burabaysky districts, Stepnogorsk and Kokshetau (Table 2). They are characterized by a high level of air pollution, inadequate provision of cultural and children's institutions, and adverse water supply and treatment conditions.

Table 1. The state of the territories according to changes in the incidence rates for individual classes by the sigmal deviation method

Name of cities and areas	Territory status according to changes in the incidence rate for individual classes				
	Blood diseases	Disease of respiratory system	Oncology	Disease of digestive system	Overall incidence
Akkolsky	SES	CEES	STES	CEES	RTES
Arshalinsky	STES	CEDS	RTES	CEDS	STES
Astrakhan	STES	CEES	RTES	CEDS	RTES
Atbasar	SES	CEDS	STES	CEDS	CEES
Bulandinsky	SES	CEES	STES	STES	SES
Egindykolsky	RTES	CEDS	STES	CEDS	STES
Enbekshildersky	SES	CEES	RTES	STES	RTES
Yereimentau	RTES	SES	RTES	CEES	RTES
Esilsky	STES	CEES	SES	CEDS	RTES
Zhaksynsky	STES	CEDS	STES	CEDS	CEES
Zharkainsky	SES	CEES	RTES	CEDS	RTES
Zerendinsky	RTES	CEES	RTES	CEDS	RTES
Korgalzhyn	CEES	RTES	STES	CEDS	SES
Sandyktau	CEES	SES	SES	SES	SES
Tselinogradsky	CEES	CEDS	CEES	RTES	RTES
Shortandinsky	STES	CEES	SES	CEDS	RTES
Burabay	CEES	CEDS	RTES	CEDS	CEES
Stepnogorsk	RTES	CEDS	CEES	CEDS	STES
Kokshetau	CEES	CEDS	CEES	CEDS	CEDS

The Akkol, Astrakhan, Enbekshilders, Zhaksynsky, Zharkaysky, Tselinogradsky districts have significant shortcomings in these sections. The most prosperous regions according to the total assessment are Sandyktau, Yereymentau, Korgalzhyn, Bulandinsky, Esilsky, Zerendinsky and Shortandinsky districts.

Table 2. Ecological situation in the territories of Akmola region with respect to the relative risk of morbidity by the method of sigmal deviations

Ecological situation of the territory	Territory
satisfactory	Sandyktausky
relatively tense	Yereimentausky, Korgalzhyn
substantially tense	Bulandinsky, Esilsky, Zerendinsky, Shortandinsky
critical or emergency	Akkolsky, Astrakhan, Enbekshildersky, Zhaksynsky, Zharkainsky, Tselinogradsky
catastrophic or environmental disaster situation	Arshalinsky, Atbasarsky, Egindykolsky (water), Burabaysky, Stepnogorsk, Kokshetau

4. Conclusion

The work by the example of Akmola region over a ten-year period allowed us to assess the environmental risks associated with environmental pollution and analyze the incidence rates in the regions. The integrated environmental assessment used by the sigmal deviation method allowed ranking the territory of Akmola region and identifying crisis zones requiring constant monitoring and additional environmental decisions that would reduce the risk of population morbidity. High incidence rates for diseases caused by environmental risks (respiratory diseases, oncopathologies, etc.) prove the increased influence of environmental factors on the health of the population in Akmola region. This circumstance indicates the need for a more detailed study of the ecological state of the environment in the region.

In Akmola region, the number of existing air condition monitoring is insufficient for an objective assessment of the quality of atmospheric air. Poor technical and instrumentation of the posts does not permit monitoring of all hazardous pollutants present in the air basin. An increase in the number of observation posts in the region is required; it is necessary to re-equip existing posts in order to more effectively assess the quality of atmospheric air.

For more effective risk management, it is necessary to implement a centralized information program for processing and analyzing a large amount of data on the state of the environment and public health and identifying cause-effect relationships in the "environment - public health" system for risk analysis.

Environmental pollution as a risk factor for health is an extremely urgent problem. The application of the assessment principles and risk management allow more efficient use of financial resources for environmental improvement, ensuring the sanitary-epidemiological and environmental well-being of the population. At the same time, it is important to consider the occurring negative changes in the habitat using new methods and assessment technologies. Health risk assessment is an internationally recognized scientific tool for developing optimal solutions for managing environmental quality and public health.

5. Acknowledgments

The article was prepared in the framework of the project: "Enhancing competences of sustainable waste management in Russian and Kazakh HEIs / EduEnvi" with the financial support of the program "Erasmus+".

References

[1] Kurmanbaeva A C, Russel A, Althonayan A, Ali M, Zhumabaeva C E, Gazdieva B A, Akhmetov K K and Mukanov E 2016 Air pollution in Akmola region and environmental risk assessment for public health *News of NAS RK, A series of biological and medical* **6** 23–29

[2] Russell A, Ghalaieny M, Gazdiyeva B, Zhumabayeva S, Kurmanbayeva A S, Akhmetov K K, Mukanov Y, McCann M, Ali M, Tucker M, Vitolo C and Althonayan A 2018 A Spatial Survey of Environmental Indicators for Kazakhstan: An Examination of Current Conditions and Future Needs *International Journal of Environmental Research* **12 (5)** 735-748

[3] Tulebaev R K, Slazheneva T I, Kenesariyev U I, Belonog A A and Korchevsky A A 2004 *Hygiene risk assessment in industrial regions of the Republic of Kazakhstan (*Almaty, Iskander)

[4] Sraubaev E N, Serik B 2013 Development of technologies for managing the health of the population of Kazakhstan, based on an integrated assessment of the combined impact of environmental factors *Hygiene and sanitation* **2** 12 - 16

[5] Guidelines A 1997 No. 2510 / 5716-97-32 *Comprehensive assessment of the degree of tension in the medical and environmental situation in various territories due to toxicants polluting the living environment of the population*, available at: http://www.consultant.ru/ search /? Q

[6] Ministry of National Economy of the Republic of Kazakhstan. Statistics committee, available at: http://stat.gov.kz/

[7] Prusakov V M and Prusakova A V 2013 Criteria for assessing the medical and environmental situation based on the sigmal deviation method *Hygiene and sanitation* **1** 72–76.

[8] Prusakov V M and Prusakova A V 2013 Risk of disease as a criterion of medical-environmental conditions of the quality of life *Acta Biomedica Scientifica* **3(2)** 120-124

[9] Ed. Musina E E 2014 Statistical indicators of socio-economic development of the Akmola region *Monthly informational and analytical magazine*

Ecological and economical assessment of efficiency of application of alternative energy sources to achieve the goals of the Climate Strategy of St. Petersburg

V A Savoskula, O I Sergienko and A S Pavlova

ITMO University, 9, Lomonosova st., St. Petersburg, 191002, Russia

E-mail: violettasavoskula@gmail.com

Abstract: The paper considers the procedure and methodology for assessing the environmental and economic efficiency of using alternative energy sources installed at industrial enterprises, which can significantly reduce climate risks by reducing greenhouse gas emissions in replaced sources. Based on the proposed methodology, the costs of introducing a hybrid solar installation at an electronic equipment manufacturing enterprise were determined, resource efficiency indicators of efficiency and environmental impact, as well as the cost of generated energy, were calculated. It is shown that the application of the methodology will contribute to the selection of the best available technologies for alternative energy and the achievement of the goals of the Climate Strategy of St. Petersburg.

1. Introduction

Assessing the effectiveness of possible impacts in order to reduce and prevent climate change is one of the most pressing problems on the agenda of the international community, the solution of which is provided for in the 2015 Paris Agreement and subsequent documents [1].

In the framework of this task, the need for Russia's participation in the global system of scientific and technological exchange of knowledge and cooperation is not only important for the import substitution of equipment, but also for the development of assessment methods and the selection of the most effective solutions, as these ensures timely response to the requirements of the international market, and therefore will increase the production of green energy from alternative sources in Russia by 2035.

In St. Petersburg, the Climate Strategy developed until 2030 aims to create conditions for assessing and mobilizing development potential in changing climate conditions through the timely implementation of adaptation measures aimed at reducing and (or) preventing climate risks in relation to urban sectors, as well as facilities of urban infrastructure, ecosystems and public health. Among possible measures, consideration should be given to replacing traditional sources of hydrocarbon fuel energy with alternative energy sources (AIE) [2].

The aim of this work is to develop a procedure and methodology for assessing the environmental and economic efficiency of using alternative energy sources installed at industrial enterprises, which can significantly reduce climate risks by reducing greenhouse gas emissions in replaced sources. The objectives of the work are: 1) to verify the feasibility of the technical implementation of the installation of a solar electric power generator and to reduce the environmental load when introducing electronic equipment at one of the enterprises of St. Petersburg, and 2) to evaluate the investment and operating costs associated with the introduction of equipment in the equipment's life cycle.

The relevance and timeliness of the research topic is confirmed by the need to solve the pressing problems of the energy sector development, to determine changes in economic parameters in accordance with the Energy Strategy of Russia for the period up to 2035, on the one hand, and the need to invest in the production of alternative energy, on the other.

Based on environmental and economic assessment, one can choose the most suitable alternative energy sources or combinations of these for the St. Petersburg region, and choose the best option for implementation in a particular cluster.

Such assessment is important for the implementation of relevant projects for the creation of wind farms, solar power plants, devices capable of receiving geothermal energy, hydropower and biomass energy. Its application will contribute to the achievement of the all-Russian strategic indicator for the development of renewable electricity by 2024 at the level of 4.5% of the total energy production [3].

2. Materials and methods

The research methodology was based on the methods of statistical analysis, the problem-target approach, reverse forecasting with the participation of stakeholders, resource efficiency analysis (ARE), environmental life cycle impact assessment (LCA) and life cycle cost assessment (LCA), which all together allowed developing a multi-step procedure for environmental and economic assessment of the choice of the best available technology for an alternative energy source [4].

A review of the main sources of greenhouse gas (GHG) emissions into the atmosphere of St. Petersburg shows that the largest mass of GHG emissions is characteristic of the areas where the heat power plant is located. Emissions from thermal power plants, namely soot and CO_2 emissions lead to the greenhouse effect, and nitrogen dioxide emissions to smog in the city and, consequently, affect climate change.

At the first stage, the climatic parameters of the region were assessed, which determine the effectiveness of using an alternative energy source: a solar battery.

The standard solar system for autonomous power supply consists of solar panels with a total capacity of 3.4 kW [5] and can reliably meet the electricity needs of most owners of private residential buildings. The use of alternative energy sources is currently being considered at a number of large industrial enterprises, ensuring the achievement of market leadership and sustainable development in the long term.

At the next stage, the main stakeholders were analyzed, as well as constraining and stimulating factors in the development of the alternative energy market.

Among the stakeholders with the greatest degree of influence, as shown in the authors' work [4], there are the authorities, shareholders, creditors, suppliers and employees of the company.

Among the factors hindering the development of alternative power sources, an abundance of traditional fuels and the lack of a regulatory framework to support green energy were highlighted. As a stimulating factor, one can note the rapid development of technologies and the widespread use of alternative energy sources abroad, which inevitably should lead to the development of the production of equipment and technologies for obtaining renewable and alternative energy in Russia, since in the future this will eliminate the dependence on foreign energy technologies.

Formulating a vision of the future and determining the necessary steps to achieve the desired goals (development scenario), called the method of reverse forecasting, were used in the early 2000s in relation to the development of energy efficiency at the state level [5,6] and later at the corporate management level [7,4].

At the company level, independent scenarios for the implementation of AES can be carried out based on an analysis of the current situation in the technology market, setting specific goals to increase energy efficiency and reduce the carbon footprint, and creating an "ideal" picture of the company's development in the future.

During the next stage of "Scenario Analysis", the main indicators of environmental assessment of selected alternative energy sources in their life cycle are determined: environmental impact indicators of output flows (emissions, discharges and waste generation) by categories of impacts and input flows of consumed resources (abiotic, biotic, water and air).

"Scenario analysis" was carried out according to the method proposed by the authors for assessing the environmental and economic efficiency of alternative energy sources, which was tested by the example of a solar power plant [8] at an electronic equipment manufacturing enterprise in St. Petersburg.

The main steps of the methodology include: assessing technical feasibility based on available technologies, performing a resource efficiency analysis (ARE), a life cycle assessment (LCA) and costs of life cycle (LCC).

3. Results and discussion

The suggested methodology was applied to assess the cost-efficiency and environmental results of implementation of solar power plant at the electronic equipment manufacturing enterprise.

The main components of a photovoltaic solar power plant are: photovoltaic modules, a controller, a battery, and an inverter. However, to achieve reliable and efficient operation of any solar system, it is necessary to connect a diesel generator (DG).

The boundaries of the production system are determined by the following stages of the life cycle: transportation of the solar battery to assembly at the consumer, transportation of batteries to assembly at the consumer, transportation of the controller to assembly at the consumer, transportation of the inverter to assembly at the consumer, transportation of the diesel generator to assembly at the consumer, temporary storage of components of the solar battery in the organization's warehouse, the process of assembling the solar battery at the consumer, the operation stage, the transportation process to the place of disposal in the end of life.

Based on the initial data and the performed calculations, material inputs or MI indicators were determined by the types of resources consumed. Analysis of resource efficiency (ARE) showed that the maximum impact is exerted by the operation stage - 86%; transportation by rail from China to St. Petersburg - 12% and by road to the organization – the assembler of an inverter - 2%. The remaining stages, the contribution of which is less than 1%, can be neglected.

At the stage of production of the solar battery, the biggest share of resources is consumed such as water - 88% and abiotic resources - 12%. The only material input at the operation stage is diesel fuel, which contributes to the consumption of abiotic resources and water. At the stage of transportation of solar modules, controller, battery and diesel engine by railway, the largest consumption share is water - 88% and abiotic resources - 12% due to the use of diesel fuel.

The calculation of the environmental impact of the output flows of the solar battery in the life cycle was performed based on the SimaPro software [9]. For a solar battery, by categories of impact, indicators such as global warming potential, g-eq. CO_2/ g, photochemical smog formation potential, g-equethen / g, eutrophication potential, g-equiv O_2 / g were determined.

To realize the goals of the Climate Policy, the analysis of the carbon footprint within the boundaries of the production system is of the greatest interest. The carbon footprint of the solar installation within the production system was 0.504 t-eq CO_2.

Costs in the life cycle were determined according to the methodology [10] and amounted to 4,481.35 thousand rubles. The annual duration of sunshine is an average of 1800 h / year in St. Petersburg and the Leningrad Region. With a solar battery power of 3 kW and a DG power of 10 kW, the electricity generated by the solar installation will average 75,000 kWh/ year.

4. Conclusion

According to the calculated data, the cost of the entire life cycle of the solar battery is 4,481.35 thousand rubles, and to obtain 1 kWh of energy it is necessary to spend 3.98 rubles, which is comparable to the tariff for electricity from utility networks – 4.58 rubles per one kWh.

However, with a decrease in diesel consumption at the operational stage, as well as a reduction in the cost of acquiring AIE, it is possible to obtain much cheaper electricity and ensure sufficient autonomy from the electric grid energy and a significant reduction in the carbon footprint. According to the analysis of the technical potential of solar energy in St. Petersburg, it can be concluded that part of the energy received at industrial enterprises from traditional CHP plants can be replaced by energy received from solar panels, which will reduce emissions into the atmosphere from the burning of fossil fuels, affecting climate change in St. Petersburg.

The proposed methodology of environmental and economic assessment of the choice of an alternative energy source can be used to identify the best available technology from those available on the market that have minimal negative impact on the environment and minimize greenhouse gas emissions in the installation's life cycle.

Such assessment ensures the implementation of the goals of the Climate Strategy of St. Petersburg until 2030 and the Energy Strategy of Russia for the period until 2035 to diversify the fuel and energy sector, and to increase the energy efficiency of enterprises. The cost of electricity in the life cycle can be used as an important indicator for further economic justification of investment projects for the implementation of AEI.

The practical significance of the proposed method lies in the fact that decision makers can use it to select the most environmentally friendly energy sources, both at individual enterprises and in the region as a whole.

References

[1] *Annual technical progress report of the Paris Committee on Capacity-Santiago Climate Change Conference* 2019. Available at: buildinghttps://unfccc.int/documents/200078

[2] *Climate strategy of St. Petersburg for the period until 2030*, available at: https://docplayer.ru/26790826-Proekt-klimaticheskaya-strategiya-sankt-peterburga-na-period-do-2030-goda.html

[3] Government of the Russian Federation 2009 *On the main directions of state policy in the field of increasing the energy efficiency of the electric power industry based on the use of renewable energy sources for the period up to 2024,* available at: http://consultant.ru/document/cons_doc_LAW_83805

[4] Pavlova A S and Sergienko O I 2014 Analysis of factors affecting energy efficiency and the development of corporate social and environmental responsibility of Russian companies based on the method of reverse forecasting with the participation of stakeholders *Scientific journal NRU ITMO. Series Economics and Environmental Management* **4** 340-355, available at: http://economics.ihbt.ifmo.ru/file/article/11202.pdf

[5] Mulder H A J, Biesiot W and Elgar E 1998 *Transition to a sustainable society. A backcasting approach to modeling energy and ecology* (Cheltenham) 287-304

[6] Quist J and Vergragt P 2006 Past and future of backcasting: The shift to stakeholder participation and a proposal for a methodological framework *Futures* **38** 1027-1045

[7] Green K and Vergragt P 2002 Towards sustainable households: a methodology for developing sustainable technological and social innovations *Futures* **34** 381-400

[8] *Hybrid solar power plant 3 kW. Model: SA-3000,* available at: https://www.solnechnye.ru/gotovye-resheniya/solnechnaya-elektrostanciya-avtonomnogo-pitaniya.htm

[9] *LCA software for fact-based sustainability,* available at: https://simapro.com/

[10] *Hybrid solar power plant 3 kW. Model: SA-3000,* available at: https://lib.sale/upravlenie-proektami-knigi/134-zatratyi-jiznennyiy-tsikl-metodyi-55931.html

IOP Conf. Series: Materials Science and Engineering **826** (2020) 012032 doi:10.1088/1757-899X/826/1/012032

Perspective of using the biological method for gaseous emissions cleaning in the food productions

A V Nazarova, N V Kurnikova, N R Molodkina, R F Yulmetova and O I Sergienko

ITMO University, 9, Lomonosova st., Saint-Petersburg, 191002, Russia

E-mail: naz.nastasya@gmail.com

Abstract. The article considers the problem of contamination of exhaust gases of confectionery plant because of harmful substances and unpleasant odors. A confectionery production uses a wide range of aromatic substances. Moreover, despite the presence of operating air treatment facilities the specific smell of ambient air could be sensed. To identify the most powerful sources of odor emissions the organoleptic study was conducted. Also, with the help of olfactometric study of the sources of odor, the composition of the aromatic mixture with a predominant content of terpenes and benzaldehyde was revealed. Several methods of treatment of gases-air emissions were studied including biological, thermal and thermocatalytic. Among the biological ways of air treatment, three types have been considered: traditional biofilters, trickling biofilters and bio scrubbers. Bacteria that helps to decompose aromatic substances like benzaldehyde or some of the terpene series have been found. Additionally, the mechanism of transformation of odorous substances has been studied.

1. Introduction

The enterprises of the confectionery industry aren't the main pollutants of the atmosphere, however, being within the city boundaries, they influence the quality of aambient air due to emissions of smelling substances [1]. The feature of the considered gas-air emissions is the presence of a specific smell of aromatic compounds used at the enterprise, despite the presence of treatment facilities. It is known that when several pollutants are in the ambient air that cause unpleasant sensations, the person's sense of smell perceives the presence of only one gas, which has the strongest odor, as it maskes the remaining gases. The presence of even minor concentrations of these substances in the air causes dizziness, headaches and even neuralgia. At high concentrations, they affect the kidneys, causing diseases such as nephritis, pyelonephritis [2].

In this work, we studied the possibility of using the biological method for the purification of gas-air emissions from a confectionery plant from volatile flavoring compounds (VFC).

To achieve this goal, the following tasks were considered: determining the concentration of odor, studying the composition of the gas-air mixture of a confectionery enterprise, considering the possibility of using the biological method to purify air from aromatic compounds.

2. Materials and methods

There are various methods to neutralize odors from the confectionery industry exhaust gases, namely: thermal and thermocatalytic, absorption and adsorption methods, gas-phase processing. Thermal and thermocatalytic methods are based on the processes of destruction and oxidation of pollutants atmospheric oxygen at high temperatures using special catalysts [3]. During gas-phase processing, ozone is introduced into gas-air emissions or special substances that neutralize odors [4]. In the absorption method, gases are flushed with liquid absorbers, for example, aqueous solutions of alkalis, acids, oxidizing agents [5]. During adsorption, unpleasant smelling substances are absorbed by solid sorbents [6]. However, these methods require the presence of certain chemicals and catalysts. And besides, as a result of such purification, products appear that become sources of secondary pollution [7]. In this regard, biological treatment has great prospects.

The principle of biological treatment of exhaust gases is based on the fact that microorganisms act on the corresponding organic substances with their transformation into harmless compounds [8]. In this work, the possibility of using biological method for the additional treatment of gas-air emissions of the confectionery plant from volatile flavoring components (VFC) was considered.

At the first stage, it was necessary to identify the most powerful sources of odor emissions, which was determined using organoleptic examination of the sources.

To determine the level of negative effects of odor, an olfactometric study of the sources of odor was performed. Smell samples were taken in sampling bags. Bags for the vacuum sampler are made of polyethylene terephthalate (PET, nalofan). This material is suitable for odor sampling, since it practically does not adsorb organic substances. Connecting and sampling tubes are made of polytetrafluoroethylene.

Samples taken from the studied sources (from 1 to 14) were subjected to olfactometric analysis to determine the odor concentration and instrumental analysis using a gas analyzer to measure the concentration of volatile flavorings compounds (VFC).

3. Results of the study of perspective of the biological treatment of gas emissions from VFC

The experimental results of the study of perspective of the biological treatment of gas emissions from VFC are given in table 1.

The values of odor concentration and VFC are the averaged values for three samples, show the concentration of VFC used in the confectionery industry, taken from the design documentation.

Table 1. Results of olfactometric determination of odor concentration in samples taken from sources of odorous substances

Number of sources	Source Name	Concentration of odour, UO / m^3	Concentration of the amount of VFC, rel. units	VFC
1	Glazing gild № 1	1076	13.3	1.0
2	Glazing gild №1	1322	8.4	0.97
3	Internal warehouse of flavouring additives	5114	23.9	2.3
4	Glazing gild №1	4894	16.1	2.3
5	Glazing drum №5	412	-	1.3
	Glazing drum №8	2021	10.8	
6	Mixing and rolling gild №1	7174	100.3	0.88
7	Glazing shop number 2	1502	8.4	1.3
8	Candy production line	3143	12.8	1.4
9	Glazing gild №2	2313	5.1	1.3
10	Dragee production	1307	2.7	2.2
11	Mixing and rolling gild №2	23716	62.8	1.2
12	Cooling Cameras	9759	30.1	0.6
	Chewing gum settling rooms	5349	19.7	
13	Chewing gum settling rooms (tempering)	1087	7.8	2.2
	Chewing gum production line	3161	75.3	
14	Candy production	4363	32.1	5.9

According to the results of pollutant dispersion calculations, three sources making the most contribution to air pollution at the border of the sanitary protection and residential zone have been identified (No. 11, No. 12, and No. 14). The total contribution of these sources is about 70% (about 40% for No. 11, about 20% for No. 12, and about 10% for No. 14). The contribution of other sources varies from 0.4% to 6.7%.

The next task was to determine the composition of the aromatic compositions has been analyzed. The results of chromatography-mass spectrometry analysis of flavors, conducted in accordance with the "Methodological recommendations on analysis unknown composition objects via gas-liquid chromatography, gas chromatography-mass spectrometry, high performance gas chromatography, mass spectrometry with inductively coupled plasma atomic emission spectrometry inductively coupled plasma methods by Mendeleev VNIIM" No. 01-07, show a significant content of terpenes (limonene, menthol) and benzaldehyde.

Considering the possibility of using a biological purification method, microorganisms capable of transforming the volatile components of flavours have been found, and the mechanism of transformation of these substances has been studied. The results are presented in table 2.

Table 2. Types of microorganisms and the mechanism of biotransformation of volatile flavoring components

Type of microorganism	The name of the flavoring component	The mechanism of biotransformation
Rhodococcus opacus	d- limonene	Biomass from the culture of the glucose-toluene chemostat converts d- limonene to trans-carveol, which is then oxidized to carvone trans-carveol by dehydrogenase [9]
Pseudomonas putida	Limonene	Converts limonene to p-ment-1-ene-6,8- diol and perillyl alcohol [10]
Rhodococcus erythropolis DCL14	Limonene	Epoxide hydrolase from Rhodococcus erythropolis DCL14 catalyzes the hydrolysis of limonene-1,2-epoxide to limonene-1,2-diol [11,12]
Thauera terpenica 21 Mol	Menthol	Thauera terpenica 21 Mol converts menthol as a source of sole carbon. The mechanism of degradation involves oxidation reactions leading to ment-2-enone, followed by hydration and an additional oxidation stage [13]
Castellaniella defragrans	Limonene	As an initial stage, oxygen-independent hydroxylation of the methyl group of limonene to perillyl alcohol occurs, followed by oxidation to perillic acid. [14]
Saccharomyces cerevisiae	Benzaldehyde	Biotransformation of benzaldehyde into free cells L - phenylacetylcarbinolbi - yeast effect from β-cyclodextrin [15]

Three existing biotechnological solutions for biological gas treatment have been considered: traditional biofilters, trickling biofilter and bio scrubbers.

Biofilters are open bulk sites filled with natural charging material (peat, sawdust, compost, pieces of bark, etc.) [16]. The formation of the microbial community proceeds naturally on the filler particles. However, the load of natural origin is staked over time, which leads to deterioration of the efficiency of purification, and can also serve as a habitat for microorganisms that are antagonists of active biomass [17]. In addition, such a biofilter requires humidification of the incoming air for the most complete extraction of pollutants by bacteria, which is solved by the introduction of a gas scrubber- humidifier. Among the weakness of traditional biofilters, the authors [18] also note the lack of a constant supply of fluid to the body of the biofilter and conclude that such a biofilter is most effective for cleaning gases from hydrophobic and poorly water-soluble compounds. It also indicates a relatively small amount of pollutant extraction (compared to other methods).

The configuration of the trickling biofilter reactor is generally similar to the traditional one. One of the main differences is the continuous supply of the liquid droplet phase and the use of inert feed materials - fillers. In trickling biofilters, polluted air is passed through a column with a filler, in which the liquid circulates [19]. The contaminant is dissolved in the liquid film, and then captured by microorganisms growing on the surface of the filler. The fluid provides moisture, nutrients, pH control in the biofilm and allows to remove inhibitory products. Ideally, excess biomass is also removed by liquid droplets, thereby achieving equilibrium and stable operation of the biofilter. However, in practice, the circulating fluid does not cope with excessive growth of the microbial mass [20], which leads to an increase in hydraulic resistance, a reduction in the residence time and a drop in productivity. In extreme cases, the biofilter reactor is closed and cleaned. Some authors [17,18] suggest other options to control the growth of microbial populations while maintaining microbial activity. These ideas include selecting of the size and structure of the filler, limiting the supply of nutrients and adding inhibitors. The possibility of using microbial predators, such as protozoa and nematodes, is also indicated.

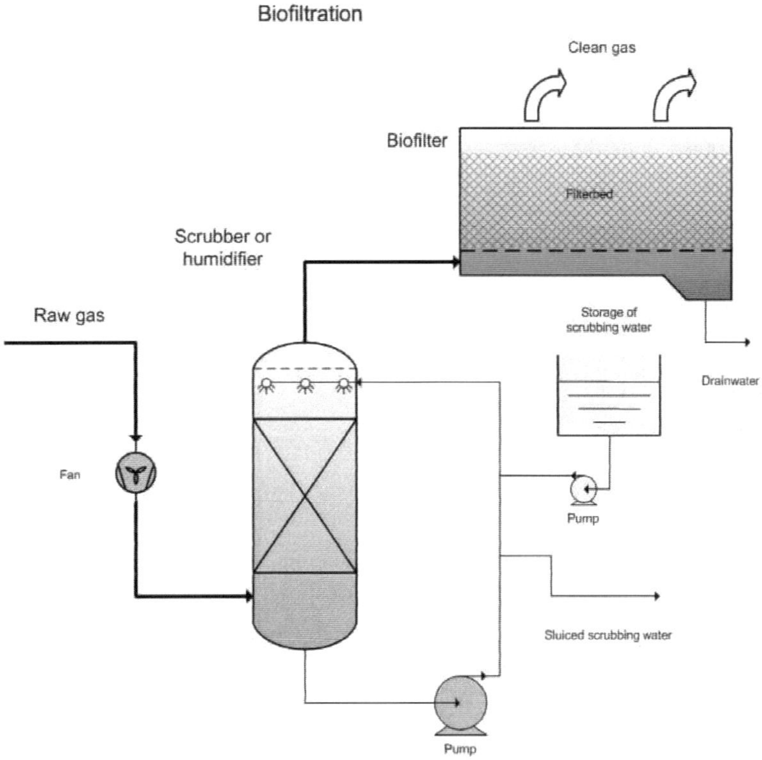

Figure 1. Schematic representation of bio-scrubber

Bio-scrubbers for gas purification are characterized by the release of two processes: the absorption of volatile compounds into water and the biological purification of water in a bioreactor. Exhaust gas is served in an absorbent column with a layer of filler and a countercurrent of gas and water, in which pollutants are converted into a liquid form. Gas-liquid absorbent columns are designed to facilitate mass transfer from the air in the liquid phase, thereby reducing the amount of equipment [21] and maintaining the hydraulic resistance.

Water then enters the bioreactor - a tank with an aeration system containing activated sludge, which is involved in the process of neutralizing pollutants. The volume of the reactor is much larger than the volume of the absorbing column (Figure 1). To maintain microbial activity, nutrients should be added to the water bioreactor. Additionally, titrants can be added to control the pH of the aqueous phase.

Advantages of bio-scrubbers compared to biofilters are: less equipment, better pH control, more reliable and predictable results, no problems with clogging of packaging materials, lower occurrence of toxic concentrations in the liquid phase [21].

The last two points are, at the same time, advantages over trickling biofilters. Among the weaknesses, the need to remove waste sludge and prevent periods of downtime are noted.

4. Conclusion

As a result of the conducted research, sources of volatile aromatic compounds of confectionery production making the most contribution to air pollution by terpenes and benzaldehyde have been identified. Additionally, the strains of microorganisms that can transform the detected aromatic compounds in the process of biotechnological air purification have been selected. And constructive and technological features of biotechnological installations for cleaning the gas-air mixture are considered.

The obtained data allows to consider the method of biological transformation of residual volatile components of flavoring with the help of special strains of microorganisms as promising and reasonable in terms of efficiency, safety and economically feasible.

References

[1] Podchufarov E 2003 *Sweet life* **3** 91-95
[2] Lazarev N V and Levina E N 1977 *Harmful substances in industry* 623
[3] Guoliang Liua, Manxuan Xiaob, Xingxing Zhangc 2017 *Sustainable Cities and Society* **32** 375–396
[4] Hao Z G, Su X M 2018 Determination and Distribution of Indoor Off-odor Pollutants, *Chemical Engineering transactions* **68** 475-480
[5] Zukeran A, Chang J S, Berezin A A amd Ito 1997 *Journal of Aerosol Science* **28**
[6] Xiao R, Mo J, Zhang Y, Gao D 2018 *Indoor air* **28(2)** 266-275
[7] Ren H, Koshy P, Chen W, Qi S and Sorrell C C 2017 *Journal of Hazardous Materials* **325** 340-366
[8] Asadi S, Hassan M, Nadiri A and Dyla H 2014 *Environmental Science and Pollution Research* **21(14)** 8847-8857
[9] W A Duetz, A H M Fjällman, S Ren, C Jourdat, and B Witholt 2001 *Appl. Environ. Microbiol* **67** 2829-2832
[10] Chatterjee T and Bhattacharyya D K 2001 *Appl. Microbiol Biotechnol.* **5** 541-546
[11] Van der Werf M J, Orru R V A, Overkamp K M, Swarts H J, Osprian I, Steinreiber A et al. 1999a *Appl. Microbiol. Biotechnol.* **52** 380–385
[12] Van der Werf M J, Swarts H J and De Bont J A M 1999b *Appl. Environ. Microbiol.* **65** 2092–2102
[13] Foss S and Harder J 1998 *Syst. Appl. Microbiol.* **21** 365–373
[14] Foss S, Heyen U and Harder J 1998 *Syst. Appl. Microbiol.* **21** 237–244
[15] Mahmoud W M, El Sayed A H M M and Coughlin R W 1990 *Biotechnol. Bioeng.* **36(3)** 256-262
[16] Mitin A K, Nikolaikina N E and Zagustina N A 2016 *Journal of TGTU* **22(1)** 84-87
[17] Zagorskis A and Baltrenas P 2010 *Ekologija* **56(1–2)** 72–78
[18] Kennes C Rene and Eldon R 2009 *Journal of Chemical Technology & Biotechnology* 1419–1436
[19] Kennes C and Veiga M C (Ed.) 2001 *Springer Dordrecht* 312
[20] Shareefdeen Z and Singh A (Ed.) 2005 *Springer-Verlag Berlin Heidelberg* 409
[21] Wang, Lawrence K (Ed.) 2009 *Humana Press* 833

Investigation of the influence of distribution of heavy metals from construction waste disposal on plant associations of adjacent ecosystems

N V Dinkelaker[1], E A Ovsyuk[1], N A Sinelnikova[1], I A Muraviev[1], N F J Dinkelaker[2] and T S Semenova[1]

[1] ITMO University, 9, Lomonosova st., St. Petersburg, 191002, Russia
[2] St. Petersburg forest engineering University. S. M. Kirova, 5, Institutskii per., St. Petersburg, 194028, Russia

E–mail: nvdinkelaker@mail.ru

Abstarct. The question of influence of distribution of pollution of soils by heavy metals from the landfill on the soil–plant complex of natural ecosystems in the zone of influence of object is investigated. The presence of increased bioaccumulation of heavy metals of certain specialized ecological groups of plants – submerged macrophytes and weed–ruderal plants was revealed. Specificity of accumulation of heavy metals is not a property of plant life forms. However, within plants of one life form there are species with an increased ability to selective accumulation of individual heavy metals, which is especially pronounced in mosses. The results obtained can be used to improve the efficiency of bioengineering protection of the natural environment through environmental planning of landscapes in the areas contaminated with heavy metals around landfills and other industrial facilities.

1. Introduction

Against the background of the acute problem of disposal of construction waste in modern Russia and the prevalence of such type of disposal as landfill disposal, the issues of environmental safety of such facilities for adjacent ecosystems and their components, including humans, are of particular importance. Construction waste landfills are characterized by soil contamination of adjacent territories with heavy metals. These pollutants in concentrations higher than natural are toxic to all groups of living organisms. They cause stress reactions both at the level of ecosystems and communities of living organisms, and at the organizational level. Long–term exposure to high concentrations of HM can lead to irreversible changes in ecosystems (change of communities, decrease in the number and production of organisms, extinction of species). The increase in the area of landfills in Russia inevitably leads to an increase in the area of various ecosystems located in the zone of soil and surface water pollution by heavy metals. The main danger for animals and humans in the spread of heavy metals in soils around landfills is the inclusion of these toxicants in the food chains of local ecosystems in the zone of pollution and their transition through food chains to larger ecosystems, located outside the territory of the direct impact of the landfill. Such areas often contain both valuable natural ecosystems and residential areas.

The first and main barrier for inclusion of HM in food chains of local ecosystems is the possibility of their accumulation in plant biomass. Plants are able to extract metals from the soil from sedentary forms [1]. At the same time, with the help of physiological barriers, they can limit the transition of xenobiotics from the root system to the aboveground part [2]. It is known that the ability to accumulate HM varies in different species of plants. However, the relationship of ecological and systematic characteristics of plants with the ability to accumulate heavy metals from soils is not currently established due to the small number of data, which makes it difficult to develop and apply environmentally effective methods of reclamation of contaminated areas around landfills of construction waste.

2. Materials and methods

This study is aimed at studying the differences in the ability to bioaccumulate HM in different ecological groups and life forms of wild species of higher plants.

The research was carried out in 2017–2019 in the territories adjacent to the construction waste landfill "North Samara" (61 hectares), located in Vsevolozhsk district of the Leningrad region. This landfill is active and has existed since 1974. The amount of waste disposed from 1974 to 2016 was 22,354,000 m^3 and 4,350,000 m^3 in a compacted state [3]. To date, it is almost completely filled and in the coming years is subject to reclamation. The facility is surrounded by natural ecosystems.

We conducted field studies in June – July 2017 and 2018, during which samples of soils and major plant species (135 species in total) were selected from different biotopes in the impact zone of the landfill. The study of HM content in soils and leaves of plants was carried out by x–ray fluorescence method [4] using the spectrometer "SPECTROSCAN Max – G" after grinding using a disk eraser and drying the material to an air-dry state.

3. Results and discussions

In the zone of impact of the landfill there are moist pine–green moss, pine trees mixed with birch, meadow areas, grasslands, wet floodplain meadows. The territories remote from the landfill for more than 1.5 km are mainly represented by old–aged and middle–aged pine forests, blueberry–grass, as well as orchards. 137 species of higher plants from different ecological groups and having different life forms were studied.

The study of soil pollution by heavy metals shows a mosaic distribution of pollutants within areas not related to distance from the landfill in the zone of influence. Excess of the content of heavy metals in the soil in the zone of influence of the landfill (800 m) relative to the maximum permissible concentrations (MPC) established by hygienic standards (GN 2.1.7.2041–06) was observed for arsenic (7 times on average) and zinc (3.5 times on average). At the same time, soil pollution is mosaic and in some points there were significant (up to 100 times) exceedances of MPC. Statistical single–factor analysis of ANOVA with a significance level of $p=0.1$ showed that almost all metals have no significant relationship between their concentration in the soil and the type of ecosystem. At the same time, it is shown that strontium accumulation in soils significantly differs among local ecosystems: moist pine forests, grassland swamp (average 99 mg/kg); ridge, quarry, old forest, stream (average 151 mg/kg); gardens (average 196 mg/kg). The principal component method showed that strontium and zinc make the main contribution to the difference between the spectrums of heavy metals from different ecosystems of the polygon zone of influence.

All studied plant species showed increased relative to the background (for the region) values of zinc and strontium. To analyze HM accumulation in different plant species, the following parameters were used: local ecosystem; plant life form (according to Serebryakov [5]), ecological group. The ratio of HM in plants from different local ecosystems showed high variability, significant links between the accumulation of HM and the type of habitat has not been established. The study of the relationship of the accumulation of individual HM with such parameters as belonging to an ecological group and the life form of plants. For this purpose, series of groups of each trait were constructed in order of decreasing the average content of each studied heavy metal in plants of each group and life form (table 1).

Table 1. Series of groups in order of decreasing metal concentration

Sr	Environmental group	Submerged macrophytes>ruderal > > forest=meadow=coastal–aquatic
	Form	aquatic grasses=terrestrial grasses>woody = mosses>shrubby
Pb	Environmental group	submerged=weed>forest=coastal–water>meadow
	Form	mosses>aquatic grasses>woody > terrestrial grasses>shrubby
As	Environmental group	forest>weed=submerged > meadow>coastal–water
	Form	mosses>terrestrial grasses>aquatic grasses=woody=shrubby
Zn	Environmental group	weeds> > submerged>forest>meadow>coastal–water
	Form	woody > > terrestrial grasses>mosses>aquatic grasses> > shrubby
Ni	Environmental group	weeds>submerged>forest>meadow>coastal–water
	Form	aquatic grasses>woody > terrestrial grasses>mosses>shrubby
Fe^{3+}	Environmental group	Submerged> > coastal–water>forest>weed>meadow
	Form	aquatic grasses> > mosses> > terrestrial grasses>woody>shrubby
Mn^{2+}	Environmental group	submerged> > coastal–aquatic> > forest=meadow> > weed
	Form	aquatic grasses> > woody>shrub>terrestrial grasses>mosses
Cr	Environmental group	submerged>forest=coastal = water = weed=meadow
	Form	shrub>aquatic grasses=mosses>woody = terrestrial grasses
Ti^{4+}	Environmental group	submerged>forest>weed>coastal–water=meadow

Analysis of bioaccumulation of heavy metals in plants of different ecological groups showed species–specific accumulation of these toxicants in different species. Nevertheless, the general regularities of significantly increased content of heavy metals in groups of aquatic submerged and weeds were revealed in comparison with the rest of the studied groups.

Strontium accumulation is significantly higher (on average 2 times) in groups of submerged aquatic plants and plants of weed associations. At the same time, the average strontium content in forest, meadow and coastal water plants is generally lower and does not differ between groups. Increased average zinc content is observed in plants of ruderal associations. It is in this group that almost all the high rates of zinc accumulation are observed (sagebrush (1091 mg / kg), wormwood (792 mg / kg), pink hybrid clover (643 mg / kg), bird Highlander (320 mg / kg)). The group of submerged plants on a par with weed plants has twice-increased accumulation of titanium, iron, chromium, strontium, lead and manganese in comparison with other groups.

Analysis of the accumulation of heavy metals in plants of different life forms shows the highest accumulation of zinc in woody species. Mosses accumulate arsenic, lead and titanium more strongly, and water grasses – strontium, nickel, iron, manganese.

The intensity of biological accumulation was estimated using the biological accumulation coefficient (BAC) [6]:

$$BAC = C_p/C_n$$

where C_n is the gross content of the element in the soil, C_p is the concentration of the element in the plant.

The value of biological accumulation was determined by the gradation of A I Perelman [7]:

Group I – "Vigorously accumulated" (CBN 10–100)

Group II – "Highly accumulated" (CBN 1–10)

Group III – "Weak accumulation or medium capture" (CBN 0.1–1.0)

Group IV – "Weak capture" (CBN 0,01–0,1)

Group V – "Very weak capture" (CBN 0.001–0.01)

Category III was further divided into two subcategories: III b (above–average accumulation) – 0.1 to 0.5 and Iii (increased accumulation) – 0.5 to 1.0. As the gross content of the element in the soil, the average values of concentrations from the growing places of the studied plants were taken. Table 2 presents the results of the study of the distribution of ecological groups and plant life forms to the groups identified by bioaccumulation of heavy metals.

Table 2. Value of biological accumulation of plant groups

	Sr, mg/kg	Pb, mg/kg	As, mg/kg	Zn, mg/kg	Ni, mg/kg	Fe2O3, %	Cr, mg/kg	TiO2, %
Environmental group								
Forest vegetation	**IIIa**	IIIb	IIIb	**II**	IIIb	IV	**II**	IIIb
Medow plants	IIIb	IIIb	IIIb	**IIIa**	IIIb	V	**II**	IIIb
Submersed plants	**IIIa**	IIIb	IIIb	**II**	**II**	**II**	**IIIa**	**IIIa**
Helophytes	**IIIa**	IIIb	IIIb	**IIIa**	IIIb	IIIb	**II**	IIIb
Ruderal plants	**IIIa**	IIIb	IIIb	**II**	IIIb	IV	**IIIa**	IIIb
Life form								
Water grasses	**IIIa**	IIIb	IIIb	**II**	**IIIa**	IIIb	**II**	IIIb
Wood	**IIIa**	IIIb	IIIb	**II**	IIIb	V	**II**	IIIb
Shrubby	**IIIa**	IIIb	**II**	IIIb	IIIb	V	**IIIa**	IIIb
Ground grasses	**IIIa**	IIIb	IIIb	**II**	IIIb	V	**II**	IIIb
Mosses	IIIb	IIIb	IIIb	**IIIa**	IIIb	IIIb	**II**	IIIb

The value of biological accumulation in the majority of studied plants of different ecological groups and life forms belongs to groups IIIa and IIIB. However, when considering the characteristics of the accumulation of individual heavy metals, the following features can be noted:

– submerged plants are the strongest bioaccumulators of heavy metals,

– among heavy metals, the greatest bioaccumulation in plants of various ecological groups and forms is manifested in zinc and chromium;

– very low bioaccumulation capacity in plants was observed in vanadium and copper.

Differences in the ability to accumulate heavy metals in different plant life forms are not expressed, and to a greater extent determined by the specifics of the species. The analysis showed relatively low average bioaccumulation of the studied heavy metals except zinc and chromium in mosses. However, species–specific increased accumulation of individual metals in certain species was noted: Polytrichum commune – on the accumulation of zinc (II category, CBN = 1,4), Dicranum polysetum Sw. ex Mlchx. and Pleurozium schreberi (Brid.) Mitt – on the accumulation of strontium – it has a weak accumulation of all studied metals, except strontium. None of the 16 species of mosses studied have any special abilities to accumulate heavy metals and their biological storage capacity is estimated on average as "weak accumulation or average capture" and below.

4. Conclusion

The obtained results indicate the existence of a relationship between the belonging of plants to an ecological group or life form and the ability to accumulate heavy metals in some of the most specialized ecological groups of plants–in submerged macrophytes and ruderal vegetation. At the same time, it was found that within the ecological groups of plants that do not have on average a pronounced increased or decreased ability to accumulate, there are species that have specificity in relation to the accumulation of a certain heavy metal. This feature is most pronounced in mosses. The results obtained testify to the ecochemical aspect of the importance of preserving the species diversity of plant ecosystems under the influence of industrial facilities, in particular, landfills. Also, reasonable use of the differences in accumulative capabilities of individual ecological groups of plants and the properties of certain plant species can improve the efficiency of protective zones around landfills and other industrial enterprises.

References

[1] Vorobyeva L A 1998 *Chemical analysis of soils* (Moscow: MSU)
[2] Kabata-Pendias A, Pendias H 2001 *Trace elements in soils and plants* (London: CRC Press)
[3] Materials of public hearings on the project Reconstruction of solid waste landfill "North Samara" 2016 (St. Petersburg)
[4] Federal Register 1.31.2018.32143 D*etermination of elements and element oxides in soil and sediment samples*
[5] Serebryakov I G 1962 *Ecological morphology of plants* (Moscow)
[6] Popova L F, Nakvasina E N 2014 *Normalization of urban soil quality and organization of soil–chemical monitoring* (Arkhangelsk)
[7] Perelman A I, Kasimov N S 1999 *Geochemistry of landscape (*Moscow: MSU)

IOP Conf. Series: Materials Science and Engineering **826** (2020) 012034 doi:10.1088/1757-899X/826/1/012034

Study of the content of heavy metals in the berries of the Vaccinium genus in the background areas of the North-West of Russia

N V Dinkelaker [1], T S Semenova[1], N B Ulyanov[1], A E Didikov[1], O V Petrova[1], E N Moiseenko[1], N F J Dinkelaker[2] and N A Sinelnikova[1]

[1] ITMO University, St. Petersburg, 9, Lomonosov st., 191002, Russia
[2] St. Petersburg forest engineering University S. M. Kirova, St. Petersburg, 5, Institutskii per., 194028 Russia

E–mail: nvdinkelaker@mail.ru

Abstract. The study of heavy metal contamination of products of wild berries of the genus Vaccinium (blueberry, cowberry, cranberry), growing in intact forests in four regions of North-West Russia- Leningrad, Pskov, Novgorod and Vologda regions is carried out. It is established that the greatest accumulation in berries of all studied species is observed for zinc, chromium and manganese. At the same time, the excess of the maximum permissible concentrations was noted for other metals - lead and arsenic, the level of pollution was moderate. In relation to these pollutants, the most sensitive genus is cranberry. The most likely source is roads and logging activities. The obtained results indicate the need to control chemical contamination of food products of wild berries and to expand the range of controlled parameters, in particular, the establishment of maximum permissible concentrations for chromium and nickel due to their significant bioaccumulation in berries.

1. Introduction

The development of the consumer market of wild forest food products is one of the key opportunities to improve the efficiency of the use of forest resources of the boreal zone of Russia, which is currently not fully implemented. On the one hand, artificial cultivation of wild berry crops in Russia is poorly developed in comparison with foreign countries to lower nutritional qualities and contamination of products from plantations because of the inevitable use of pesticides and herbicides in compacted industrial plantings [1]. On the other hand, it is related to the abundance of wild berries in the North-West of Russia. The greatest industrial value is represented by berries of the genus Vaccinium – blueberries, cranberries, cranberries [2]. Wild berries are widely distributed in the boreal forests of Russia, have stable yields and are practically not susceptible to diseases, unlike those grown on plantations. Their products are superior to industrial ones in terms of accumulation of biologically active substances, in food and pharmacological value [3]. In addition to the use of berries in the food industry, the procurement of medicinal raw materials from various parts of these plants is of great importance. Like all plants, Vaccinium species can also receive pollutants from the soil along with nutrients [4]. The usual soil pollutants in forests are heavy metals, which creates a risk of penetration into the production of berries and pharmacological raw materials. This type of pollution is widespread in forest areas due to the development of the road system and industrial enterprises. In the North-West of Russia, this problem is of great importance as in the course of economic development of territories, there is an increasing fragmentation of forests by linear objects. The increase in the area of contaminated forest land entails the need for environmental wild berry products control. Despite the urgency of the problem of contamination of forest food resources with heavy metals, the modern information about the contamination of berries and leaves of Vaccinium species is not sufficient for assessment of the forest berry products safety for the food and pharmaceutical use in the regions of the North-West of the Russian Federation

2. Methods and materials

The aim of this study was to assess the state of production of Vaccinium berries from different regions of the North-West of Russia, collected in the background areas – that is, in areas with minimal risk of contamination. For this purpose, intact forests located near specially protected natural areas (PNA) of the Leningrad, Novgorod, Pskov and Vologda regions were selected. The collection of material was carried out in 2017 – 2019 outside the protected areas, in places of traditional harvesting offorest berries.

The study of the content of heavy metals was performed using an X-ray fluorescence method [5] for the following heavy metals: strontium, lead, zinc, copper, vanadium, chromium, as well as iron, titanium and manganese oxides. Preparation of the material for the study involved the transportation of the material from the collection sites for 2 days, then - the selection of undiluted ripe berries without damage and drying them in a thermostat with strong ventilation at 20°C [6]. After no less than 40 hours of drying by weight control, the material was ready for further grinding using a disk eraser. The prepared samples were examined using X-ray fluorescence spectrometer «SPECTROSCAN-Max G».

As a result of the study it was established that the content of heavy metals in wild berries of shrubs of the genus Vaccinium (cranberries, blueberries and cranberries marsh) from the collection regions within the North-West of Russia had some differences, on average not more than 30-40% in different species (table 1).

There was a moderate excess of maximum permissible concentrations of heavy metals for arcenic and mercury. They are established by normative documents of the Russian Federation.

Such excesses are more common in the Leningrad region than in other regions studied. Contamination of berry products with lead and arsenic is more pronounced in cranberries. Cases of exceeding the content of heavy metals were observed in all studied species of shrubs, despite the difference in the areas of growth. The data obtained indicate the presence of local sources of pollution, given the absence of industrial enterprises near the studied forests, the most likely source of pollution can be considered roads and timber harvesting.

The greatest accumulation in the berries of the studied species have such metals as strontium, zinc and chromium (total) (Figure 1), while cobalt, vanadium and titanium oxide are virtually absent.

3. Results and discussion

High content compared to other metals is observed for manganese oxide (Figure 2), it is 10 – 100 times higher than strontium, lead, arsenic, zinc, nickel and chromium (table 1). Strontium, zinc and chromium have the highest accumulation in berries of all studied species in all regions (Figure 1).

Table 1. The content of TM in berries of the genus blueberries, cranberries and cranberries marsh different regions of the North-West of Russia (bold text indicates values that exceed the maximum permissible concentrations [7], [8]).

Genus	Gathering place	Heavy metal content, mg / kg						
		Sr	Pb	As	Zn	Ni	MnO	Cr
Cowberry	Novgorod region	29±3	7±2,5	2±0,7	46±5,2	2±0,9	554±36	37±4
	Vologda region	40±5,1	4±1,9	5±1,4	35±3,7	2±0,5	577±20,1	35±2
	Pskov region	36±4	2±0,8	5±1	38±3,1	1±0,2	657±40,8	36±4
	Leningrad region	25±5,3	3±1,2	4±1,3	36±4	2±0,2	418±32	36±3
Blueberry	Novgorod region	31±4,8	-	3±1	53±6,8	2±0,4	759±61,2	40±2
	Vologda region	43±2,2	6±1,8	2±0,4	31±4,5	3±0,1	699±23	38±3
	Pskov region	39±3	5±1,2	4±0,9	39±4,2	1±0,2	577±35	36±5
	Leningrad region	31±4,6	8±2,3	3±0,8	34±1,5	1±0,1	697±61	35±4
Cranberry	Novgorod region	34±5,2	3±1,5	3±0,4	39±5,4	1±0,1	501±25,3	30±5
	Vologda region	20±2,4	4±1,3	2±0,5	41±6	1±0,1	501±25,3	30±5
	Pskov region	31±6,1	3±1	3±0,8	40±5,2	1±0,1	612±54,4	34±6
	Leningrad region	27±8	5±1,7	4±0,4	43±3,5	2±0,1	605±27	40±5

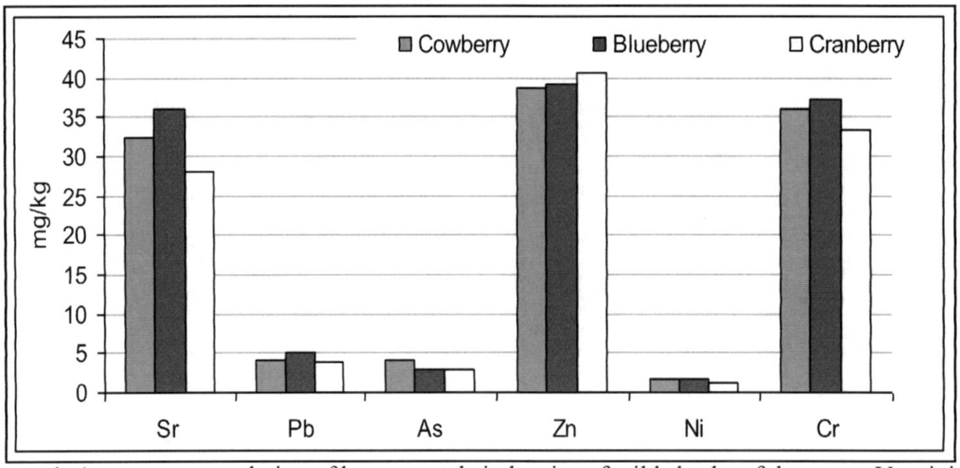

Figure 1. Average accumulation of heavy metals in berries of wild shrubs of the genus Vaccinium

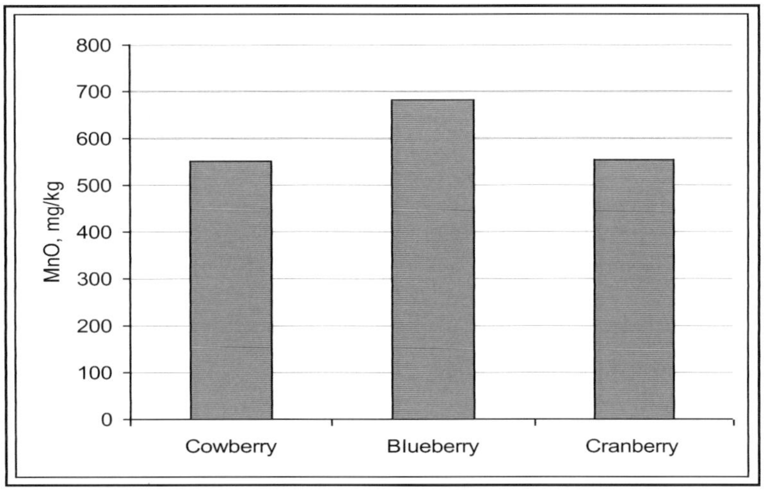

Figure 2. Average accumulation of manganese oxide in berries of wild shrubs of the genus Vaccinium

4. Conclusion

The studied species of wild berry bushes are confined to different forest biotopes; their habitats practically do not intersect. However, the accumulation of heavy metals in the studied species has a similar character: manganese oxide, strontium, zinc and chromium are preferably accumulated. High values of zinc content are typical for plant material in the norm, as it is required in large quantities for the biochemical processes of the plant. Strontium, zinc and chromium are present in low concentrations and reflect the geochemical characteristics of the soils of the collection sites of plant material. The accumulation of zinc is increased in comparison with other metals due to the high physiological need of plants for it. Moderate excess established in Russia maximum permissible concentrations was noted only for lead and arsenic However, other heavy metals can be considered as potentially dangerous for raw berries due to local contamination of biotopes and should be subject to control, for the effectiveness of which it is necessary to introduce standards of content, especially for nickel and chromium, intensively accumulating in berries.

References

[1] Cherkasov A F, Tyak G and Makeev V A 1999 *Guide to technology and agrotechnics of plantation cultivation of cranberries, cranberries and blueberries*(VNIISPK) 481 – 492
[2] Trehane J 2009 *Blueberries, cranberries, and other vacciniums* (Portland: Timber Press)
[3] Zhidkin V I and Samusev A M 2014 Food contamination by nitrates, pesticides and heavy metals *Entrepreneurship* **5** 190 – 198
[4] Efremov A A, Shatalina N V, Strizheva E N and Pervyshina G G 2002 Influence of ecological factors on chemical composition of some wild plants of Krasnoyarsk Krai *Chemistry of plant raw materials* **3** 53– 56
[5] Determination of elements and element oxides in soil and sediment samples. FRA.1.31.2018.32143.
[6] 1992 *Guidelines for the determination of heavy metals in soils of farmland and crop production* (Moscow: tsinao)
[7] Eurasian Economic Commission 2010 *Unified sanitary-epidemiological and hygienic requirements for goods subject to sanitary-epidemiological supervision (control)*, available at: http//tsouz.ru/KTS/KTS17/Pages/P2_299.aspx

[8] Chief state sanitary doctor of the USSR 1986 *Maximum permissible concentrations of heavy metals and arsenic in food raw materials and food products* N 4089-86
[9] Vorobyeva L A 1998 *Chemical analysis of soils* (Moscow: MSU)
[10] Kabata-Pendias A and Pendias H 2001 *Trace elements in soils and plants* (London: CRC Press)
[11] Lyanguzova I, Yarmishko V and Gorshkov V 2018 *Impact of Heavy Metals on Forest Ecosystems of the European North of Russia,* available at: https://www.intechopen.com/books/heavy-metals/impact-of-heavy-metals-on-forest-ecosystems-of-the-european-north-of-russia
[12] Serebryakov I G 1962 *Ecological morphology of plants* (Moscow: *Vyschaya Shkola*)
[13] Perelman A I and Kasimov N S 1999 *Geochemistry of landscape* (Moscow: MSU)

Extreme vibration increase of the heat transfer coefficient of short heat pipes

AV Seryakov and AV Konkin

LLC Rudetransservice, 64, Saint Petersburg Str., Veliky Novgorod, 173003, Russia

e-mail: seryakovav@yandex.ru

Abstract. The results of the heat transfer coefficient studies of the short linear heat pipes (HP's) with a formed by the capillary-porous insert vapour channel in the Laval nozzle-liked form under high thermal loads and external vibrations are presented. Under high heat power loads the vaporization process in the capillary-porous evaporator and vapour propagation in the vapour channel acquires a pulsating character with a frequency of 400-500 Hz and the amplitude up to 10^3 Pa. For the first time it was experimentally detected that under external longitudinal vibration, attached to the HP's body with an limited amplitude of 0.15 mm and frequency equal to the frequency of arising internal vapour pulsations, the HP's heat transfer coefficient shows the resonance nature and increases by up to 20%. The increase in the heat transfer coefficient is due to the boiling process intensification in the grid HP's capillary-porous evaporator.

1. Introduction

The problem of the heat transfer coefficient increasing of the short linear HP's is an important technological task in the development of the spacecraft with strict takeoff mass regulation. The use of the short linear HP's is justified in the case of constructive impossibility of the closed loop HP's placing, as well as to improve the long-term reliability and stability of passive cooling systems without any distributed supply and discharge lines of loop HP [1-3].

It is well-known that in the linear HP under high heat power loads (high temperature head at the evaporator) in the vapour channel arise pressure and velocity pulsations [4-5]. These pulsating phenomena lead to the complication of vapour flows in the Laval-liked vapour channel and to the intensification of evaporation and condensation and increase the heat transfer coefficient in such HP [5].

The problem of the external vibrations influence on the heat transfer systems, especially on the short linear HP, is of great interest in terms of controlling inhomogeneous hydrodynamic flows of two-phase working fluid in the capillary-porous structure, and in the analysis of vapour flows in the HP vapour channel.

Alaei et al. [6-7] published the results of studies of the low-frequency longitudinal vibrations influence with amplitude up to 2 mm on the heat transfer parameters of vertically oriented HP. Authors measured the HP's thermal resistance under various heat power loads, the coefficients of the working fluid filling and the vibration frequencies, they recorded a decrease in the thermal resistance of HP's up to 0.064 K/W at the external longitudinal vibrations frequency of 30 Hz.

Chen et al. [8] investigated the horizontal vibrations influence and variable condensation temperature on the heat transfer coefficient in a cylindrical copper HP with longitudinal risks on the inner surface for the working fluid return to the evaporator. Longitudinal horizontal vibrations with frequencies of 3-9 Hz and amplitudes of 2-25 mm, which caused vibration acceleration within the range of (0.1-1) g, were applied in the HP's experimental study. It is shown that when longitudinal vibrations are applied to a horizontally oriented HP, the heat transfer coefficient increases in the direct ratio to the vibrational energy in the interval up to the value of 500 mm^2 Hz^2. When this limit is exceeded, the simple proportional relationship between the vibration energy and the HP's heat transfer coefficient is broken, the coefficient value sharply reduced. In addition, the influence of the HP condensation surface temperature on the heat transfer coefficient value is significantly greater, than the influence of vibrations.

In Asias'a, Shusser'a et al. [9] detailed study of the heat transfer coefficient of linear HP's at high axial accelerations up to 12g very interesting results were obtained. Under such mechanical loads, the HP's accelerating effect exceeds the capillary pressure developed by the porous system of the HP's insert, and a result of such accelerating is the drainage of the evaporator, when the external acceleration acts opposite to the working fluid flow direction through the insert. The effect of acceleration on the working fluid flow direction is complex phenomenon. However, already at the acceleration of 6 g, the HP's instability operation begins, manifested in the occurrence and development of high-amplitude temperatures oscillations of the evaporator and heater, associated with the onset of fragmented film boiling, drying and overheating of the evaporator due to insufficient working fluid flow. Instability increases at high accelerations, the average heater and evaporator temperatures systematically increase along with the growth of temperature oscillations. After switching off the mechanical accelerations, the HP's normal functioning is restored.

In the short linear HP's with a Laval-liked vapour channel and with large amount of the working fluid in the insert (wick) pores and at high temperature head on the evaporator and the internal pulsations arisen, the influence of longitudinal external vibrations is particularly significant. To study this effect of longitudinal external vibrations on the HP's heat transfer coefficient, previously manufactured short stainless steel HP's were used, the vapour channel in which is formed along the entire length of the heat transfer by means of a grid capillary-porous insert connected to a grid flat evaporator. The internal vapour channel is made in the form of Laval-liked nozzle [4-5], measuring capacitive sensors [10] are flush mounted with the flat top cover, and for the actual heat transfer coefficient measurements, stationary and vibrating calorimetric measuring stands designed and manufactured, a brief description of which is given below.

2. Experimental setup

A schematic diagram of the experimental test setup is shown in Figure 1. In this study was applied stainless steel HP's with shaped vapour channel in the converging-diverging Laval-liked nozzle form with nonlinear varying diameter and with flat top and bottom covers.

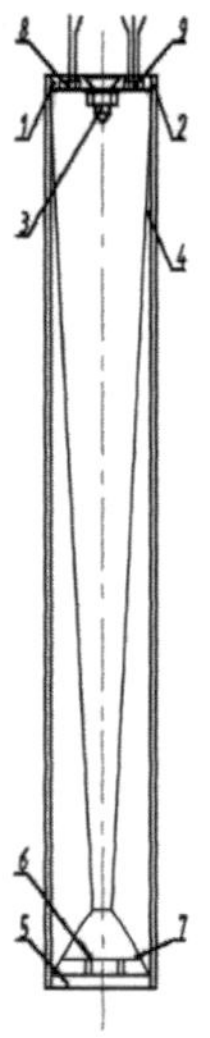

Figure 1. HP`s diagram: 1 – top cover; 2 – cylinder body of the HP`s; 3 – cone-shaped turbulator; 4 – capillary-porous insert; 5 – bottom cover; 6 – capillary injector channels, 7– bottom flat capillary-porous insert-evaporator. There are capacitance sensors 8, 9, installed inside the top cover [4 – 5], one of which is intended for a condensate film thickness measurement, while another one has a sensing element of CT3-19 thermistor mounted on its electrodes to measure the film temperature.

The diethyl ether $C_4H_{10}O$ is used as the working fluid, which has the boiling temperature under the atmospheric pressure of T_B =308.65K (35.5°C), freezing temperature T_F = 156.95K (−116.2°C) and critical parameters T_C = 466.55K (193.4°C), P_C =3.61MPa. The diethyl ether mass in the pore system of a metal mesh capillary-porous insert and evaporator is not less than $12.28\cdot10^{-3}$kg. The developed sensors enables us to perform measurements of local characteristics of the film thickness and temperature, without making any major disturbances in the flow. The HP's dimensions are chosen in such a way that it is possible to establish capacitive sensors [10] for film thickness and temperature measuring.

All heat transfer coefficient measurements in stationary mode were carried out using two HP's, measuring one and reference one [10]. The main HP, called measuring, is filled with diethyl ether and the reference one, which is completely identical to the main HP, is filled with dehumidified air at a pressure of 1 bar with dew point temperature lower than 233.15K (-40°C). The heat transfer coefficient K_{HP2} of the second HP does not exceed 0.15% from the first one (measuring one) K_{HP} and is not taken into account. The second HP, completely identical to the first one, performing the reference function in measurements with the help of high-frequency generators the condensate film thickness in the first HP.

The pulsation characteristics of HP's in stationary mode have been investigated previously [4–5] and the experimental values of the pulsation frequencies are shown in table 1. Measuring inaccuracy does not exceed 3-5 Hz.

Table 1. Pulsation rate values

Overheating of the evaporator, δT_{ev}, K	Pulsation frequency f_{puls}, Hz
9.05	386±5
10.1	396±5
11.03	426±5
12.15	450±5
13.0	456±5
14.07	474±5
15.03	474±5
16.0	478±5
17.1	490±5
18.06	491±5
19.02	495±5
20.12	502±5

3. Vibration heat transfer coefficient measurements

To measure the HP's heat transfer coefficient with external longitudinal vibration exposure, an experimental stand was created, including the previous measuring HP's and a specially designed flow microcalorimeter. The whole experimental stand, including a flat resistive heater with polished heating surfaces, HP and a vortex flow microcalorimeter 1, shown in Figure 2, was installed on a movable paper diffuser of a high power sound electrodynamics loudspeaker type 100 GDN-3-8, the operating frequency range of which lies within 31.5 Hz–1000 Hz. The average vibration diffuser amplitude at low frequencies (up to 50 Hz) in the free state reaches $\xi = 7$ mm, and decreases sharply with an increase in frequency above 300 Hz. When HP together with a water-filled flow microcalorimeter and a resistive heater installing, the vibration amplitude at a frequency of 500 Hz does not exceed $\xi_0 \leq 0.1$– 0.15 mm, $\xi = \xi_0 \sin(2\pi f_{vibr})$. The rate of HP movement in the vertical direction at such vibrations is equal to $\xi = f_{vibr}\, \xi_0 \cdot \cos(2\pi f_{vibr}\tau) = 0.15$ mm $\cdot 500$ Hz $= 7.5 \cdot 10^{-2}$ m/s, the magnitude of the overload arising in the capillary-porous structure (CPS) due to vibration acceleration reaches of 0.15 mm$\cdot(500\ s^{-1})^2 \sim 37.5\ m^2 / s \sim 3.8$ g, which is less than the available pressure developed by the HP's capillary-porous insert structure ~ 5.5 g.

The flow microcalorimeter design for the HP's heat transfer coefficient K_{HP} measuring under vibrations is a difficult task associated with the small size and weight of the microcalorimeter when it is necessary to achieve a heat exchange conditions that is not inferior to α_{cal} in a large stationary vortical flow calorimeter. The general layout of the measuring vibration microcalorimetric stand is shown in Figure 2.

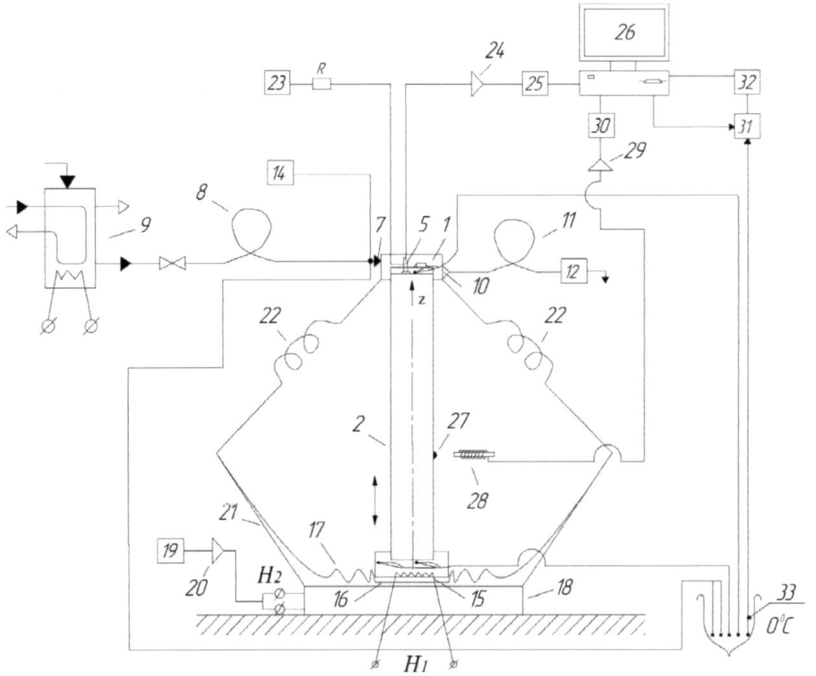

Figure 2. Scheme of vibration measurements of the HP's heat transfer coefficient: 1–microcalorimeter; 2 – HP; 3–flange; 4 – generator; 5 – capacitive sensor; 6–channels for wires; 7–inlet fitting; 8 – pressure hose; 9–constant pressure vessel of flowing water; 10–outlet fitting; 11 – drain hose; 12–flowmeter; 13 – cover; 14–bubble generator; 15 – heater; 16 –support disk; 17 – loudspeaker diffuser; 18 –sound loudspeaker; 19 – generator; 20 –amplifier; 21 – diffuser holder; 22 – centering springs; 23 –generator; 24 –amplifier; 25 – oscilloscope; 26 – computer; 27 – small permanent magnet; 28 – measuring coil; 29 –vibration control device; 30 –frequency meter; 31 –switch; 32 –voltmeter; 33 –Dewar vessel with zero point 0 °C.

Figure 3 shows the resonance curves of the HP's heat transfer coefficient K_{HP} dependence on the longitudinal external vibrations frequency at three different evaporator temperature head δT_{ev} and pressure pulsations fixed in the Laval-liked vapour channel.

1: Evaporator temperature head $\delta T_{ev} = T_{ev} - T_B = 20.1$K, the frequency of internal pulsations $f_{puls} = 502 \pm 5$Hz, stationary vapour flow inside the HP channel $G_{mix} \sim 3.147 \cdot 10^{-4}$ kg/s, the average vapour channel pressure is $6.13 \cdot 10^5$ Pa, the increase in the K_{HP} coefficient at resonance with superimposed external longitudinal vibrations with amplitude $\xi_{vibr} = 0.15$ mm reaches 20 %;

2: Evaporator temperature head $\delta T_{ev} = T_{ev} - T_B = 15.02$K, the frequency of internal pulsations $f_{puls} = 474 \pm 5$Hz, stationary vapour flow inside the HP channel $G_{mix} \sim 2.56 \cdot 10^{-4}$ kg/s, the average vapour channel pressure is $5.25 \cdot 10^5$ Pa, the increase in the K_{HP} coefficient at resonance with superimposed external longitudinal vibrations with amplitude $\xi_{vibr} = 0.15$ mm reaches 18 %;

3: Evaporator temperature head $\delta T_{ev} = T_{ev} - T_B = 10.1$K, the frequency of internal pulsations $f_{puls} = 396 \pm 5$Hz, stationary vapour flow inside the HP channel $G_{mix} \sim 1.71 \cdot 10^{-4}$ kg/s, the average vapour channel pressure is $4.53 \cdot 10^5$ Pa, the increase in the K_{HP} coefficient at resonance with superimposed external longitudinal vibrations with amplitude $\xi_{vibr} = 0.15$ mm reaches 14 %.

Presented in Figure 3 the values of the HP's heat transfer coefficients depending on the temperature head on the evaporator at three different values of the vibration parameters show extreme resonant vibration behavior. Deviations of experimental points from approximating curves do not exceed 3.5 %.

At different frequencies of external vibrations, when $f_{vibr} < f_{puls}$ or $f_{vibr} > f_{puls}$ begins to occur misalignment and partial oscillations damping, there are slowing antiphase movements of the liquid layer, leading to a decrease in the oscillations amplitude of the vapour bubbles and to a decrease in the resonance contribution to the HP's heat transfer coefficient.

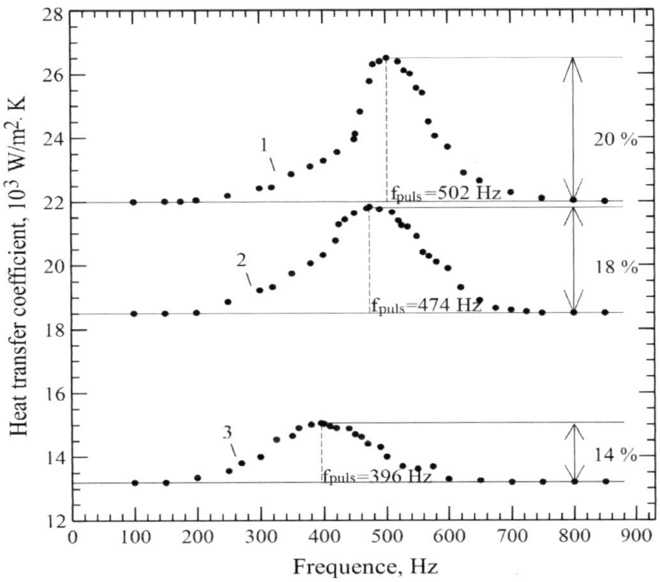

Figure 3. The results of the Laval-liked nozzle vapour channel HP's heat transfer coefficient K_{HP} measurements with a depending on the longitudinal external vibrations frequency.

The obtained results mean the vaporization intensity increase in the capillary-porous evaporator and the growth rate increase of the vapour bubbles, compared to the known results of the stationary growth rate of the vapour bubbles in a large volume of superheated liquid [11-13]:

$$r_{bub} \simeq r_{min} + \frac{4\,\Delta T_s}{r(T,P)\rho_{vp}^{mix}}\sqrt{\lambda_l C_{pl}\rho_l}\cdot\sqrt{\tau}\,. \quad (1)$$

The exponent 0.5 for the time parameter τ, s, in equation (1) reflects the stationary growth of vapour bubbles over time in a large volume of superheated liquid without taking into account the relative motion of the phases. However, in reality, due to the difference in densities, there is a convective movement-the bubbles pop up in the gravitational force field 1g, and in the experimental data obtained on the dependence of the bubble growth rate on time, the degree index is somewhat large, reaching a value of ~ (0.6 – 0.7) [13].

During the vapour bubble movement along the grid layers with a constant temperature T_{ev} (z=const) to the injection channels in the capillary-porous HP's evaporator $\tau = R / u_{bub} \sim 1\cdot10^{-2}\text{m} / 5\cdot10^{-3}\text{m/s} = 2\text{s}$, the total length of the perpendicular oscillatory bubble displacements at the harmonic oscillations with frequency f_{vibr} = 500 Hz and amplitude $2\xi_{l\Sigma} = 3\cdot10^{-4}$ m reaches $\ell_\Sigma = f_{vibr}\cdot 2\,\xi_{l\Sigma}\cdot 2\text{s} \sim 0.3$ m, while the growth rate of its radius $\dot{r}_{bub}$ can be considered close to constant. As a result, an additional contribution appears in the functional dependence of the bubble radius on time, as shown in equation (2):

$$r_{bub} \simeq r_{min} + \frac{2\,\Delta T_s}{r(T,P)\rho_{vp}^{mix}}\sqrt{\lambda_l C_p^l \rho_l}\cdot\sqrt{\tau} + b f_{vibr}\, 2\xi_{l\Sigma}\tau^n \quad (2)$$

where the exponent of n can be tried to estimate from the experimental results, and the coefficient b does not exceed two.

The exponent n can be estimated from the assumption that the resonant increase in the heat transfer coefficient K_{HP} is due to a vapour bubbles faster growth at harmonic vibrations in a capillary-porous evaporator as follows:

$$b f_{puls}\, \xi_{l\Sigma}\tau^n = 1.2\cdot\frac{2\,\Delta T_s}{r(T,P)\rho_{vp}^{mix}}\sqrt{\lambda_l C_p^l \rho_l}\cdot\tau^{0.5}\,. \quad (3)$$

And after the transformation of the equation (3) we obtain:

$$\tau^{n-0.5} = 1.2 \cdot \frac{2\,\Delta T_s}{r(T,P)\rho_{vp}^{mix}} \sqrt{\lambda_l C_p^l \rho_l} \Big/ b f_{puls}\, \xi_{l\Sigma}\ ;\ n - 0.5 = \ln\left(1.2 \cdot \frac{2\,\Delta T_s}{r(T,P)\rho_{vp}^{mix}} \sqrt{\lambda_l C_p^l \rho_l} \Big/ b f_{puls}\, \xi_{l\Sigma}\right). \tag{4}$$

We substitute in equation (4) all the necessary parameters [14-15], including the thermophysical properties of diethyl ether $r(T,P) = 345$ kJ/kg at evaporator temperature T_{ev} =328.75K(55.6°C) and pressure $P = 6.13\cdot10^5$ Pa ; $\rho_{vp}^{mix} = 3.3$ kg/m^3 ; $\lambda_l = 0.13$W/m·K; $C_p^l = 2.33$ kJ/kg ; $\rho_L = 703$ kg/m^3; ΔT_s =20 K; $b \sim 2$; $f_{puls} = 5\cdot10^2$ 1/s; $\xi_{l\Sigma} = 3\cdot10^{-4}$ m and after calculations obtain the following value of the exponent n:

$$n - 0.5 = \ln(2.050757) = 0.718\ ;\ n = 1.2\ \pm 0.1\,. \tag{5}$$

The exponent n value is greater than one in equation (2), $n > 1$ for the rate of vapour bubbles growth as a function of time τ, s, and this means the manifestation of a strong influence of the relative motion of vapour bubbles in the diethyl ether boiling layer under external influence.

The faster vapour bubbles growth due to their forced relative motion in the perpendicular direction to the main flow with constant conditions of the grid evaporator with periodic cell structure leads to more intensive vaporization at a constant temperature head value on the evaporator δT_{ev}, thereby increasing the heat transfer coefficient α_{ev} in the evaporator. In addition, with a small thickness of the liquid layer on the HP's lower cover surface, the introduced change by the vibration effect in the initial growth rate of a vapour bubbles is relatively large and the vibration further contributes to the liquid redistribution over the heating surface into the superheated injector capillary channels. This increases the flow rate of liquid and vapour in the injection channels and thus the heat transfer coefficient of the evaporator.

4. Conclusion

The resulting resonant increase in the heat transfer coefficient K_{HP} at the coincidence of the frequencies of internal pulsations in the vapour channel and the HP's evaporator and longitudinal external vibrations of the HP's body, indicating an accelerated growth of vapor bubbles in the boiling layer of diethyl ether on the surface of the bottom cover, depends on the value of the temperature head on the evaporator $\delta T_{ev} = T_{ev} - T_B$ and increases with its increase, but it is limited by the evaporator drying.

References

[1] Akachi H. 1990 Structure of Heat Pipe US patent 4921041
[2] Tong B, Wong T, Ooi K 2001 *Appl. Ther. Eng.* Closed-loop pulsating heat pipe **21** 1845
[3] Naik R, Varadarajan V, Pundarika G and Narasimha K 2013 *J. App. Fluid Mech.* Experimental investigation and performance evaluation of a closed loop pulsating heat pipe **6** 267
[4] Seryakov A V 2014 *Int. J. Heat Mass Tran. Theory Appl.* Pulsation flow in the vapour channel of short low temperature range heat pipes **2** 40
[5] Seryakov A V 2016 *Russian J. Appl. Mech. Tech. Phys.* Characteristics of low temperature short heat pipes with a nozzle-shaped vapour channel **57** 69
[6] Alaei A, Kafshgari M, Atashi H 2012 *Int. J. Heat Mass Tran.* A new designed heat pipe: an experimental study of the thermal performance in the presence of low-frequency vibrations **48** 719
[7] Alaei A, Kafshgari M, Rahimi S 2013 *Int. J. Heat Mass Tran* A vertical heat pipe: an experimental and statistical study of the thermal performance in the presence of low-frequency vibrations **49** 285
[8] Chen R, Lin Y, Lai C 2013 Heat Tr. Eng. The influence of horizontal longitudinal vibrations and the condensation section temperature on the heat transfer performance of a heat pipe **34** 45
[9] Asias A, Shusser M, Leitner A, Nabi A, Grossman G 2007 *J. Heat Tr.* Instability of heat pipe performance at large axial accelerations **129** 137
[10] Seryakov AV 2018 *Am. J. Modern Phys.* Intensification of heat transfer processes in the low temperature short heat pipes with Laval nozzle formed vapour channel **7** 48
[11] Nigmatulin R 1987 *Dynamics of multiphase media* **2** (Moscow: Nauka)
[12] Kutateladze S 1979 *Fundamentals of the heat transfer theory*. Ed. 5 (M: Atomizdat)
[13] Labuntsov D, Yagov V 2000 *Mechanics of two-phase systems* (M: Publishing MEI)
[14] *Tables of physical values* 1976 Guide under the editorship of Kikoin I (Moscow: Atomizdat)
[15] Vargaftic N 1972 *Spravochnick po teplophizicheskim svoistvam gasov i zhidkostey* (Moscow: Publishing house of Physics and Mathematical literature)

Development of a fast method for forming Braille on the surface of steels with IR ns-pulsed 50 W fiber laser

J V Amiaga[1], S A Vologzhanina[1], B S Ermakov[2] and S G Gorny[3]

[1] Saint Petersburg National Research University of Information Technologies, Mechanics and Optics University ITMO, 49, Kronverksky Pr., Saint Petersburg, 197101, Russia
[2] Peter the Great St. Petersburg Polytechnic University, 29, Politekhnicheskaya St., Saint Petersburg, 195251, Russia
[3] Laser Center LLC, 25, Piskarevsky ave., Saint Petersburg, 195176, Russia

E-mail: Joooneg@yandex.ru

Abstract. Convex linear structures of various configurations with a height of up to 200 microns and a depth of up to 100 microns, point structures with a height of up to 500 microns were obtained on the surface of the steel using a 50 W 1.064 μm wavelength fiber laser source. No additives were used. The structures were formed from the material of the irradiated steel by controlling the dynamics of the melt. The possibility of forming Braille on a steel surface using this technology is demonstrated. The results are encouraging for a number of industry sectors where electron beam machines and high-powered laser sources are cost prohibitive, and one need a high rate of obtaining relief.

1. Introduction

In recent years, the research trend has been gradually shifting toward studying the capabilities of lasers with short and ultrashort pulses (pico- and femtosecond). There are a number of works devoted to laser ablation performance or material removal rate [1,2]. Leitz et al. compared micro-, nano-, pico- and femto-second sources of laser radiation in light of the rate of removal of material in the receiving holes process [1]. Authors reported that among the investigated laser systems the nanosecond one allows highest ablation efficiency per energy. Traditionally, the presence of splashes of the liquid phase and burr at the edges of the holes in the ablation process is undesirable. This reduces the efficiency of the process and also the quality of the holes or other technological operations. Generally, increasing ablation rate leads to increasing the amount of liquid phase in its turn decreasing quality of the microstructuring.

In [3,4] liquid phase control is included in the process. Moreover, the presence of a melt is an advantage [4]. In [3] the authors obtained surface diffraction structures using radiation of nanosecond duration with a wavelength of 1.064 μm. It was achieved by melt control in micron size areas. In [4], the authors obtained various stable structures on the surface of steel using a 200 W CW Yb-fibre laser with a height of the order of 1 mm without the use of filler materials. Moreover, authors used their own trademark process called Surfi-Sculpt®.

The aim of the work was to find out the optimal strategy for steel surface laser treatment to obtain a stable relief with a maximum height difference (engraving depth, surface structure heights, total difference) using a low-power laser source (50 W) with a wavelength of 1.064 μm. Also we demonstrate the capabilities of a steel treatment strategy with pulses of a nanosecond laser, according to which the liquid phase is not unwanted, but useful.

2. Experimental details

2.1. Material and system

The normal quality of structural carbon steel of grade St3ps (Russian grade abbreviation) was used as a material for investigation. The chemical composition in % for grade St3ps is given in table 1.

Table 1. Chemical composition in % for grade St3ps

C	Si	Mn	Ni	S	P	Cr	N	Cu	As
0.14 - 0.22	0.05 - 0.15	0.4 - 0.65	max 0.3	max 0.05	max 0.04	max 0.3	max 0.008	max 0.3	max 0.08

A 1-5 mm thickness sheet of leads of the St3ps (GOST 3778-77) were used as samples for investigation.

Tthe steel sheet was processed by 1.064 μm laser irradiation. The laser source was YPLN-1-100-50-M (IPG Photonics). The laser source was integrated in the laser system (TurboMarker, Laser Center LLC) with a scanner head. The general specifications of the laser system is listed in table 2. The scanner head was equipped with an f-theta lens. The calculated beam diameter was 75 μm. The system could reach up to 15000 mm/s of scan speed (v). The laser could generate up to 50 W of average power (P_{avg}) and could be operated between 2-200 kHz of the pulse repetition rate (PRR) range. The appearance and structure of the laser system are shown in figure 1.

Table 2. General specifications of the used laser system

Brand and model	YPLN-1-100-50-M
Architecture	Fibre, Q-switched
λ	1064 nm
Max P_{avg}	50 W
PRR	2-200 kHz
Max. E	1 μJ
τ	100 ns
beam quality, M^2	<2
d0	75μm

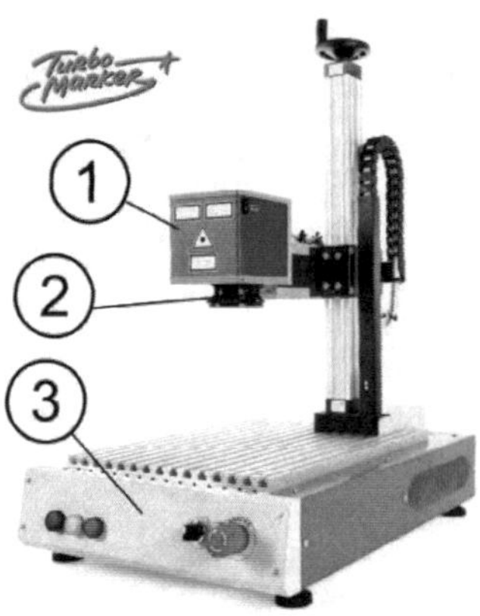

Figure 1. The appearance of the laser system including: (1) – scanner head, (2) – f-theta lens, (3) – control system and laser source

The relief was inspected through optical microscopy and relative measurements were taken using the images obtained via a digital camera.

2.2. Experimental plan

The first step was to determine how the melt behaves near the rectilinear groove if the beam path does not coincide with the direction of the groove. It was necessary to determine the optimal regimes of laser radiation in which it is possible to effectively control the melt during laser exposure to steel. For the laser source YPLN-1-100-50-M, the maximum pulse energy is 1 mJ. This energy is possessed by pulses which are emitted at a frequency of 50 kHz. These values of these parameters are widely used in obtaining fast relief using engraving in the evaporative mode using this laser emitter. These two parameters were unchanged.

Then the speed of the laser beam on the surface is the first parameter in the problem. It is widely known that when performing welds with electric arc welding, transverse vibrations of the electrode along a certain path are used, performed with a constant frequency and amplitude and combined with movement along the seam. Such movements make it possible to control the melt and the degree of heating of the welding areas, and to obtain a weld of the required width and quality. To control the melt in the problem of the article, the rectilinear motion of the beam was replaced with motion along a curve called a trochoid. There are two parameters for this curve: the cyclic frequency and the width of the curve. For convenience, we will study not the cyclic frequency, but the distance between the loops of the trochoid. We have three parameters for the study of obtaining linear convex structures on the surface of steel: the speed of the beam (**V**), the width (or height) of the trochoid (**H**), the distance between the loops (**D**).

After finding the optimal parameters of the trochoid and the speed of the beam, it was necessary to study how the shape of the guide curve affects the result. Drawing an analogy with electric arc welding, it can be called a curve responsible for the shape of the weld. It was a second step in the study. Figure 2 shows a diagram on which there are all the parameters of the task.

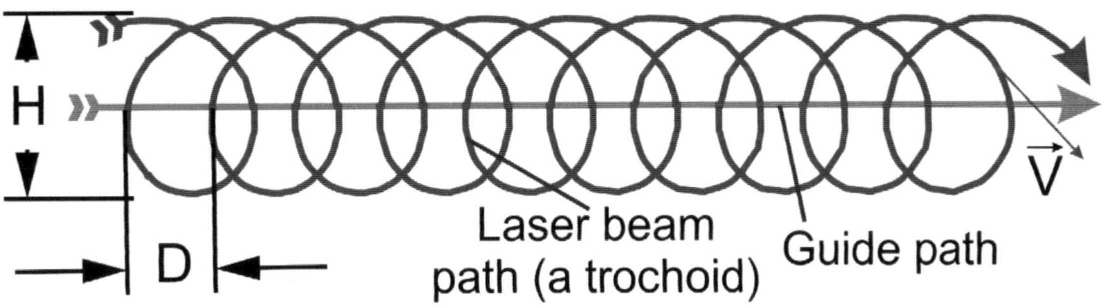

Figure 2. Diagram with parameters of the task and beam motion pattern, where D – the distance between the loops, V – the speed of the beam, H – the width (or height) of the trochoid

3. The study of the 1-dimensional structures on the steel surface

It was discovered that the trochoid movement of the laser beam on the steel surface leads to the formation of the groove and a side bead consisted of re-solidified steel. For a more detailed analysis of the relief, metallurgical grindings were produced. The cross section of these linear structures has an N-shape as can be seen in figure 3. Further, studies were conducted on the influence of the task parameters on the shape and stability of the N-structure. For better understanding of the results of the experiments, side light was used and N-structures were additionally treated with a special cleaning treatment on the same laser system. Side light gives a shadow by which one can estimate the depth of the groove, height of the beads and their stability. The cleaning treatment allows one to remove oxides and slightly melt the surface to a metallic luster. Figures 4-6 contain a special sign indicating the light source position.

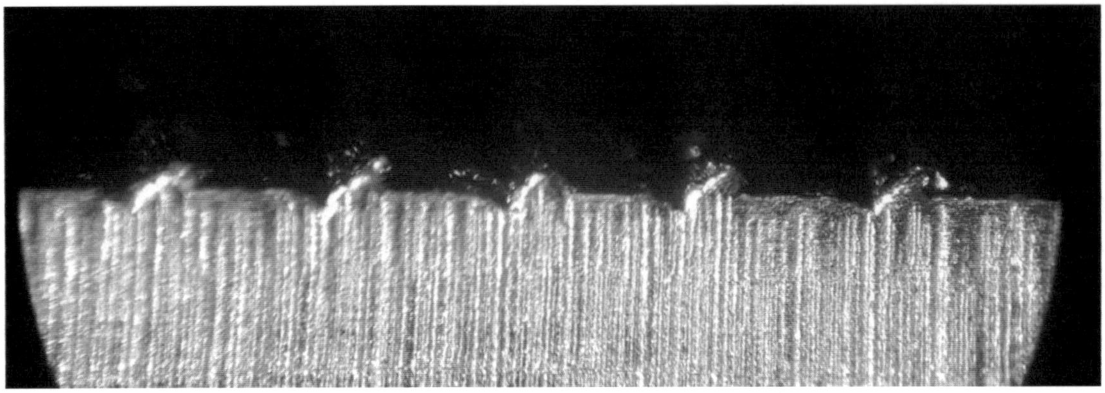

Figure 3. N-shape grooves on the steel surface. Cross section view

Decreasing the beam speed from 300 to 150 mm/s leads to increasing of the groove depth and bead height, decreasing the width of the groove. When the speed is equal to 100 mm/s, the bead moves closer to the center of the groove as the groove collapses. Moreover, the height of the bead decreases, as can be seen from figure 4. Increasing the distance between the loops from 5 μm to 15 μm leads to increasing the groove depth and bead height. Increasing the distance between the loops from 15 μm to 25 μm leads to decreasing the groove depth and collapsing the groove as can be seen in figure 5. Increasing the width (or height) of the trochoid from 0.3 mm to 0.4 mm leads to increasing the groove depth and bead height. Increasing the width from 0.4 mm to 0.5 mm leads to decreasing the groove depth. When the width is equal to 0.5, the bead becomes unstable and breaks up into individual drops, the groove collapses as can be seen in figure 6.

Based on the results obtained, one can conclude that the optimal parameter values are V=150 mm/s, D=15 μm, H=0.4 mm.

Combining the resulted structures, one can obtain other structures schematically depicted in figure 7. It is necessary to know where the melt is directed during the laser process. Figure 7 shows it. The appearance and cross section of Λ-, π- and M-structures are given in figure 8. These structures can be used to convex text marking on a steel surface.

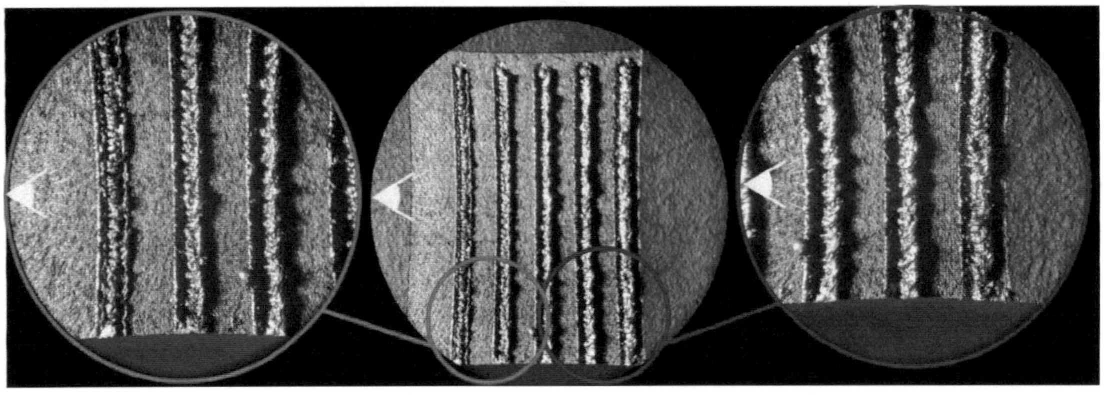

Figure 4. N-structures on the steel surface. Top view. Speed of the beam is reduced from left to right

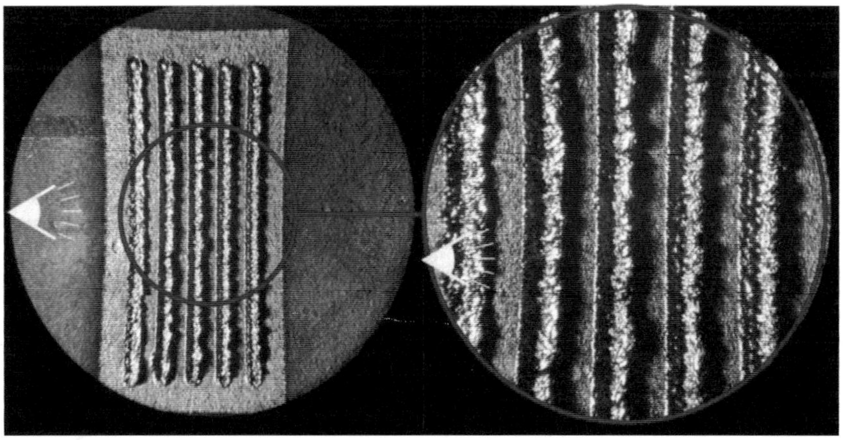

Figure 5. N-structures on the steel surface. Top view. The distance between the loops increases from left to right

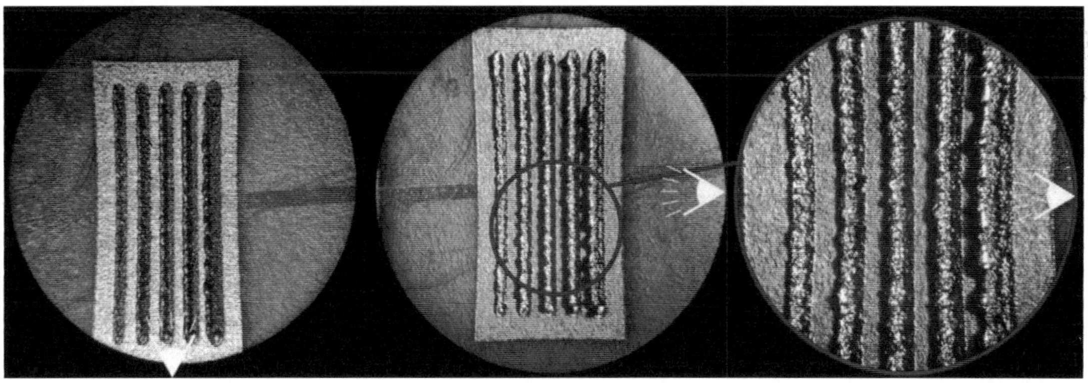

Figure 6. N-structures on the steel surface. Top view. The width (or height) of the trochoid increases from left to right

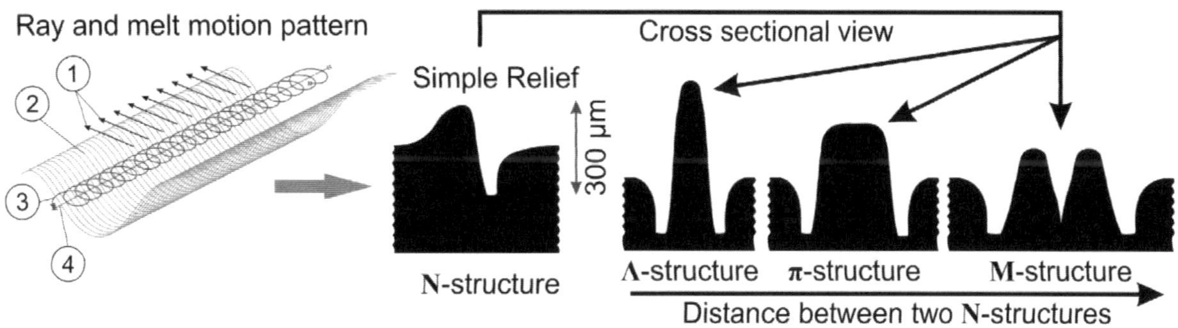

Figure 7. Scheme for obtaining various one-dimensional structures. Ray and melt motion pattern, where (1) – direction of melt travel, (2) – groove on the steel surface, (3) – beam path (a trochoid), (4) – guide path

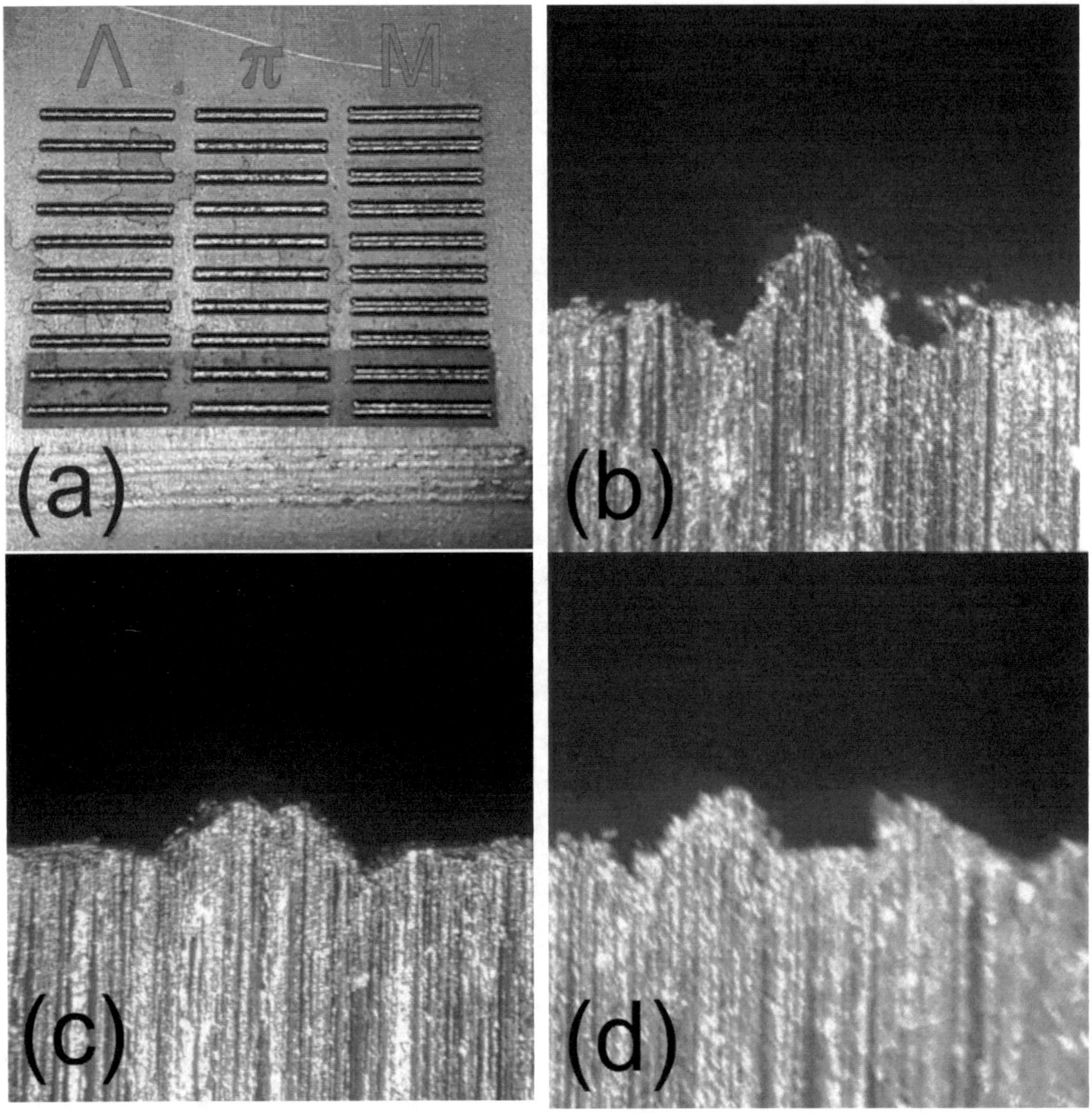

Figure 8. View of the various one-dimensional structures. (a) – Several lines of all three structures, top view. Cross section views: (b) – Λ-structure, (c) – π-structure, (d) – M-structure

4. The study of the point structures on the steel surface. Braille

It was found that if the Archimedes spiral is used as a guide path as shown in figure 9, it is possible to obtain point structures shown in figure 10. To be more precise, an additional circle was needed for the guide path. These point structures are easily tangible with the fingertips and can be used as Braille on the steel surface as shown in figure 11. The height of the point structures is 500 μm. The time required for drawing the text string "HELLO WORLD" was 4.5 min, for Braille text – 5 min. Point structures were additionally treated with a special cleaning mode on the same laser system. Linear structures were additionally machined manually to a metallic luster.

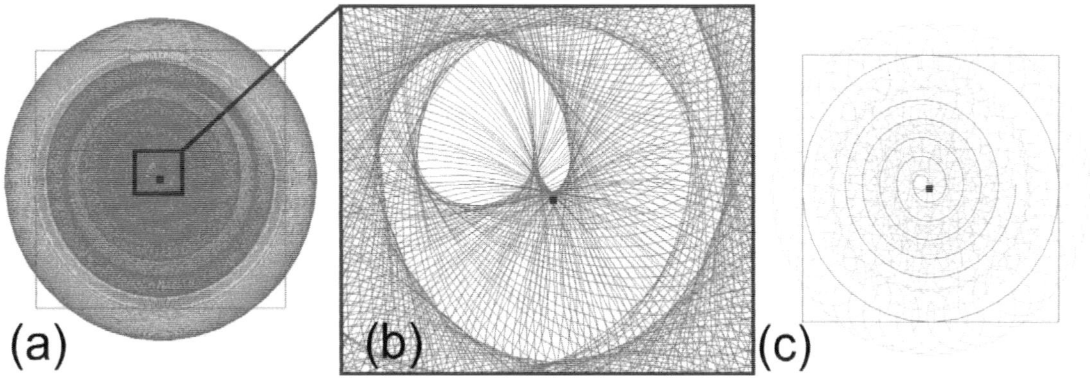

Figure 9. Beam path for obtaining the point structure from the control program window of the laser system: (a) – true beam path, (b) – enlarged area, (c) – simplified path scheme, where circle and Archimedean spiral are a guide path, the transparent line is a beam path.

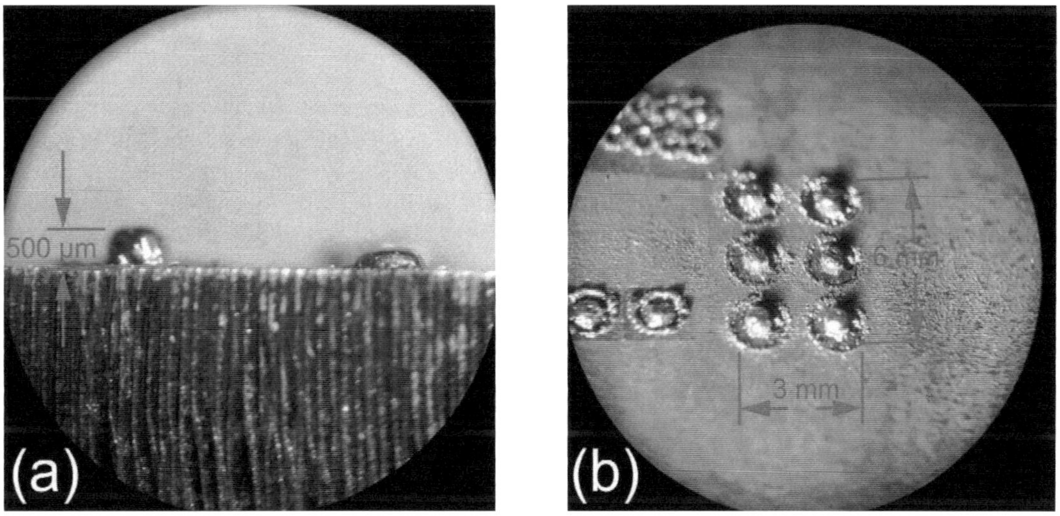

Figure 10. View of point structures: (a) – side view of point structure, (b) top view of several point structures arranged in the shape of braille font ligature *«for»*.

Figure 11. View of the convex text and braille text with the same characters.

5. Conclusion

Basic results:

1. The possibility of controlling the melt and obtaining convex stable structures when processing the surface of a steel with low-power nanosecond laser radiation was investigated.
2. It was clarified how the main processing parameters affect the structure of convex linear structures. The optimal parameters were found.
3. Point structures were received on the steel surface. These structures satisfy the requirements of Braille.
4. The maximum value of the height difference was 300 μm for linear structures and 500 μm for point structures.

6. Acknowledgments

The authors wish to thank Mr. S. G. Gorny from Laser Center LLC for putting the laser system at our disposal.

References

[1] Leitz K H, Redlingshöer B, Reg Y, Otto A and Schmidt M 2011 Metal ablation with short and ultrashort laser pulses *Physics Procedia* **12 (2)** 230–238

[2] Neuenschwander B, Jaeggi B and Schmid M 2013 From fs to sub-ns: Dependence of the material removal rate on the pulse duration for metals *Physics Procedia* **41** 794–801

[3] Wlodarczyk K L, Ardron M, Waddie A J, Dunn A, Kidd M D, Weston N J and Hand D P 2015 Laser microsculpting for the generation of robust diffractive security markings on the surface of metals *Journal of Materials Processing Technology* **222** 206–218

[4] Blackburn J and Hilton P 2011 Producing surface features with a 200 W Yb-fibre laser and the Surfi-Sculpt® process *Physics Procedia* **12 (1)** 529–536

Technological approaches to the quality management of bakery products

V A Fedotov

Orenburg State University, 13, Prospekt Pobedy, Orenburg, 460018, Russia

E-mail: vital_asm@mail.ru

Abstract. The article studies the influence of the physicochemical properties of wheat grain on the quality indicators of bakery products. Methodological approaches to managing the quality of bakery products through the development of production solutions have been formed by varying the technological parameters of the use of wheat flour (moisture of the leaven when the leaven method of dough preparation is used) taking into account its physicochemical properties (hardness of wheat grain). Wheat grain in a wide range of hardness - from low-hard to ultra-high-hard was studied. Regulated indicators of bread quality were the volume yield of bread and the integral characteristic of the organoleptic evaluation of bread on a 100-point scale. It is shown that the increase of grain hardness of flour and gain in moisture of starter dough (the average yield by weight of bread increases by 1% when the moisture of starter dough is 8 %) lead to increasing bread's yield by weight. To simplify the model of forming the quality of bread depending on the grain hardness and humidity of the used starter dough, a constant value of the specific work when dough kneading was used. Regression equations for the formation of white bread quality indicators with a high degree of confidence (determination coefficient 0.72-0.89) and corresponding bread quality nomograms are constructed.

1. Introduction

The problem of forming the quality of bakery products is closely related to the low quality of the used grain raw materials [1]. The technological qualities of grain and products of its processing are closely related to its structural and mechanical properties, described by indicators of strength or hardness [2]. Strength is characterized by yield strength, creep strength. Strength can be estimated by the value of the breaking force or stress at a certain type of deformation, as well as by the energy consumption per unit of the newly formed surface [3]. As a comprehensive indicator of technological properties, an indicator of grain hardness can be used, or hardness grain applied to wheat, since it is closely related to the physicochemical properties of grain [4].

The aim of the research was to study the influence of the physicochemical parameters of wheat grain on the formation of the quality of bakery products and, based on the patterns found, to develop methodological approaches to the quality management of bakery products.

In accordance with the goal, the following research objectives were formulated:

- on the basis of the analysis of literature data to determine the most significant factors in the formation of the quality of bakery products on the example of white bread produced by leaven method from wheat flour of the first grade;

- assess the effect of the physicochemical properties of wheat grain and the technological parameters of white bread production on its quality indicators;
- to formulate approaches to improving the quality of white bread by varying the technological parameters of production (leaven moisture) taking into account the physicochemical qualities of wheat (grain hardness).

2. Materials and Methods

The objects of the research were samples of 13 most common wheat varieties grown in the Orenburg region—Varyag, Orenburg 13, Orenburg 10, Uchitel, Prokhor, Bezenchukskaya Yantar, Step 3 and others [5]. The flour used as raw material was produced from grain, characterized by a wide range of hardness (from 10 to 28 kg/mm^2). The flour produced by grinding in a Nagema laboratory mill corresponded to first grade flour [10]. The samples of white bread were made from wheat flour of the first grade.

Grain hardness was defined by its microhardness—the ability of grain to resist indentation—on a PMT-3 microhardness tester. The ranges of hardness values were as follows: low hardness—less than 10 kg/mm^2, medium hardness—from 10 to 15 kg/mm^2, high hardness—from 15 to 20 kg/mm^2, superhigh hardness—more than 20 kg/mm^2.

The following were analyzed as adjustable indicators of bread quality: the volumetric yield of bread, which varied in the range from 200 to 600 cm^3 per 100 g of flour, and the integral characteristic of the organoleptic evaluation of bread on a 100-point scale, which varied in the range from 40 to 80 points according to the method developed at the department bakery production technology methods of Moscow State University of Food Production. This technique allows organoleptically evaluating the main quality indicators of products made from flour, given the significance of each of the indicators in a 5-point system [6]. Volume and weight yield of bread, its shape stability was determined according to GOST 27669-88 "Baking wheat flour. Laboratory test baking method".

3. Methodology for researching quality of bakery products

The studied process of forming the quality of bakery products is conditionally reflected by the parametric scheme in Figure 1.

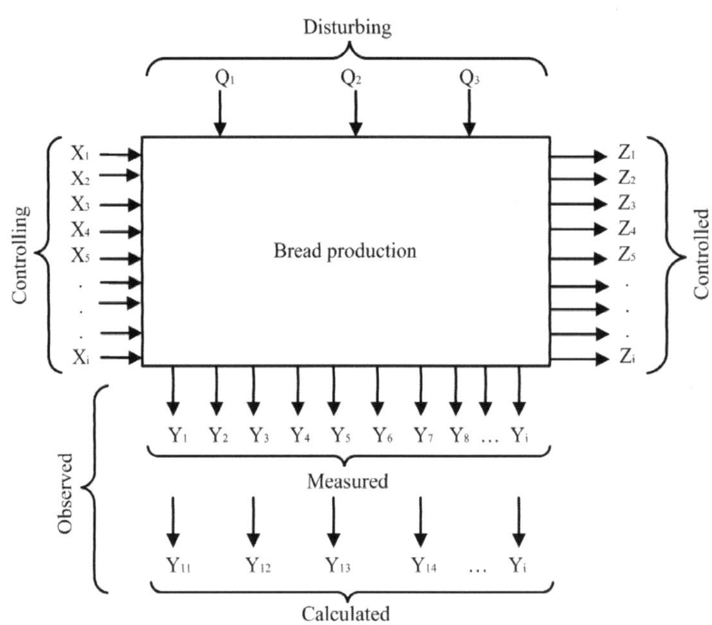

Figure 1. Parametric diagram of bread production processes

IOP Conf. Series: Materials Science and Engineering **826** (2020) 012037 doi:10.1088/1757-899X/826/1/012037

Following the generally accepted catagorizing of factors and parameters, the following factors of forming the quality of bakery products were taken into account in the studies.

The disturbing parameters include:

Q_1 – the external environment parameters of (pressure, humidity and temperature, other)

Q_2 – chemical composition and quality of raw materials (including floor grain size distribution);

Q_3 – technological deviations (mass of kneaded dough, temperature of components, humidity of raw materials and others);

The controlling parameters include:

X_1 – grain hardness, kg/mm²;

X_2 – amount of flour crude gluten, %;

X_3 – quality of flour crude gluten, units of FDM;

X_4 – moisture content of raw materials (flour), %;

X_5 – flour ash content, % (whiteness indicator, flour grade);

X_6 – falling number, s;

The controlled parameters include:

Z_1 – dough moisture content, %;

Z_2 – dough temperature, °C;

Z_3 – dough kneading time, s;

Z_4 – specific dough kneading intensity, kW/kg;

Z_5 – specific mechanical work for dough kneading, kW*s/kg;

Z_6 – proofing time, min;

Z_7 – amount of raw materials (wheat, water) for optimal weight of kneaded dough, kg;

The observed (measured or calculated) parameters include:

Y_1 – moisture content of ready products, %;

Y_2 – acidity of ready products, deg;

Y_3 – porosity, % (specific volume in cm³ per 100 g of flour);

Y_4 – form stability;

Y_5 – COS (complex organoleptic score), point;

Y_6 – shelf life, days;

Y_7 – weight yield, %.

Y_8 – structural and mechanical properties of crumb.

Since the leaven method is inherently more technologically flexible compared to the non-leaven method, the bread was produced according to the recipe for white bread from wheat flour of the first grade according to GOST 26987-86 "White bread from wheat flour of the highest, first and second grades" pan and hearth leaven method [7]. Preparation of raw materials for bread production, calculation of the main and auxiliary components were carried out according to the collected book "Technological instructions for bakery products".

The following requirements are imposed on pan bread: crumb moisture should be not more than 45.0%; crumb acidity should be not more than 3.0 degrees; the crumb porosity should be not less than 68.0%.

IOP Conf. Series: Materials Science and Engineering **826** (2020) 012037 doi:10.1088/1757-899X/826/1/012037

The flour was produced from the studied wheat grain samples in the Nagema laboratory mill, then laboratory baking was performed from the studied flour samples. Samples of leaven were prepared with the moisture content varying from 41% to 72% in 1% increments to cover the entire possible range of humidity used in the production of jars. Based on the quality indicators of the produced bread samples (volumetric yield, shape stability, overall score, weight yield of bread), the optimal moisture content of the dough was established. The dough was kneaded in the MTVK-80 dough mixing machine, later experiments were repeated on factory dough mixing machines.

The minimum weight yield of white bread from flour of the first grade with a pan mass of 0.75 kg was 136.1%, that of bread with a hearth mass of 0.5 kg was 131%. The calculated value of the weight yield should correspond to the planned one or exceed 1-2%. The calculation of the amount of water for kneading dough is based on this planned weight output.

To simplify the model for the formation of the quality of bread depending on the hardness of the grain and the moisture content of the used dough, a constant value of the specific work was used when mixing the dough for all bread samples. The duration of the kneading test was set based on the optimal specific work of the kneading test (40 J/g).

4. Results

The obtained results include equations of the dependences of bread quality indicators on the structural and mechanical properties of the feedstock and the technological parameters of the dough preparation—the moisture content of the leaven selected for the production (Figures 2, 3 and 4):

V is volumetric yield, cm^3 per 100 g of flour;

H/D is form stability;

Est is total score;

B is weight yield, %;

X_1 is grain hardness, kg/mm^2;

X_2 is leaven moisture content, %.

Studies have shown that flour from low-hardness grain (with hardness up to 14 kg/mm^2) allows producing bread with the highest possible volumetric yield with a minimum moisture content of leaven of 40%; with an increase in hardness to obtain maximum volumetric yield of bread, leaven with a humidity of 50-55% should be used. However, the maximum possible weight yield of bread from the flour of the initial grain hardness of up to 14 kg/mm^2 is achieved by using leaven with higher moisture content (up to 70%).

The regression equations of the quality indicators of bread show the features of their formation under the influence of hardness of grain and moisture content of the leaven.

$$V = 79{,}518 \cdot X_1 - 15{,}313 \cdot X_2 - 3{,}489 \cdot X_1^2 + 0{,}0015 \cdot X_2^2 + \\ +0{,}852 \cdot X_1 \cdot \tilde{O}_2 + 224{,}099$$

$$V = 79{,}519 \cdot X_1 - 15{,}153 \cdot X_2 - 3{,}489 \cdot X_1^2 + 0{,}852 \cdot X_1 \cdot \tilde{O}_2 + 220{,}063$$

$$H/D = 0{,}099 \cdot X_1 - 0{,}0179 \cdot X_2 - 0{,}0041 \cdot X_1^2 + 0{,}0012 \cdot X_1 \cdot X_2 + 0{,}1813$$

$$Est = 14{,}124 \cdot X_1 - 2{,}657 \cdot X_2 - 0{,}620 \cdot X_1^2 + 0{,}157 \cdot X_1 \cdot X_2 + 23{,}185$$

$$B = 0{,}251 \cdot X_1 + 0{,}107 \cdot X_2 + 128{,}713$$

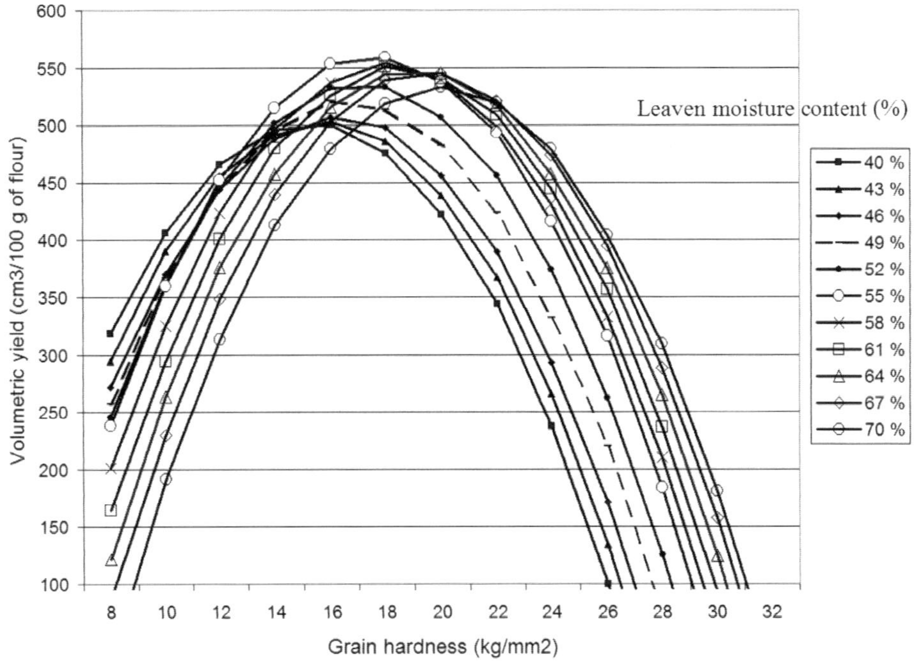

Figure 2. Nomogram of the specific volumetric yield of white bread from wheat flour of the first grade

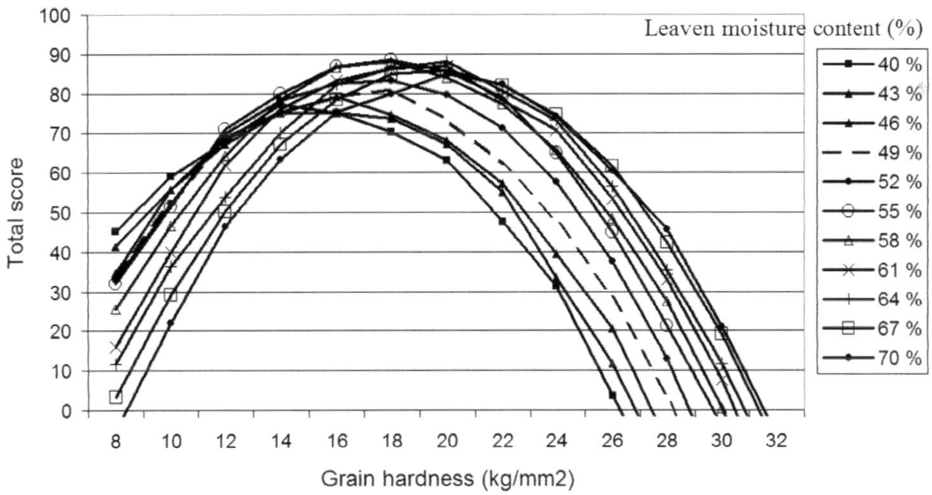

Figure 3. Nomogram of the total score for white bread made from wheat flour of the first grade

To simultaneously maintain high product quality (which can be estimated by the high volumetric yield) and its profitability (which can be estimated by the high weight yield), it is recommended that a compromise be reached between them when choosing leaven moisture. Such moisture content will allow high-quality bread with a sufficiently high weight yield to be produced. The situation changes when using grain flour with a hardness of 14-15 kg/mm^2: the optimal leaven moisture content for bread with the best quality indicators (volumetric yield, overall score) varies from 50 to 55%. For flour from high-hardness grain (above 22 kg/mm²) it is preferable to use leaven with a moisture content of above 55%, this will also provide the maximum volume and weight yield of bread. However, the quality of the resulting bread is low compared to that produced from medium-hardness grain, which is why it is preferable to use such flour as an improver.

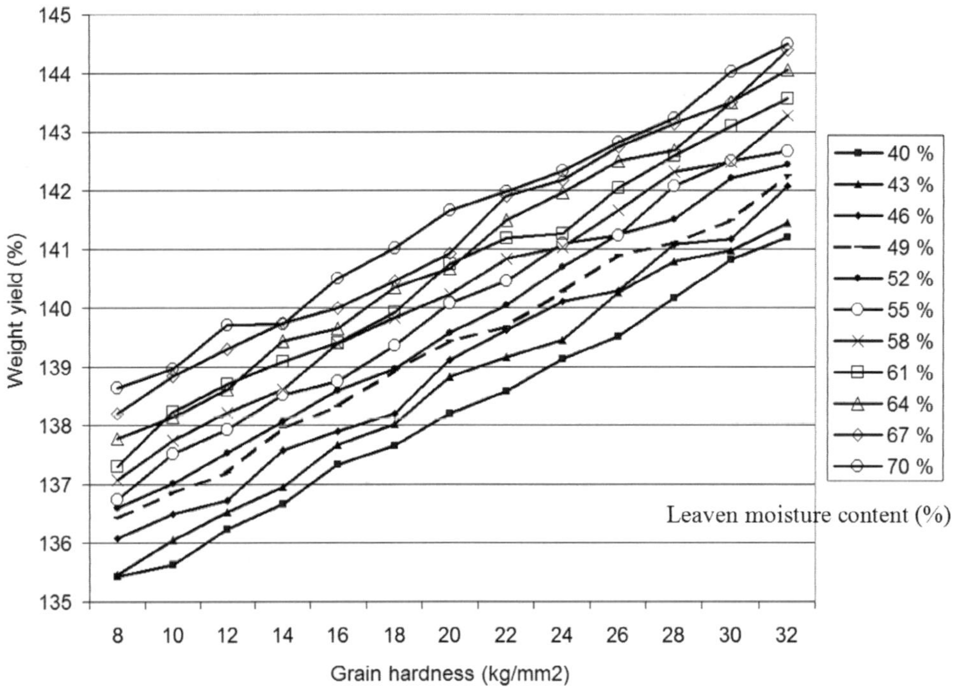

Figure 4. Nomogram of the weight yield of white bread from wheat flour of the first grade

The weight yield of bread increases linearly with increasing hard grain from which the flour was produced, and also linearly increases with increasing moisture content of the leaven (on average by 1% with an increase in moisture by 8%).

Diagrams of changes in bread quality indicators demonstrate the preference for leaven moisture depending on the hardness of the original grain. The maximum possible values of volumetric yield and weight yield for flour from grain with a given hardness in most cases are achieved by using leavens with different humidity. For example, for grain flour with a hardness of 13 kg/mm², the maximum volumetric yield is 475 cm³/100 g of flour when using a leaven with 40% of moisture; maximum weight yield of 139.8% is reached when using a leaven with 70% of humidity.

5. Conclusion

Judging by the results of the studies, the main indicators of the quality of bakery products (volumetric yield, shape stability, overall score, weight yield of bread) can be predicted based on the physical and chemical properties of wheat grain (grain hardness) and production parameters (leaven moisture).

The regression equations for the formation of indicators of the quality of white bread are characterized by high values of the determination coefficient (0.72-0.89), which indicates their reliability. Based on them, for clarity and simplicity of decision making on the preferred moisture content of the leaven and the forecasting of the quality indicators of the bakery products, corresponding nomograms were plotted. Quality control of bakery products is possible by choosing the moisture of the leaven with the leaven method of dough preparation, taking into account information about the hardness of grain, which served as a raw material for the flour.

IOP Conf. Series: Materials Science and Engineering **826** (2020) 012037 doi:10.1088/1757-899X/826/1/012037

References

[1] Maningat C C, Seib P A, Bassi S D, Woo K S and Lasater G D 2009 Wheat starch: production, properties, modification and uses *Starch Chem. Technol* (Whistler 3rd ed. Academic Press New York) 441-510

[2] Nuttall J G, O'Leary G J, Panozzo J F, Walker C K, Barlow K M and Fitzgerald G J 2017 Models of grain quality in wheat *Field Crop Res.* **202** 136-145

[3] Katyal M, Singh N, Chopra N and Kaur A 2019 Hard, medium-hard and extraordinarily soft wheat varieties: Comparison and relationship between various starch properties *International Journal of Biological Macromolecules* **123** 1143-1149

[4] Huang M, Wang Q G, Zhu Q B, Qin J W and Huang G 2015 Review of seed quality and safety tests using optical sensing technologies *Seed Sci. Technol.* **43(3)** 337-366

[5] Tanabata T, Shibaya T, Hori K, Ebana K and Yano M 2012 Smart Grain: high-throughput phenotyping software for measuring seed shape through image analysis. *Plant Phys-iol.* **160(4)** 1871-1880

[6] Mo X S, Huang W and Chen J W 2008 Test on the new national standard of wheat hardness index *Journal of flour* **6** 22-23

[7] Souza E J, Martin J M, Guttieri M J, O'Brien K, Habernicht D K, Lanning S P, Carlson G R and Talbert L E 2004 Influence of genotype, environment, and nitrogen management on spring wheat quality *Crop Sci.* **44** 425-432

Comparative analysis of elasticity and rigidity of morphological parts of wheat grain

V A Fedotov and S Yu Solovykh

Orenburg State University, Prospekt Pobedy, 13, Orenburg, 460018, Russia

E-mail: vital_asm@mail.ru

Abstract. The article presents the results of a comparative assessment of the elasticity and rigidity of various morphological elements of wheat grain depending on the nature of the grain (normal, frosted, after drying, after scouring machines). The dependences of the rigidity of wheat grain on the structure, chemical composition and humidity are described. A close correlation was revealed between the structural and mechanical properties of wheat, assessed by microhardness, strength index, and the results of laboratory and industrial grinding of wheat grains of various consistencies. For different consistency of wheat grain, microhardness was determined using the microhardometer "Miniload", and the strength index was determined using the Brabender's hardness tester. At relatively low stresses, the shells have flexural elasticity. IT was detected that with increasing humidity, the rigidity of the shells during bending decreases. It was found that for hard and hard-grained varieties of soft wheat, the compression force is required 2-4 times more than for soft-grained varieties. It is shown that the microhardness of glassy grains of hard-grained and soft-grained wheat varieties is basically at the same level. At the same time, with increasing humidity, the microhardness of the grain decreases regardless of the variety, area of growth and vitrescence of wheat.

1. Introduction

The most important role in the formation of physicochemical characteristics of grain is assigned to structural and mechanical properties. The species (hard and soft), type, variety, region and growing conditions of wheat have a huge impact on the latter. The heterogeneity of the microscopic structure and chemical composition of the individual anatomical parts of the grain (primarily the shells and endosperm) causes differences in their physical and mechanical properties, which must be taken into account when preparing wheat for grinding and processing.

Structural and mechanical properties of grain are characterized by strength, hardness and rheological characteristics. Strength is characterized by tensile strength, yield strength, creep strength. Strength can be estimated by the value of the breaking force or stress at a certain type of deformation (compressive tensile, shear, shear, chipping, bending, impact, abrasion), as well as the energy consumption per unit of the newly formed surface.

The term "relaxation" is understood to mean a gradual decrease in the acting stresses to zero or some value with a constant deformation in time. Studies by many scientists have made it possible to determine the magnitude of the force required to break down grain under compression, shear and bending strains. Scientists noted the effect on the strength properties of grain type and variety of wheat, size and shape, moisture and vitreousness [1].

Shells are less studied than other parts of the grain. Almost all published works on the study of shells concern only their strength and completely do not touch upon such issues as elasticity and rigidity. Meanwhile, the study of these mechanical properties of grain will significantly improve the quality of the products produced.

By "elasticity", as you know, they understand the property of bodies subjected to deformation to return to their original (undeformed) state after the termination of the forces that caused this deformation. In other words, elasticity is the body's ability to experience reversible deformations. “Elastic limit” refers to the maximum stress to which practically no residual deformations are obtained.

The bending stiffness of a beam is numerically expressed as the product of the elastic modulus under linear deformation and the moment of inertia of the beam section relative to the neutral axis. Thus, the magnitude of the elastic modulus is not related to the dimensions of the body and characterizes only the elasticity of the material from which it is made, while the stiffness of the body simultaneously takes into account both the elastic properties of the material and the shape and size of the cross section of this particular body.

This paper presents the results of a study of the elasticity and stiffness of the shells of grain depending on its moisture content and vitreousness. The tests were carried out on a specially designed device. The device consisted of the following main parts: a tripod, on which a disk with two transparent disks was mounted, handles with an arrow to rotate the axis in the center of the disk, axis, spring clip and arrow. The test shell specimen was mounted between the upper and lower clamps. Turning the knob clockwise, the shell sample was bent under the action of the weight of the arrow [2, 3].

2. Materials and Methods

The objects of research were grain samples of hard (Kharkovskaya 3) and soft (Yugo-Vostochnaya 3, Saratovskaya 42, Orenburgskaya 13 and Varyag) wheat of the Orenburg region.

The grain was preliminarily classified in size and vitreousness. For the study, grain screenings from a screen with mesh size of 2.4×20 mm were used. The grain thus selected was placed in containers with water. After the endosperm was so softened that it could be separated from the membranes, they were removed. The grain was placed with a groove up, with grain hair down and cut with a razor into two halves. Endosperm was completely removed from each half, after which the membranes were placed in separate boxes and then their moisture was determined. If the moisture of the shells exceeded the set value, they were dried in a desiccator over water. If the moisture of the shells was within the set limits, they were tested.

The shell was removed from the box, cut in the longitudinal direction from two sides in order to obtain a strip 2 mm wide and 3 mm long, and then clamped with ends in spring clips, one of which was located on the axis of the device and the other on the end of the loading arrow. After fixing the shell in the clamps, the axis of the device was inserted into the hole of the disk so that the angle of rotation indicator and the arrow with the shell fixed were set to zero. It should be noted that when moistened the grain (in particular in the shell) experiences some leaching of water-soluble substances, and some biochemical changes are possible. Nevertheless, this research method provides quite reliable and comparable results.

The microhardness of various parts of the grain was evaluated using a Miniload micropenetrometer, and the index of tensile strength using a Brabender hardness tester.

3. Results

With increasing moisture, the rigidity of the shells during bending decreases. Thus, at a moisture content of 11.7 to 12.5%, the shells of soft wheat from the eastern regions (the studied varieties) are characterized by higher hardness compared to the shells of wheat from the central regions. With increasing moisture of the shells to 48.0%, their rigidity decreases (the shells become more malleable).

Interestingly, this is more characteristic of wheat from the eastern regions than from the central regions.

Hard wheat shells at a moisture content of 11.7 to 12.5% are more rigid than soft wheat shells; with an increase in moisture of up to 47.5%, the resistance of the shells to a change in shape decreases, though to a lesser extent, than that of the shells of soft wheat. Thus, the bending tests of the shells showed that with an increase in their moisture, the elastic modulus decreases.

Since the elastic modulus characterizes the elasticity of a substance, it can be said that with increasing moisture, the shells become less elastic. The elastic modulus for durum wheat is higher than for soft wheat, and that for wheat in the eastern regions (studied varieties) is slightly higher than for wheat in the central regions. Consequently, at the same moisture, the shells of durum wheat have the highest elasticity, followed by the shells of wheat in the eastern regions, while the lowest moisture content is characteristic of wheat in the central regions.

Table 1 shows data on the stiffness and modulus of elasticity of the shells of frosted grain, grain after drying, grain processed in scouring machines with an abrasive cylinder, and grain with subzero temperature. From the data given in the table it follows that the shells of grains damaged to some extent by frost after drying or processing in scouring machines with an abrasive cylinder, as well as grains with subzero temperatures compared with shells of normal grains, have a lower resistance to bending and less elasticity [4, 5].

Table 1. Rigidity and elastic modulus of grain

Wheat variety	Grain nature	Bending stiffness, [kg mm^2]	Elastic modulus [kg/mm^2]
Saratovskaya 42	Normal	0.145	21.8
	Frosted	0.118	18.7
	After drying	0.113	18.5
	After abrasive scouring machine	0.101	10.2
Orenburgskaya 13	Normal	0.135	22.6
	Frosted	0.132	16.5
	After drying	0.120	14.9
	After abrasive scouring machine	0.110	8.3
Varyag	Normal	0.145	29.2
	Frosted	0.128	16.4
	After drying	0.132	14.9
	After abrasive scouring machine	0.114	8.5

Frosted grain is known to usually differ from normal one by its reticular surface, while the grain itself is more or less deformed. The shells of such grains are easily deformed.

When processing grain in scouring machines with an abrasive surface as a result of impacts of grain on an acutely rough surface, cuts, scratches, cracks and other damage are formed in the shells, which leads to a significant decrease in their rigidity and elasticity.

With a decrease in grain temperature, especially during early frosts, when the grain temperature drops to -1 °C or lower, the water between the macromolecules expands and increases the distance between them, weakening the intermolecular adhesion forces. In addition, the free moisture in the shells of the grain, turning into ice, increases the volume of macropores, and this leads to a complication of the stress state of the whole grain.

Thus, the shells experience a load not only from the pressure forces of the increasing volume of the endosperm, but also from the forces resulting from the freezing of water in the pores or capillaries of the grain. Both of these cases can lead to the formation of microcracks in the shells and, therefore, not only to their reduced strength, but also to reduced elasticity and stiffness [6].

There are several hypotheses regarding the causes of cracks on the surface of the material during drying. For example, when wet grain is dried (especially if the latter is conducted at improper temperature conditions), the moisture in the pores of the endosperm and shells turns into steam. The latter, by virtue of its elasticity, exerts pressure on the endosperm and the shell, which leads to the appearance of microcracks on the surface of the latter and a decrease in its resistance to the load.

When the shell is bent, the latter experiences only or predominantly elastic deformations. During this period, due to the elasticity of the shells, they significantly resist bending. After removing the load, the shell sample restores to its original size and shape.

With a further increase in load, the bending of the shell increases, and on the convex side, individual layers of the shell stretch, and on the concave side they are compressed. Since the elongation and shortening of individual layers of the shell are accompanied by transverse compression and expansion, as a result, a relative shift of the individual layers of the shells occurs on the concave and convex sides.

At the same time, a gradual increase in plastic deformation occurs, and the angle measured on the scale decreases. If during the deformation of the shells any section is weaker than others, then the pattern of development of the deformation will be violated. In a weak cross section and in neighboring cross sections, the stresses reach maximum values, which leads to an increase in residual strains and deformation of the sample.

The elasticity, rigidity, strength of the shells of wheat grain depends on the structure, chemical composition and humidity. Dry shells (with humidity from 11.7 to 12.5%) are characterized by the highest rigidity. With increasing humidity of the shells, their bending stiffness decreases, and their resistance to grinding increases. The presence on the surface of the grain of cracks, scratches and other damage leads to a decrease in resistance and, consequently, to easy destruction of the shells.

In addition, the correlation of the structural and mechanical properties of grains of various consistencies with humidity, temperature, and loading speed was studied. With an increase in the moisture content of grain of the Kharkovskaya 3 variety to 16.0%, the destructive force drops by 18.5%, by 22.7% for Southeast 3 wheat and by 27.7% for Saratovskaya 42 wheat. The destruction work reaches its minimum when the moisture content of Kharkovskaya 3 wheat grain is from 15.0 to 16.0% and Saratovskaya 42 moisture content is from 13.0 to 14.2%. With an increase in the loading rate, the work on the destruction of grain decreases, since the proportion of plastic deformations sharply decreases [7].

Flour-milling technologists came to the conclusion that, depending on humidity, three phases of the grain state can be distinguished, characterized by special deformation properties. At a moisture content of up to 11.0%, the grain is a viscoelastic body; at a moisture content of 11.0 to 13.5%, it is a viscoelastic-plastic body; above 17.0%, it is an elastoplastic body.

Of practical interest are the strength properties of the main parts of the grains—shells and endosperm. Dry shells (grain moisture from 11.7 to 12.5%) of durum wheat (Kharkovskaya 3) are more rigid than soft shells (hard-grain Yugo-Vostochnaya 3 and soft-grain Saratovskaya 42). Similar ratios were revealed for the shell elasticity index. With increasing humidity, the stiffness and elastic modulus of the shells decrease.

In the study of the strength of pure endosperm, it was found that for durum and hard-grain varieties of soft wheat, the compressive force is 2 to 4 times greater than that for soft-grain varieties (23.2, 15.1 and 12.5 kg/mm^2, respectively). In connection with the anisotropic microscopic structure and the heterogeneous chemical composition of the endosperm of the caryopsis, it is necessary to study the strength properties of its individual sections.

Flour milling technologists studied the topography of the microhardness of the endosperm and the shells of grains of vitreous and mealy fractions of wheat varieties of various consistencies. The microhardness value of the endosperm and the shells of durum and hard-grain soft wheat kernels is usually higher than the similar indicators of soft-grain wheat kernels.

The microhardness of vitreous grains of hard-grain and soft-grain wheat varieties is basically on the same level. The same is observed for mealy grains. With increasing humidity, the microhardness of grain decreases regardless of the variety, growth area and vitreousness of wheat.

When studying slices of three endosperm zones of durum and soft durum wheat, it was found that the microhardness values increase from the center to the periphery of the grain (from 24.1 to 25.5; from 15.5 to 17.8; from 10.7 to 12.5 kg/mm^2, respectively). For soft-grain varieties of white wheat, the microhardness values are slightly lower.

The microhardness of individual fragments of an intermediate protein and starch granules was studied. The moisture content of the grain ranged from 9.0 to 11.0%. The microhardness values of vitreous grains are on average two times higher than the that of mealy ones. The microhardness of the vitreous fraction of grains of soft wheat varieties is similar to the microhardness of semi-vitreous grains of durum wheat. The same ratio was observed for the mealy fraction of hard-grain varieties and the semi-vitreous fraction of soft-grain wheat (Table 2).

Table 2. Microhardness of grain of hard and soft wheat of various vitreousness

Vitreousness range [%]	Gran microhardness [kg/mm^2]			
	vitreous	semi-vitreous	mealy	total
	durum wheat			
80-100	20.2	16.5	12.1	19.5
	hard grain wheat			
to 40	18.5	14.2	11.5	15.4
41-60	18.2	14.1	11.2	16.2
60-80	18.9	14.0	11.0	16.8
80-100	19.1	14.3	11.0	17.1
	soft grain wheat			
to 40	13.5	11.2	10.2	11.2
40-60	13.0	11.0	10.3	10.8
60-80	12.9	10.9	9.7	10.5

The correlation coefficients of vitreousness and microhardness of durum, hard-grain and soft-grain wheat varieties were 0.71, 0.92 and 0.67, respectively.

4. Conclusion

In the course of this study, the rigidness of the shells was determined and the ultimate bending stresses were calculated. In the working area of the roller mills, the shells are deformed under conditions of complex deformation; in this case, the required power or energy consumed for their grinding is determined. In the latter case, with increasing humidity of the shells, due to an increase in their plasticity, resistance to grinding increases and, accordingly, the specific energy consumption.

At relatively low stresses, the shells exhibit bending elasticity, i.e., upon subsequent unloading of the shells, they regain their size and shape.

The modulus of elasticity decreases with increasing humidity of the shells. Consequently, under the same other conditions, dry shells are more resilient than wet ones.

The elastic modulus of the shells of durum wheat is higher than that of the shells of soft wheat, while that of the wheat of the eastern regions (studied varieties) is higher than that of the shells of wheat in the central regions. Consequently, at the same humidity, durum wheat shells are more elastic than the shells of the studied soft wheat varieties, and wheat shells of the eastern regions are more elastic than the wheat shells of the central regions.

The shells of frosted grain, grain dried after processing in scouring machines with an abrasive cylinder and grain with a temperature below 0 °C, are characterized by lower elasticity and resistance to bending compared to shells of normal grains.

A close correlation was found between the structural and mechanical properties of wheat, evaluated by microhardness (according to the Miniload microhardness tester), strength index (by the Brabender

IOP Conf. Series: Materials Science and Engineering **826** (2020) 012038 doi:10.1088/1757-899X/826/1/012038

hardness tester), and the results of laboratory and production grindings of wheat grains of various consistencies.

References

[1] Souza E J, Martin J M, Guttieri M J, O'Brien K, Habernicht D K, Lanning S P, Carlson G R and Talbert L E 2004 Influence of genotype, environment, and nitrogen management on spring wheat quality *Crop Sci.* **44** 425-432

[2] Caporaso N, Whitworth M and Fisk I 2016 Application of calibrations to hyperspectral images of food grains: example of wheat *Falling Number Journal of Spectral Imaging* **1** 46-52

[3] Huang M, Wang Q G, Zhu Q B, Qin J W and Huang G 2015 Review of seed quality and safety tests using optical sensing technologies *Seed Sci. Technol.* **43(3)** 337-366

[4] Tanabata T, Shibaya T, Hori K, Ebana K and Yano M 2012 Smart Grain: high-throughput phenotyping software for measuring seed shape through image analysis *Plant Phys-iol.* **160(4)** 1871-1880

[5] Mo X S, Huang W and Chen J W 2008 Test on the new national standard of wheat hardness index *Journal of flour* **6** 22-23

[6] Nuttall J G, O'Leary G J, Panozzo J F, Walker C K, Barlow K M and Fitzgerald G J 2017 Models of grain quality in wheat *Field Crop Res.* **202** 136-145

[7] Katyal M, Singh N, Chopra N and Kaur A 2019 Hard, medium-hard and extraordinarily soft wheat varieties: Comparison and relationship between various starch properties *International Journal of Biological Macromolecules* **123** 1143-1149

IOP Conf. Series: Materials Science and Engineering **826** (2020) 012039 doi:10.1088/1757-899X/826/1/012039

Thermodynamic cycle of lithium bromide absorption chillers with two-level absorption and generation processes and with associated mass flow

O S Malinina, A V Kasyanov and A V Baranenko

Department of Low Temperature Energy, ITMO University, 49, Kronverksky Ave., St. Petersburg, 197101, Russia

E-mail: holmash_malinina@mail.ru

Abstract. The paper presents the assessment of the energy efficiency of the thermodynamic cycle of the absorption lithium bromide refrigeration unit with two-level absorption and generation processes with the associated mass flow. The analysis of temperature effect of heating and cooling sources on thermal coefficient is carried out. The parameters of external sources when it is possible to determine the energy efficiency of the thermodynamic cycle are defined.

1. Introduction

In recent decades, there has been a growing interest to lithium bromide absorption chillers (LBAC) [1, 2, 3, 4, 5, 6, 7]. Their performance can be based on renewable energy sources (RES) and secondary energy resources (SER), which corresponds well to the world energy development tendencies aimed at increasing the share of these resources in the energy balance in order to increase energy efficiency and reduce carbon intensity of the global economy [8]. LBAC has been widely applied in cooling systems of basic technologies, comfortable and technological conditioning in chemical, oil refining, electronic, food industry, housing and communal sector and social sphere. The studies of cooling and air conditioning systems using solar-powered absorption lithium bromide refrigeration units and RES are presented in [1, 3, 7, 9, 10, 12, 13, 14, 15].

The analysis of LBAC thermodynamic cycles is considered in a number of works [11, 12, 13, 16, 17, 18, 19, 20, 21]. The authors of [17] give the classification of cycles of absorption refrigeration units thus identifying a one-level cycle, cycles with external and inner closures of devices for circulation of a solution, a coolant, heat and mass exchange. The latter two categories may include a series of simple or complex interconnected cycles. This paper briefly describes thirty-seven different absorption cycles. The one-level cycle is the most widely spread and is therefore considered the base cycle. It is most convenient to study the possibilities of other cycles and determine their advantages in comparison with a one-level base cycle. Multi-stage cycles, including cycles with multi-stage solution generation and multi-stage absorption, expand the capabilities of the base cycle. They allow using both high- and low-temperature heating sources for operation of the machine.

LBAC thermodynamic cycles involving multi-stage absorption, as well as single multi-stage absorption and generation of a solution can be carried out by externally coupled heat or mass transfer processes. These cycles are performed in a region of lower solution concentrations compared to a cycle with the two-level generation. This reduces the risk of solution crystallization during LBAC operation. All other things being equal, these cycles require a lower heating source temperature than the base cycle. Therefore, in such cycles it is possible to effectively use low temperature RES and solar energy as a heating medium. The positive qualities of these cycles indicate the relevance of their study.

The work [22] carried out the analysis of LBAC cycles with two-level absorption and two-level generation of a solution with externally connected mass flow with air cooling of absorbers and a capacitor at a coolant boiling point of 5°C and the use of low-temperature solar heat for generation of a solution in generators. The research was primarily aimed at studying cycle parameters that exclude crystallization of the solution in the machine. It was found that this cycle can be realized at a condensation temperature of not more than 53°C. Thus, the temperature of the heating source was about 80°C. The theoretical thermal coefficient made about 0.38.

2. Research methods

The authors of this paper performed the study of the thermodynamic cycle with two-level absorption and two-level generation of a solution with externally connected mass flow. In the cycle there are two circulation loops of the solution, high-temperature G_1 – A_1 and low-temperature G_2 – A_2. The externally connected mass transfer process is G_2 – A_1 The coolant steam evaporated in the low-pressure generator G_2 is absorbed into the high-pressure absorber A_1. The actual cycle is shown in Figure 1.

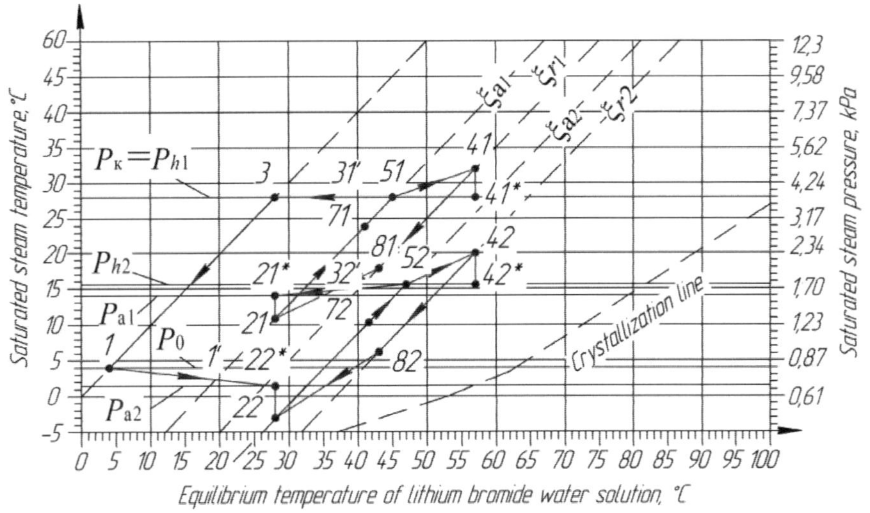

Figure 1. Thermodynamic cycle of LBAC with two-level absorption and two-level generation of a solution with externally connected mass flow

The main processes of the thermodynamic cycle are as follows: 1 – 1' – boiling of a working substance in an evaporator; 22 – 72 – heating of weak solution in a heat exchanger of the second stage solutions; 72 – 52 – adiabatic-isobaric desorption of working substance steam; 52 – 42 – solution boiling in the second stage generator; 42 – 82 – cooling of the rich solution in the heat exchanger of the second stage solutions; 82 – 22 – adiabatic-isobaric absorption of working substance steam in the second stage absorber; 21 – 71 – heating of a weak solution in a heat exchanger of the first stage solutions; 71 – 51 – adiabatic-isobaric desorption of working substance steam; 51 – 41 – solution boiling in the first stage generator; 41 – 81 – cooling of rich solution in a heat exchanger of the first stage solutions; 81 – 21 – adiabatic-isobaric absorption of working substance steam in the first stage absorber; 3' – 3 – removal of the heat of superheat and vapor condensation of a working substance in the capacitor.

The calculations were performed with the following initial data: cooled source temperature t_{s1} = 12°C, cooling source temperature made $t_{wc1} = t_{wa11} = t_{wa12}$ = (20; 25; 30; 35; 40)°C, temperature of the heating source varied within $t_{h11} = t_{h12}$ = (50 ÷ 80)°C, irreversible losses of the actual cycle were accepted according to the recommendations stated in the known literature. To calculate thermodynamic cycles depending on different parameters of external sources, a software complex was created with the help of the gfortran compiler. The program package consists of the main program and a number of subprograms. The input of initial data and the output of results are performed in separate files. The developed complex mathematical model includes the following subprograms: calculation of thermodynamic cycle point parameters; calculation of heat loads on units; determination of thermal coefficient. A comparative analysis of one-level LBAC and LBAC with two-level absorption and generation processes was conducted.

3. Results and discussion

Table 1 shows the calculation results of a one-level LBAC.

Table 1. Calculation results of a one-level LBAC

Parameter	$t_{wc1} = t_{wa1}$, °C	Cooling source temperature t_{h1}, °C					
		80	85	90	95	100	105
t_{s1}, °C	20	12	12	*[a]	*	*	*
	25	–[b]	12	12	12	*	*
	30	–	–	–	–	12	*
ξ_r, %	20	61.1	63.4	*	*	*	*
	25	–	60.5	62.7	65.1	*	*
	30	–	–	–	–	64.6	*
$\Delta\xi$, %	20	5.61	7.94	*	*	*	*
	25	–	2.1	4.27	6.69	*	*
	30	–	–	–	–	3.61	*
ζ	20	0.773	0.786	*	*	*	*
	25	–	0.652	0.728	0.753	*	*
	30	–	–	–	–	0.694	*

[a] Crystallization of the solution is possible.
[b] Cycle cannot be performed.

Figure 2 shows a diagram of thermal coefficient (ζ) dependence on heating and cooling source temperature for the LBAC cycle (Figure 1).

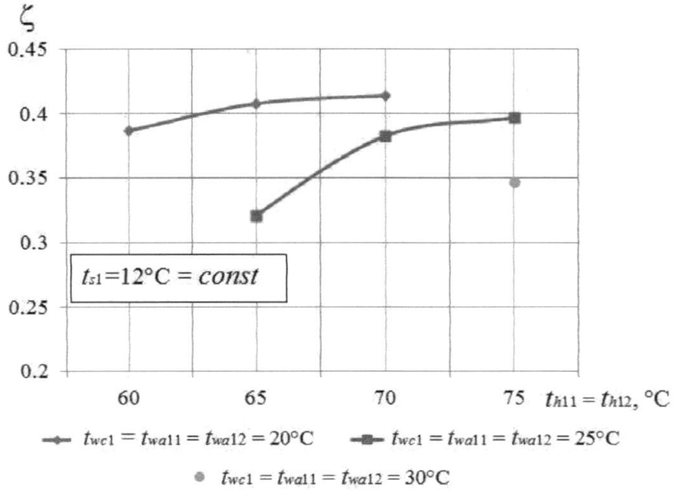

Figure 2. Diagram of thermal coefficient dependence on heating and cooling source temperature

IOP Conf. Series: Materials Science and Engineering **826** (2020) 012039 doi:10.1088/1757-899X/826/1/012039

Table 2 shows the calculation results of LBAC with two-level absorption and generation processes with associated mass flow.

Table 2. Calculation results of LBAC with two-level absorption and generation processes with associated mass flow

Parameter	$t_{wc1} = t_{wa11} = t_{wa12}$, °C	Cooling source temperature $t_{h11} = t_{h12}$, °C					
		55	60	65	70	75	80
t_{s1}, °C	20	–	12	12	12	*[a]	–[b]
	25	–	–	12	12	12	*
	30	–	–	–	–	12	*
	35	–	–	–	–	–	–
ξ_{r2}, %	20	–	58.3	61.1	63.4	*	–
	25	–	–	59.6	62.1	64.7	–
	30	–	–	–	–	63,3	–
	35	–	–	–	–	–	–
ξ_{r1}, %	20	–	51.0	53.5	56.0	*	–
	25	–	–	50.6	53.1	55.6	*
	30	–	–	–	–	52.6	*
	35	–	–	–	–	–	–
$\Delta\xi_2$, %	20	–	2.86	5.61	7.91	*	–
	25	–	–	1.21	3.70	6.30	*
	30	–	–	–	–	2.40	*
	35	–	–	–	–	–	–
$\Delta\xi_1$, %	20	–	5.24	7.77	10.31	*	–
	25	–	–	2.51	5.04	7.49	*
	30	–	–	–	–	2,72	*
	35	–	–	–	–	–	–
ζ	20	–	0.387	0.408	0.414	*	–
	25	–	–	0.321	0.383	0.397	*
	30	–	–	–	–	0.347	*
	35	–	–	–	–	–	–

[a] Crystallization of the solution is possible.
[b] Cycle cannot be performed.

4. Conclusion

The calculations made it possible to establish that in case it is possible to avoid crystallization of a solution the studied cycle can be performed at a heating source temperature of 60°C ÷ 75°C, the cooling water temperature shall be not more than 30°C. The actual thermal coefficient of the cycle is within the range of 0.35 ÷ 0.41. At the same temperatures of the cooled and cooling media, the one-level cycle makes 0.69 ÷ 0.79, but requires a heating source with a temperature of 20°C ÷ 25°C higher than that of the studied cycle.

References

[1] Palacios E, Izquierdo M, Marcos J D and Lizarte R 2009 Evaluation of mass absorption in LiBr flat-fan sheets *Appl. Energy* **86** 2574

[2] Bujedo L A, Rodriguez J and Martinez P J 2011 Experimental results of different control strategies in a solar air- conditioning system at part load *Solar Energy* **85** 1302

[3] Lizarte R, Izquierdo M, Marcos J D and Palacios E 2012 An innovative solar-driven directly air-cooled LiBr-H_2O absorption chiller prototype for residential use *Energy Build.* **47** 1

[4] Ketfi O, Merzouk M, Merzouk N K and Bourouis M 2017 Feasibility study and performance evaluation of low capacity water – LiBr absorption cooling systems functioning in different Algerian climate zones// *International Journal of Refrigeration* **82** 36

[5] Chena J F, Daia Y J, Wang H B and Wang R Z 2018 Experimental investigation on a novel air-cooled single effect LiBr-H_2O absorption chiller with adiabatic flash evaporator and adiabatic absorber for residential application *Solar Energy* **159** 579

[6] Sabbagh A A and Gómez J M 2018 Optimal control of single stage LiBr/water absorption chiller *International Journal of Refrigeration* **92** 1
[7] Shiue A, Hu S and Chiang K 2018 Effect of operating variables on performance of an absorption chiller driven by heat from municipal solid waste incineration *Sustain. Energy Technol. Assess.* **27** 134
[8] IEA 2019 *Global Energy & CO_2 Status Report 2018* (International Energy Agency)
[9] Al-Ugla A A, El-Shaarawi M A I and Said S A M 2015 Alternative designs for a 24-hours operating solarpowered LiBr-water absorption air-conditioning technology *International Journal of Refrigeration* **53** 90
[10] Li M, Xu C, Hassanien R H E, Xu Y and Zhuang B 2016 Experimental investigation on the performance of a solar powered lithium bromide – water absorption cooling system *International Journal of Refrigeration* **71** 46
[11] Lubisa A, Jeonga J, Giannettia N, Yamaguchia S, Saitoa K, Yabasea H, Alhamid M I and Nasruddin N 2018 Operation performance enhancement of single-double-effect absorption chiller *Applied Energy* **219** 299
[12] Sarabia Escriva E J, Lamas Sivila E V and Soto Frances V M 2011 Air conditioning production by a single effect absorption cooling machine directly coupled to a solar collector field. Application to Spanish climates *Solar Energy* **85** 2108
[13] Xu Z Y and Wang R Z 2017 Comparison of CPC driven solar absorption cooling systems with single, double and variable effect absorption chillers *Solar Energy* **158** 511
[14] Martínez P J, Martínez J C and Martínez P 2016 Performance comparison of solar autonomous and assisted absorption systems in Spain *International Journal of Refrigeration* **71** 85
[15] Muye J, Ayou D S, Saravanan R and Coronas A 2016 Performance study of a solar absorption power-cooling system *Applied Thermal Engine*ering **97** 59
[16] Chahartaghi M, Golmohammadi H and Shojaei A F 2019 Performance analysis and optimization of new double effect lithium bromide – water absorption chiller with series and parallel flows *International Journal of Refrigeration* **97** 73
[17] Xu Z Y and Wang R Z 2016 Absorption refrigeration cycles: Categorized based on the cycle construction *International Journal of Refrigeration* **62** 114
[18] She X, Yin Y, Xu M and Zhang X 2015 A novel low-grade heat-driven absorption refrigeration system with LiCl-H_2O and LiBr-H_2O working pairs *International Journal of Refrigeration* **58** 219
[19] Stepanov K I, Mukhin D G, Volkova O V and Baranenko A V 2016 Analysis of COP thermodynamic cycle of LBAC with two-level absorption when obtaining negative temperatures of cooling *Journal of the International Academy of Refrigeration* **1** 86
[20] Aprile M, Toppi T, Garone S and Motta M 2018 STACY – A mathematical modelling framework for steady-state simulation of absorption cycles *International Journal of Refrigeration* **88** 129
[21] Bowie D and Cruickshank C A 2017 Experimental evaluation of a triple-state sorption chiller *International Journal of Refrigeration* **81** 12
[22] Izquierdoa M, Venegas M, Rodriguez P and Lecuona A 2004 Crystallization as a limit to develop solar air-cooled LiBr – H_2O absorption systems using low-grade heat *Solar Energy Materials & Solar Cells* **81** 205

Calculation of cold accumulators with phase change materials

V Yu Zakharova, P A Kuznetsov, A V Baranenko and R O Faizullin

ITMO University, 49, Kronverksky Prospect, St. Petersburg, 197101, Russia

E-mail: viktoriasju@yandex.ru

Abstract. The paper presents an improved technique (mathematical model) for numerical calculations of solid-liquid phase transitions based on a zero-dimensional solution of the Stefan problem for cases of placing phase change materials in ball-shaped or cylindrical capsules. According to the proposed model, numerical calculations of the liquid phase volume fraction in Scilab have good compliance with experimental data for n-octadecane, and the calculation time is reduced several hundred-fold in comparison with calculations in Fluent.

1. Introduction

It is advisable to use cold accumulation at variable loads of consumers, at large difference in electricity rates during day and night, if there are technical and economic grounds for using renewable energy resources for cold energy accumulation (for example, sun energy and night radiation cooling) [1-5]. For instance, variable cold-supply capacities can be found in production refrigerators in case of large quantities of food supplies that requires cooling or freezing, in air-conditioning systems, which cold consumption during the daytime is usually higher than at night. Peak shaving by accumulated cold allows using refrigerating systems with lower capacity. The accumulation of cold by compressor refrigerators at night at double rates reduces energy costs.

Cold accumulators may use liquids, frozen water ice, phase change materials (PCM) as accumulating medium. The use of PCM reduces the capacity of accumulators by 5-14 times compared to the use of liquids [3]. The advantages of PCM also include constant phase transition temperature, chemical stability and economic efficiency. The specific heat of phase transition plays a major role in energy storage of such cold accumulators. In case of liquid-solid phase transition the accumulator is charged, and in case of reverse phase transition the solid-liquid is used [6].

The PCM can be placed in a cold storage tank in capsules in the form of balls, as well as circular or rectangular pipes. Thus, a coolant circulates in space between capsules [6]. This type of accumulators is most common. The PTS density in the accumulator in ball-shaped capsules makes about 50%, in rectangular pipes – about 90% of the volume [7]. In other cases, the PCM is placed in the accumulator tank, and the coolant circulates inside the pipes of the heat exchanger located in the tank.

2. Study of cold accumulator operation

In order to design cold accumulators with PCM, among other aspects, it is necessary to reliably calculate the processes of liquid-solid and solid-liquid phase transitions corresponding to charging and discharging of cold energy accumulators.

Many works are devoted to the study of thermal energy accumulators and thus related processes during charging and discharging [8-17]. The main calculation method in case of phase transitions is enthalpy-porous approach during numerical solution of the Navier-Stokes system of equations for PCM, which is considered as an incompressible liquid [14-17]. The phase transition boundary is not directly modeled, which is ensured by the unity of setting equations for solid and liquid phases. Thus, the volume fraction of a liquid phase β depends on temperature T as follows [14]

$$\beta = \begin{cases} 0, \text{ if } T < T_{sol} \\ 1, \text{ if } T > T_{liq} \\ \frac{T-T_{sol}}{T_{liq}-T_{sol}}, \text{ if } T_{sol} < T < T_{liq} \end{cases} \quad (1)$$

where T_{sol} – solidus temperature, T_{liq} – liquidus temperature

Change of flow characteristics during phase transition is taken into account by introducing the porous zone in the transition zone, the porosity of which varies from 0 (solid phase) to 1 (liquid phase). The constants of the porous zone are determined during the numerical experiment [18].

The idea of enthalpy approach is to replace the energy of unknown temperature in the equation by enthalpy H determined through the temperature of phase transition and the liquid phase fraction [15]

$$H = H_{ref} + \beta L + \int_{T_{ref}}^{T} c dT, \quad (2)$$

where H_{ref} – enthalpy at relative temperature T_{ref}, L – heat of phase transition, c – PCM specific heat. This approach allows the heat of the phase transition to be considered directly in the energy equation

$$\frac{\partial(\rho H)}{\partial t} + \nabla(\rho \boldsymbol{v} H) = \nabla(k \nabla H), \quad (3)$$

where v – liquid velocity, t – time, k – PCM heat conductivity, ρ – PCM density. Thermal conductivity and density are fugacity of temperature.

However, the disadvantage of this approach includes the problems with convergence of a strongly non-linear problem for numerical solution, and the large calculation time even for two-dimensional problems.

An alternative can be the approach based on the Stefan problem, where the location of a mobile phase transition boundary is unknown. The boundary conditions at the boundary of solid and liquid phases are defined here. Namely, the temperature continuity condition and the heat flux condition. Such systems of equations are solved analytically for one-dimensional cases when charging an accumulator, since it is not necessary to consider in detail the convective heat fluxes in the liquid phase. The classical solutions to Stefan problem are given in [19-21]. However, the solution in the case of accumulator discharge causes the need to consider heat fluxes in the liquid phase and the natural convection.

3. Calculation methodology

The authors of this paper propose the procedure for calculating heat fluxes during the discharge of accumulators based on a zero-dimensional solution of the Stefan problem for cases of placing PCM in ball-shaped or cylindrical capsules. The average radius of the phase transition boundary r_m and the capsule shell temperature T_c are unknown here, as shown in Figure 1. It is assumed that the phase transition occurs at a constant temperature T_m.

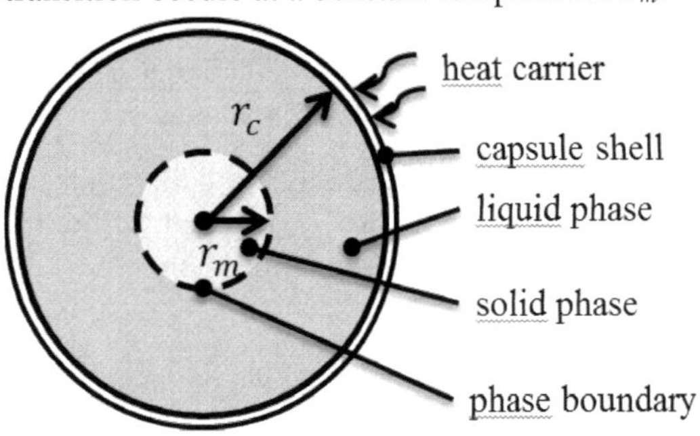

Figure 1. Thermal model of accumulator with PCM.

Equation for phase boundary

$$\frac{dr_m(t)}{dt} = -\frac{T_c(t)-T_m}{R(t)} \frac{1}{\rho_f A_m(t)[L+c_f(T_c(t)-T_m)]}, \tag{4}$$

where c_f and ρ_f – specific heat capacity and density of liquid phase, respectively, A_m – area of phase boundary calculated through the average radius $r_m(t)$, R – thermal resistance of ball or cylindrical layer of the liquid phase calculated in accordance with [22]

$$R(t) = \frac{r_c - r_m(t)}{4\pi r_c r_m(t) k_{ef}}, \tag{5}$$

$$R(t) = \frac{1}{2\pi k_{ef}} \ln\frac{r_c}{r_m(t)}. \tag{6}$$

where r_c – inner radius of the shell. The formula (5) is applicable in the case of a ball shell, (6) – for a cylindrical shell. The effective thermal conductivity coefficient k_{ef} is calculated by the formula

$$k_{ef} = \begin{cases} k_f, \ \mathrm{Pr} < 1000 \\ 0{,}18 \cdot k_f \cdot (\mathrm{Gr} \cdot \mathrm{Pr})^{0{,}25}, \end{cases} \tag{7}$$

$$\mathrm{Gr} = b_m \cdot g\left(r_c - r_m(t)\right)^3 \frac{T_c - T_m}{\nu^2}, \tag{8}$$

where ν – kinematic viscosity of the liquid phase, g – gravitational acceleration.

For $T_c(t)$, the equation is written on the assumption of a uniform shell temperature field

$$C\frac{dT_c(t)}{dt} + \frac{T_c(t)-T_m}{R(t)} + \sigma(T_c(t) - T_a) = 0, \tag{9}$$

where C – total heat capacity of the shell, σ – heat conductivity from the shell to the heat carrier with temperature T_a. At larger σ, the model turns into a model for the known temperature $T_c = T_a$ [23].

4. Results and discussion

Equations (4) – (9) are solved numerically.

In order to verify the developed method, the design values are compared with the experimental ones, which were calculated in the Fluent package using equations (1) – (3) taken from [14] for n-octadecane. In [14], a numerical solution is based on the first order approximation resulting in a computational error. The second order approximation used by the authors of this paper in the numerical solution better corresponds to experimental data. The numerical calculation according to formulas (4) – (9) is performed in Scilab package. The calculation data are shown in Figure 2.

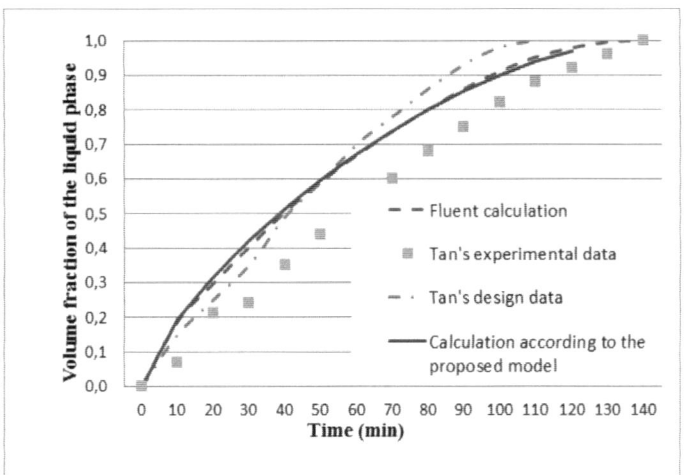

Figure 2. Comparison of calculation data.

IOP Conf. Series: Materials Science and Engineering **826** (2020) 012040 doi:10.1088/1757-899X/826/1/012040

5. Conclusion

Figure 2 shows that in fact the calculations of the liquid phase volume fraction done according to the proposed model in the Scilab package coincide with the numerical calculations in the Fluent package, but the calculation time in the latter case increases several hundred-fold. The obtained results also coincide well with experimental data, the discrepancy can be caused by the introduction of thermoelectric converters into capsules throughout the experiments, which was not taken into account in numerical calculations. This makes it possible to use the developed mathematical model in calculation and design of cold accumulators with PCM.

References

[1] Energy Storage Technology Roadmap 2014 *Technical Annex* (International Energy Agency)
[2] Rabczak S 2015 Free-Cooling in Seasonal Cold Accumulator *Int. J. of New Technology and Research* **1 8** 49–52
[3] Veerakumar C and Sreekumar A 2016 Phase change material based cold thermal energy storage: Materials, techniques and applications – A review *Int. J. Refrig.* **67** 271–289
[4] Al-Aifan Bader Parameshwaran R, Mehta Kushagra and Karunakaran R 2017 Performance evaluation of a combined variable refrigerant volume and cool thermal energy storage system for air conditioning applications *Int. J. of Refrigeration* **76** 271–295
[5] Gwi Taek Kim, Young Uk Choi, Yoong Chung, Mo Se Kim, Keon Woo Park and Min Soo Kim 2018 Experimental study on the performance of multi-split heat pump system with thermal energy storage *Int. J. of Refrigeration* **88** 523–537
[6] Gang Li, Yunho Hwang and Reinhard Radermacher 2012 Review of cold storage materials for air conditioning application *Int. J. of Refrigeration* **35** 2053–2077
[7] Iten M and Liu S 2014 A work procedure of utilising PCMs as thermal storage systems based on air-TES systems *Energy Conversion and Management* **77** 608–627
[8] Bosholm F, Lopez-Navarro A, Gamarra M, Corberan J M and Paya J 2016 Reproducibility of solidification and melting processes in a latent heat thermal storage tank *Int. J. Refrig* **62** 85–96
[9] Sharma Atul, Tyagi V V, Chen C R and Buddhi D 2009 Review on thermal energy storage with phase change materials and applications *Renewable and Sustainable Energy Rev.* **13** 318–345
[10] Zakir Khan and Zulfiqar Ahmad Khan 2017 An experimental investigation of discharge/solidification cycle of paraffin in novel shell and tube with longitudinal fins based latent heat storage system *Energy Conversion and Management* **154** 157–167
[11] Yaoa C, Konga X, Lic Y, Dud Y and Qia C 2018 Numerical and experimental research of cold storage for a novel expanded perlite-based shape-stabilized phase change material wallboard used in building *Energy Convers. Manag.* **155** 20–31
[12] Joybari M M, Seddegh S, Wang X and Haghighat F 2019 Experimental investigation of multiple tube heat transfer enhancement in a vertical cylindrical latent heat thermal energy storage system *Renewable Energy* **140** 234–244
[13] Mishra L, Sinha A and Gupta R 2019 Recent Developments in Latent Heat Energy Storage Systems Using Phase Change Materials (PCMs) – A Review *Green Buildings and Sustainable Engineering. Proceedings of GBSE 2018* pp 25–37
[14] Tan F L, Hosseinizadeh S F, Khodadadi J M and Liwu Fan 2009 Experimental and computational study of constrained melting of phase change materials (PCM) inside a spherical capsule *Int. J. of Heat and Mass Transfer* **52** 3464–3472
[15] Sattari H, Mohebbi A, Afsahi M M and Azimi Yancheshme A 2017 CFD simulation of melting process of phase change materials (PCMs) in a spherical capsule *Int. J. of Refrigeration* 2017 **73** 209–218
[16] Joybari M M, Haghighat F, Seddegh S and Al-Abidi A A 2017 Heat transfer enhancement of phase change materials by fins under simultaneous charging and discharging *Energy Conversion and Management* **152** 136–156

[17] Gulfam R, Zhu W, Xu L, Cheema I I, Sheng P, Zhao G and Deng Y 2018 Design, fabrication and numerical analysis of compact thermal management system integrated with composite phase change material and thermal bridge *Energy Conversion and Management* **156** 25–33
[18] Voller V and Prakash C 1987 A fixed grid numerical modelling methodology for convection-diffusion mushy region phase-change problems *Int. J. of Heat and Mass Transfer* **30** 1709–1719
[19] Samarsky A A, Varischevich P N 2003 *Computational heat transfer* (Moscow: URSS Editorial)
[20] Carslaw H and Jaeger J 1959 *Conduction of Heat in Solids* (USA: Oxford University Press)
[21] Meirmanov A M 1986 *Stefan problem* (Novosibirsk: Nauka: Siberian Branch)
[22] Dulnev G N, Semyashkin E M 1968 *Heat exchange in radioelectronic devices* (Leningrad: Energia)
[23] Zakharova V Y, Kuznetsov P A, Baranenko A V 2019 Mathematical model of solid-liquid phase transitions applied to thermal energy accumulators *IOP Conference Series: Materials Science and Engineering* **643** 012086

The prospects for the use of electrogasdynamic systems in dry ice-cold generators of refrigerated trucks

B S Babakin [1], M I Voronin [1], A N Suchkov[2] and S B Babakin[2]

[1]Moscow State University of Food Production, 11, Volokolamsk highway, Moscow, 125080, Russia
[2]All-Russian Scientific Research Institute of Refrigeration Industry – branch of V.M. Gorbatov Federal Research Center for Food Systems of Russian Academy of Science, 12, st. Kostyakova, Moscow, 127422, Russia

E-mail: holod@mgupp.ru

Abstract. Cold sources for cooling refrigerated vans are divided into temporary ones, ensuring the maintenance of the required temperature in the body without recharging the batteries for cold cooling for a limited period, and constant, ensuring the maintenance of the set temperature without periodically renewing the supply of cold. Temporary sources of cold include ice-salt mixture, dry ice, frozen eutectic solutions, installations with liquid nitrogen. A constant source of cold is engine cooling. Dry ice cooling is devoid of the main disadvantages of other methods. It has a relatively high refrigerating effect with a relatively small volume occupied by dry ice. Dry ice cooling can be carried out by contact and non-contact methods. In the contact method, the blocks of dry ice are not separated from the product, due to which carbon dioxide pairs wash the products. With the non-contact method, dry ice is loaded into hermetically sealed containers. The authors have proposed an environmentally friendly, reliable scheme of the process of cooling the body of a refrigerator with carbon dioxide using electrophysical methods. EGD devices with moving and stationary electrodes developed by the authors were used as pathogens for the circulation of the gas mixture in an isothermal body, which allowed to reduce the consumption of solid carbon dioxide.

1. Introduction

Using of machine-free cooling in refrigerated trucks used for transporting foodstuffs and their role in the formation of a continuous refrigeration chain in cities, especially in megacities, has recently increased significantly. With this cooling method, the isothermal bodies of automobiles (refrigerated containers) are cooled by eutectic ice, carbon dioxide and other cryogenic substances.

Carbon dioxide cooling systems are the simplest and most reliable. The advantages of their usage for cooling food products include a structural prostate, low cost of equipment, high reliability, environmental friendliness, a favorable effect on product quality, when passing from a solid state to a gaseous state, it occurs at a temperature of 78.5 °C without moisture formation. Dry ice per unit mass has almost twice as much cooling capacity as regular ice. The specific cooling capacity while maintaining the air temperature at –18 °C for solid CO_2 is approximately 1.7 times higher than that for liquid nitrogen N_2. Dry ice cooling is carried out in direct contact with the cooled object or using an intermediate coolant, most often air. In the latter case, dry ice is crushed and placed in metal containers – pockets through which air circulates. Air circulation is usually controlled by fans. The atmosphere containing carbon dioxide has a bactericidal effect.

"Cearns & Brown" manufactures installations for transporting chilled and frozen food products using dry ice. The system provides the necessary mode in the temperature range from –15 to + 5 °C – for up to 72 hours or more. Transportation of packaged products is provided. Thermo King SD-III CR refrigeration unit is designed for heavy-duty semi-trailers. Lowering the temperature to –29 °C is 3 times faster than a refrigeration unit with an autonomous diesel engine. For intercity transportation, the CO_2 Hybrid model refrigeration unit is used. The cooling capacity of the SD-III CR and CO_2 Hybrid refrigeration units is shown in Table 1.

Table 1. Cooling capacity of refrigeration units SD-III CR and «CO_2 Hybrid»

Model	Cooling capacity (kW) at an ambient temperature of 38 °C and air temperature in the truck body		
	2 °C	–18 °C	–29 °C
SD-III CR	17	17.6	17.9
CO_2 Hybrid	9.5	6.6	5.1

2. Formulation of the problem

Despite the apparent simplicity of the implementation of the above process of cooling products with carbon dioxide, the latter has a number of disadvantages, which significantly inhibit its use. These include a significant increase in carbon dioxide consumption with a decrease in cooling temperature (on average 10–12 kg/h for transportation in the city), the inability to ensure uniform circulation of the gas mixture throughout the body volume. At the same time, it should be noted that the food delivery process, especially in cities, is characterized by frequent, long stops, during which significant heat inflows enter the body through open doors, which, as a result, leads to the formation of hydroaerosols.

The noted disadvantages can be offset by the use of new modern technologies. These include technologies developed by the authors using environmentally friendly, electrophysical methods, in particular electro-gas-dynamic systems (EGD). In the general case, an EGD system is a set of technical means that ensure the use of electrosynthesis and electroconvection controlled in time and space to influence the objects which are being processed.

3. Materials and methods

We have proposed to increase the working temperature in the body, while strengthening, in order to prevent microbiological spoilage of products, the bactericidal effect of the medium. To ensure uniform circulation of the medium, it is advisable to use highly efficient EGD systems with a power of 50–500 W creating a uniform electric wind speed of 0.5–3.5 m/s. Our studies, using EGD systems specially developed for refrigerated containers, showed that corona discharge provides stable values of air flow rates up to the entire volume of the processed chamber. The total power consumption of the system did not exceed 1.5–2.0 kW, in a wide range of negative temperatures. The kinetics of inactivation of microorganisms in the medium of a mixture of CO_2 with air formed in the atmosphere was significantly higher than in the case of carbon dioxide. In figure 1 shows the survival of microorganisms in a CO_2 environment with air and CO_2 in a corona discharge field for 90 min at a medium temperature of –18 and –8 °C.

As it can be seen from Fig. 1, the bactericidal effect of using a corona discharge with increasing ambient temperature to –12 °C is higher than the use of CO_2 in air at –18 °C.

Below fig. 2 shows a diagram of the process of cooling a truck body with granular carbon dioxide using electrophysical methods.

As the causative agents of the circulation of the gas mixture in an isothermal body, we used EGD devices developed by the authors with moving and stationary electrodes operating in the corona mode, both in static electric fields and in pulsed ones.

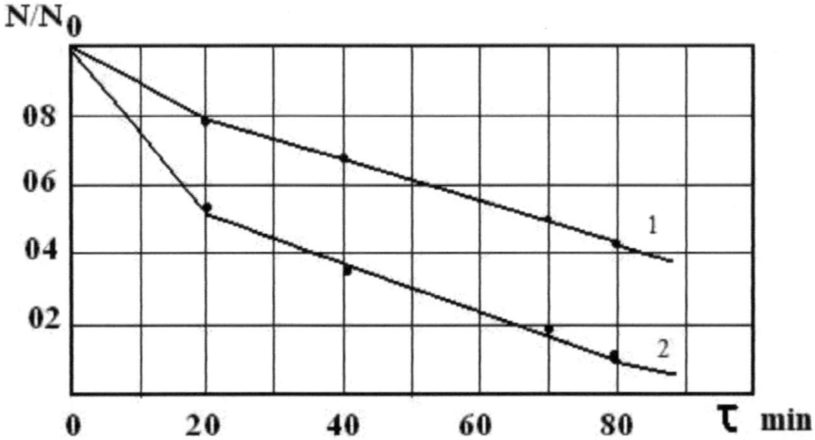

Figure 1. The survival rate of microorganisms: 1 – in a mixture of CO_2 with air at t = –18 °C; 2 – in a mixture of CO_2 with air and a corona discharge field at t = –8 °C; N0 is the initial content of microorganisms in the medium; N is the number of microorganisms in the medium after treatment

The amount of dry ice Gl (kg) required to maintain a given temperature in the body of the refrigerator is determined by the formula:

$$\text{Gl} = \frac{Q_k + Q_m \tau \cdot 3.6}{r_{di.}},$$

where $Q_{k.}$ – the amount of cold required for pre-cooling the body, kJ;

$Q_{m.}$ – hourly average cold consumption for the repayment of heat influx from the environment, W;

τ – transportation duration, h;

3.6 – conversion factor;

$r_{di.}$ – specific heat of dry ice sublimation, kJ / kg;

$r_{di.} \approx$ 620 kJ / kg.

The heat transfer surface (F, m2) is determined by the formula:

$$F = \frac{Q}{\alpha \cdot \Delta t \cdot \tau},$$

where Q – cold reserve kJ;

Δt – temperature difference between air and dry ice, °C;

τ – refrigerated truck duration, h;

α – heat transfer coefficient between heat exchange surface and air, W/(m^2·K).

Provided that the heat exchange between air and dry ice in the installation is through a metal wall, and the air velocity varies in the range from 0.2 to 1.6 m / s, the heat transfer coefficient between the heat exchange surface and air will average 20 W/(m^2·K).

The internal volume of the box for dry ice (V, m^3) is calculated by the formula:

$$V = \frac{\text{Gл}}{\rho_{di.}} K_o,$$

where $\rho_{di.}$ – dry ice density, $\rho_{с.л.}$ = 1500 kg/m^3;

$К_о$ – coefficient of increase in dry ice volume during crushing, $К_о$ = 2.5–3.

In the isothermal case of stand *1* made of reinforced polystyrene foam with a thickness of 100 mm, channels for the circulation of gaseous carbon dioxide are mounted on the inner walls. Circulating holes were made in the channel walls against which, inside the channels, EGD devices were mounted *7*. The dry ice storage unit *2* is located outside the isothermal housing and includes a movable scraper *3*, an electric motor *4* with an auger *5*, for supplying ice to the dispenser *6*. Process body cooling is carried out by feeding granular carbon dioxide into the dispenser, where, due to the operation of EGD devices, gaseous CO_2 circulates through the internal volume of the body, ensuring its cooling, disinfection and deodorization.

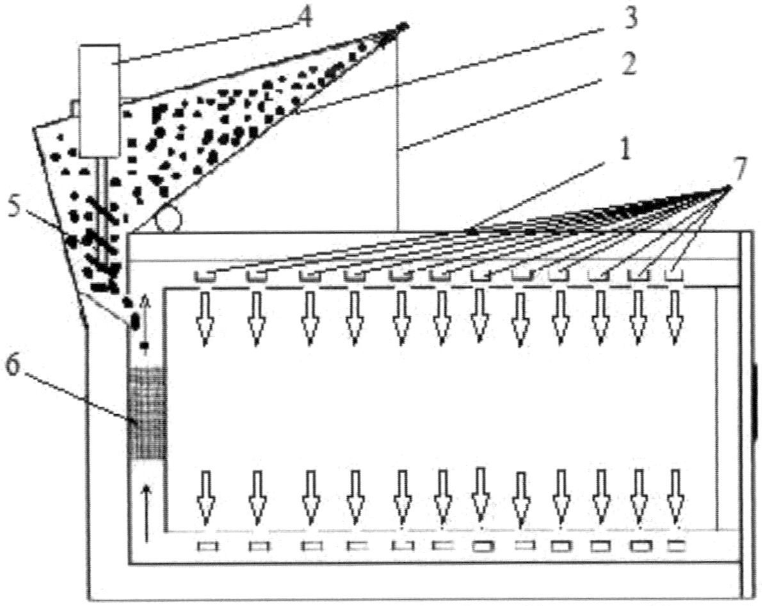

Figure 2. Diagram of the process of cooling the body of a refrigerator with granular carbon dioxide using electrophysical methods

4. Conclusion

The cooling system of the refrigerator and food products developed by the authors allows one to reduce the consumption of solid carbon dioxide up to 20 % by increasing the temperature of the working medium of the refrigerator and increasing the bactericidal effect due to the use of electrophysical methods, in particular, corona discharge.

References

[1] Voronin M I and Babakin B S 2009 Scientific – practical bases of application of electro-gazogidrodynamic system in the refrigerating technics and technology *Cryogenic technologies and equipment. Prospects of development* ed Arkharov A M (Moscow: Mir-Expo) p 44

[2] Voronin M I and Babakin B S 2010 Application of discharge elements in refrigerators *Cryogenic technologies and equipment. Prospects of development* ed Arkharov A M (Moscow: Mir-Expo) pp 66–7

[3] Voronin M I and Babakin B S 2011 *Application of electrophysical methods in food technology at cryoscopic temperatures Industrial gases* ed Arkharov A M (Moscow: Mir-Expo) pp 157–60

[4] Voronin M I and Babakin B S 2012 *Application of corona discharge in suholednykh generators of cold Cryogenic technologies and equipment. Prospects of development* ed Arkharov A M (Moscow: Mir-Expo) pp 95–8

IOP Conf. Series: Materials Science and Engineering **826** (2020) 012042 doi:10.1088/1757-899X/826/1/012042

On the issue of the usage of sealed containers for cooling the refrigerant in refrigerated trucks

B S Babakin[1], A N Suchkov[2] and M I Voronin[1]

[1]Moscow State University of Food Production, 11, Volokolamsk highway, Moscow, 125080, Russia
[2]All-Russian Scientific Research Institute of Refrigeration Industry – branch of V.M. Gorbatov Federal Research Center for Food Systems of Russian Academy of Science, 12, st. Kostyakova, Moscow, 127422, Russia

E-mail: holod@mgupp.ru

Abstract. The paper deals with the problem of using machine-free cooling in refrigerated trucks for intercity transportation in order to ensure environmental safety and to eliminate noise from a working refrigeration machine when unloading refrigerated products from refrigerated trucks to the distribution network. A machineless method for cooling the body of a refrigerator truck with cold accumulators without direct contact with a cooling medium due to the usage of frozen sealed containers is proposed. The accumulation of cold (AC) is carried out, as a rule, due to the heat of the phase transition at the melting point or crystallization of individual substances or there mixtures. The research of water-salt systems, which are widely used for temperature stabilization in refrigeration, for storage and transportation of medicines and food products, is presented. The results of the melting and freezing of three cold storage substances, which are H_2O and water-salt solutions with NaCl concentrations of 8 and 13 % are presented. According to the results of studies, the usage of pressurized containers in refrigerated trucks with machine-free cooling is beneficial and opens up great opportunities for cooling the air in various destinations.

1. Introduction

Accumulation of cold allows one to increase the cooling capacity of the existing cooling system, energy efficiency, resource and reliability of the equipment.

An approximate classification of cold accumulators is given in Fig 1 [1]. Accumulation of cold is carried out with a phase transition with direct contact of the storage substance with or without refrigerant.

Nowadays cooling devices filled with eutectic solutions are widely used for urban transport of perishable foodstuffs by refrigerated vehicles. The eutectic solution is frozen by the refrigeration machine in the parking lot, generally at night, when Q fare for electricity is applied. During the route, the desired temperature is maintained as a result of eutectic solution melting in the truck body. This type of cooling is widely used abroad (Carrier Transicold, FIC s.p.a., etc.). The use of machine-free cooling in refrigerated trucks allows to maintain an environmentally friendly environment and eliminate noise when unloading products to supermarkets with a working refrigeration machine, especially in stores which are located in residential areas, avoiding noise for residents [2].

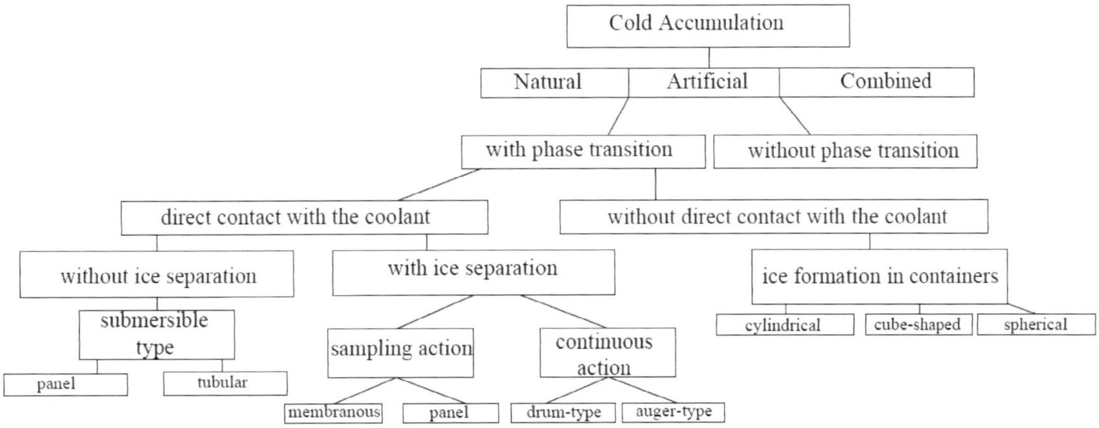

Figure 1. The classification of cold accumulators

Thermal pollution is a factor which should be avoided, in particular, the energy spent on the operation of the refrigeration system is a source of environmental pollution. In some cases, the refrigeration system is a source of carbon dioxide.

To recommend substances as cold storage materials, is necessary that they satisfy the following requirements: have the necessary melting temperature, high specific heat of phase transition, insignificant amount of supercooling during crystallization, maintaining the stability of qualities during multiple melting and crystallization cycles, insignificant toxicity, explosion safety, availability and low cost.

An analysis of the scientific and technical literature showed that many water-salt systems are potentially productive substances as AC. However, the absence of important physical and chemical characteristics for many systems precludes the possibility of using them as an AC without conducting special studies.

2. Object of research

The authors have developed a method of accumulating cold without direct contact of the cold storage substance with a cooling and cooled medium. It is possible due to the usage of sealed containers of various shapes, which are filled up to 80–90 % of their volume to ensure free expansion of the solution when it freezes. In fig. 2, a, you can see the air pocket in the capsule before freezing. Sealed containers are a system of spherical elements with diameters from 40 to 75 mm. Sealed containers can also have a different shape depending on the purpose. The tanks are filled with cold storage solution [3].

Hermetic containers are frozen in the air of the refrigerating chamber with their subsequent transfer to the truck body and their location in the supporting frame. Also, freezing of sealed elements can be carried out by immersing them in a coolant, which accelerates the process of freezing cold-storage substance in capsules.

While the process of melting a frozen solution in the truck body, an intensive absorption of heat occurs, due to which the temperature set by the technological conditions is maintained. The duration of the action of cold accumulators depends on the diameter of the sealed elements, which in turn affects the intensity of heat absorption. While reducing the diameter of the balls, using the same volume of eutectic solution, the area of the heat transfer surface between the cold storage substance and the cooled medium increases.

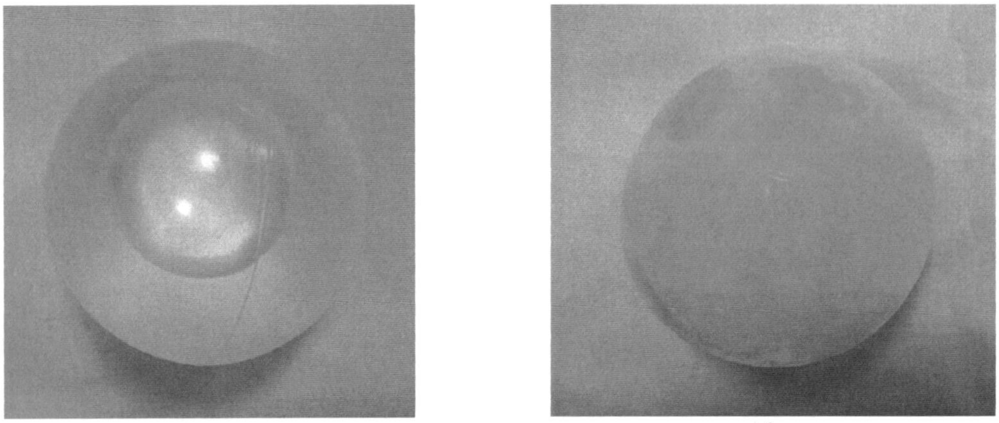

a b

Figure 2. Capsules of the experimental installation for cold accumulation: a – before freezing; b – after freezing

3. The results of the research

The results of experimental studies of the process of charging balls are presented in Fig. 3.

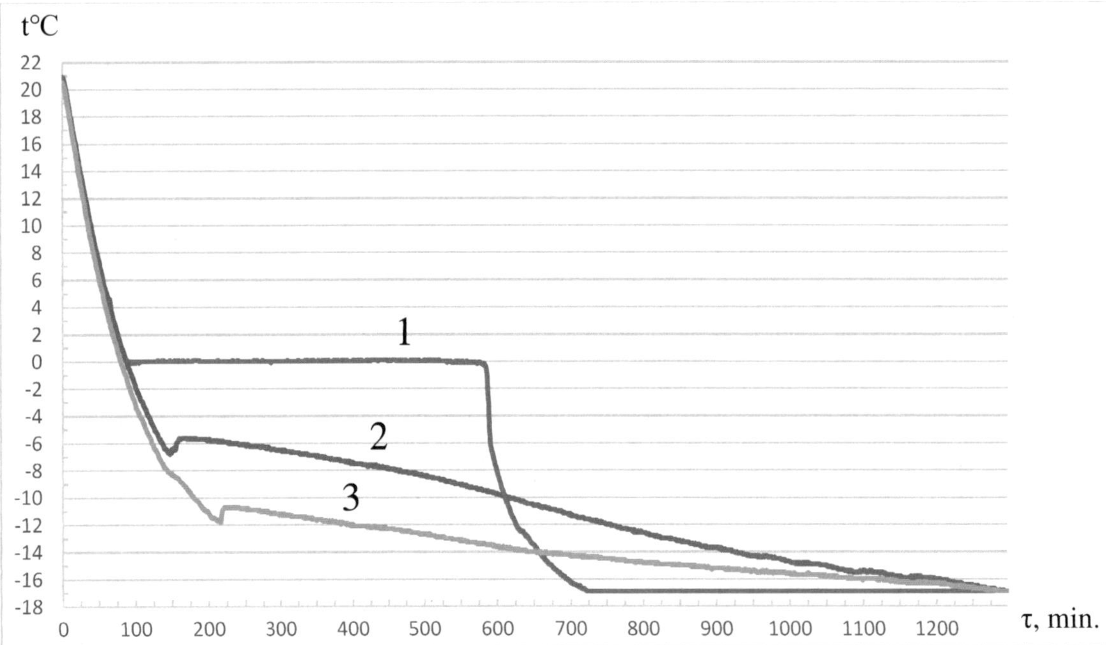

Figure 3. Temperature changing of a cold storage substance inside a sealed container during charging: 1 – H_2O; 2 – 8 % NaCl solution, 13 % NaCl solution

Distilled water (H_2O) and saline solutions (H_2O + NaCl) are presented as cold storage substances. As can be seen from fig. 3 the batteries were charged to a temperature of –18 °C.

While freezing water, the water was cooled for 1.5 hours and then 8 hours and a phase transition was observed at 0 °C, then the ice was cooled for 2 hours to an air temperature of –18 °C in the chamber.

When freezing an 8 % NaCl solution (curve 2), the liquid cooling process lasts 2 hours, after which we observe a phase transition, which begins at a solution temperature of –6 °C.

When freezing a 13 % NaCl solution (curve 3), the liquid cooling process lasts 4 hours, after which we also observe the beginning of a phase transition, but at a lower temperature of –11 °C.

The freezing process of solutions differs from crystallization of water. After the first ice crystals are formed in the solution, the strength of the solution increases and the temperature of the phase transition decreases. The process occurs until the solution becomes saturated and the whole freezes. Therefore, the charging of cold accumulators with saline solutions extends over a large temperature range. A frozen salt solution consists of ice crystals and salt, which can be seen by examining the frozen element of the cold accumulator through a transparent sealed shell.

It should also be noted a slight hypothermia of 0.5– 1 °C of saline solutions before the start of the phase transition.

Figure 4 shows the temperature change of the cooled medium, by the same elements of the cold accumulator filled with various cold storage substances H_2O, 8 % aqueous solution of NaCl and 13 % aqueous solution of NaCl.

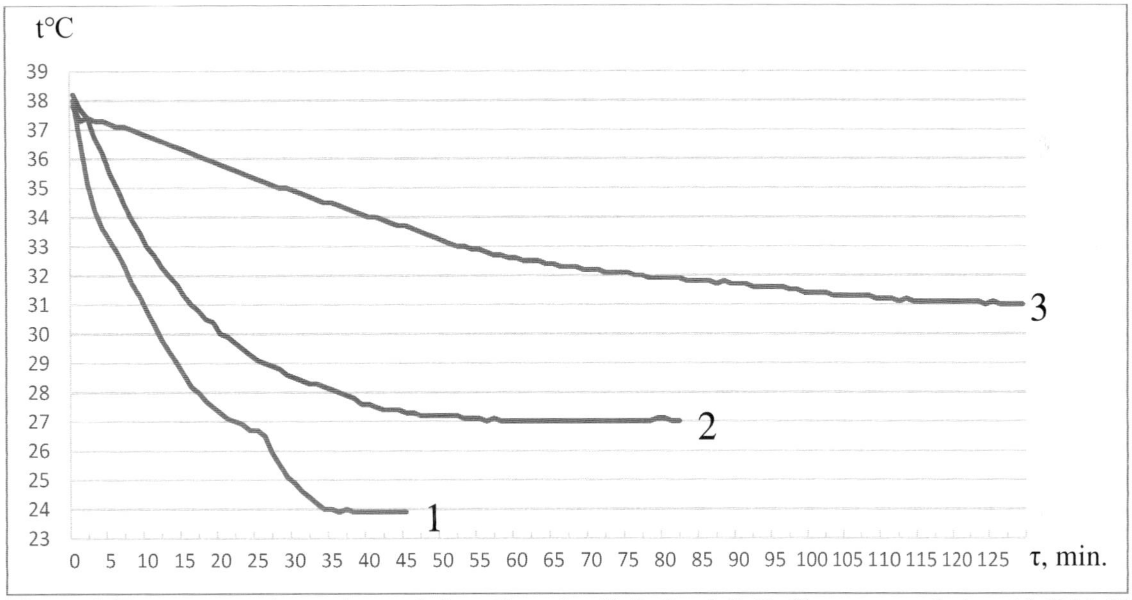

Figure 4. Temperature change by cooled medium: 1 – H_2O; 2 – 8 % NaCl aqueous solution, 13 % NaCl aqueous solution

As it can be seen from fig. 4 when using H_2O, the temperature of the medium decreases by 14 °C in 35–45 minutes, when using an 8 % solution of H_2O + NaCl, the temperature of the medium decreases by 11 °C in 55–80 minutes, when using a 13 % solution of H_2O + NaCl, the temperature of the cooled medium changes by 8 °C for 120–130 min.

Using H_2O under these conditions is more efficient than using 8 and 13 % aqueous NaCl solutions. This is because the cooling effect is based on the usage of latent heat of fusion of eutectic ice. When the eutectic melts, the temperature remains constant, in contrast to the melting of ice when it is mixed with salt. And, since the cryohydrate point for the ice-salt mixture with NaCl is –21.2 °C, at a NaCl concentration in the solution of 23.1 %, to obtain a solid eutectic NaCl solution, it is necessary to charge the elements of the cold accumulator with a NaCl concentration in the solution of 23.1 % at temperature below cryohydrate, which is not always possible.

4. Conclusion

As the operating experience of machine cooling systems of refrigerated trucks, they have a number of drawbacks, in particular, the environment is polluted with carbon dioxide emissions, when the refrigeration machine is operating while food delivery to retail chains located in residential neighborhoods, there is increased noise that adversely affects people's health.

The results of the experimental studies show the benefits of using sealed containers of various diameters with various cold storage substances for cooling of working environments. When using H_2O, an effective decrease in the temperature of the medium occurs in a shorter period of time than when using aqueous 8 % and aqueous 13 % NaCl solutions frozen to a temperature above the eutectic temperature.

References

[1] Babakin B S, Suchkov A N and Voronin M I 2019 Cooling of coolant in cold accumulators *Dairy Industry* **7** 55–6
[2] Belozerov G A and Babakin B S 2010 *Refrigerated transport and containers* (Ryazan: Ryazan Regional Printing House)
[3] Babakin B S, Voronin M I, Suchkov A N and Aitikeev R B 2019 The study of the heat load in industrial accumulators of a cold Kazakhstan-Refrigeration 2019: *Proc. of the Conf.* ed Kulazhanova T K (Almaty: ATU) pp 26–8

IOP Conf. Series: Materials Science and Engineering **826** (2020) 012043 doi:10.1088/1757-899X/826/1/012043

Modeling the crystallization process of an eutectic solution in cold accumulators of a refrigerated truck

G A Belozerov[1], B S Babakin [2,4], A N Suchkov[1], B A Makarov[3] and S B Babakin[1]

[1]All-Russian Scientific Research Institute of Refrigeration Industry – branch of V.M. Gorbatov Federal Research Center for Food Systems of Russian Academy of Science, 12, Kostyakov St., Moscow, 127422, Russia
[2]Moscow State University of Food Production, 11, Volokolamskoe shosse, Moscow, 125080, Russia
[3]Open stock corporation Central Scientific Research Institute "Kurs", 34-a, Kirpichnaya St., Moscow, 105187, Russia

E-mail: holod@mgupp.ru

Abstract. The duration of crystallization process of an eutectic solution of an infinite plate may be presented as a sum of three components: duration of the cooling process of the eutectic solution of liquid phase $\tau_1$from initial temperature t_I to the crystallization beginning; duration of crystallization process τ_2 and duration of the precooling process of solid phase τ_3 to certain final temperature t_f in the plate center.

1. Introduction

At present the use of refrigerated trucks with cold accumulators designed for interurban transportation is a perspective mode of cooling with no moving parts. The melting heat of the eutectic solution after freezing is used in accumulator systems of refrigeration. The appliance of cold accumulators in refrigerated trucks ensures maintaining environment safety.

Abroad such a kind of cooling is in widespread use. In particular Carrier Transicold (France) offers a series of "Vanta" units with machine and accumulator refrigeration for isothermal motor vans at capacity from 4 to 23 m^3 at minus 32° eutectic solution freezing temperature. The units are intended for transportation of ice cream and quick frozen products. They are designed for 14 hours of continuous operation at 30° ambient air temperature.

The world leader in eutectic plate production FIT s.p.a. from Italy offers a series of EBS and EFR models of eutectic plates in Russian market.

2. The solution of the problem

Below a mathematical model to define the duration of the process of eutectic solution freezing in accumulator cooling devices is offered (Fig.1). An approximate definition of duration of the process of eutectic solution crystallization of the infinite plate is accepted under boundary conditions of the third kind.

The duration of crystallization process of an eutectic solution of an infinite plate may be presented as a sum of three components: duration of the cooling process of the eutectic solution of liquid phase $\tau_1$from initial temperature t_I to crystallization beginning; duration of crystallization process τ_2 and duration of the process of solid phase τ_3 precooling to a certain final temperature t_f in the infinite plate center.

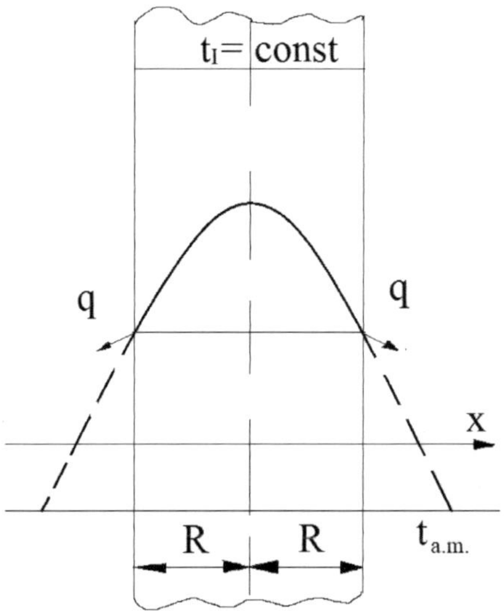

Figure 1. The temperature field of the infinite flat plate at cooling

Neglecting convective heat exchange, the duration of the process of liquid phase cooling τ_1 from initial temperature t_I to crystallization temperature t_C of the eutectic solution on the outer surface can be obtained by solution of the one-dimensional non-stationary equation of heat conductivity:

$$\frac{\partial t(x\tau)}{\partial \tau} = a_L \frac{\partial^2 t(x,\tau)}{\partial \tau} x^2, \qquad (\tau > 0; -R \le x \le R) \tag{1}$$

Under initial boundary conditions:

$$q = -\lambda_L \frac{\partial t(R,\tau)}{\partial x} = \alpha[t(R,\tau) - t_i]$$

$$\frac{\partial t(o,\tau)}{\partial x} = 0$$

$$(x,o) = t_I = const \tag{2}$$

where a_L, λ_L are the coefficients of temperature conductivity and heat conductivity of liquid phase respectively; α is the coefficient of heat emission on outer surfaces of the infinite plate; $t_{a.m.}$ is the temperature of ambient medium, °C; q is the heat flow, W/m^2.

The solution of equation (1) under boundary conditions (2) is obtained by:

$$\theta = \frac{t_I - t(x,t)}{t_I - t_i} = 1 - \sum_{n-1}^{\infty} A_n cos\mu_n \frac{x}{R} \exp\left(-\mu_n^2 Fo_L\right), \tag{3}$$

where $A_n = \frac{2sin\mu_n}{\mu_n + sin\mu_n cos\mu_n} = (-1)^{n+1} \frac{2Bi_L(Bi_L^2+\mu_n^2)^{1/2}}{\mu_n\left(Bi_L^2+Bi_L+\mu_n^2\right)}$; μ_n is the shadow of the characteristic equation root , $ctg\mu = \frac{1}{Bi_L}\mu$; $Fo_L = \frac{a_L\tau}{R^2}$ is the Fourier number; $Bi_L = \frac{\alpha R}{\lambda_L}$ is a Bio criterion for the eutectic solution liquid phase.

For large Fourier numbers it is possible to be limited to the first term in series (3), and then, the duration of the process of liquid phase cooling can be approximately obtained by the next dependence:

$$\tau_1 = \frac{R^2}{a_L\mu_1^2} ln\left[\frac{A_1\mu_1}{Bi_L^2 + \mu_1^2(1-\theta_3)^{1/2}}\right], \text{где } \theta_3 = \frac{t_I + t_C}{t_I - t_{a.m.}}$$

The accurate value τ_1 can be found using a numerical solution of equation (3) relative to the Fourier number Fo_{L} at

$$\theta_3 = \frac{t_{\mathrm{I}} + t_C}{t_{\mathrm{I}} - t_{\mathrm{a.m.}}} \text{ и } x = R$$

The definition of the duration of the process τ_2 of the eutectic solution crystallization is the most difficult.

To determine approximately the duration of the eutectic solution crystallization process we use the L.S. Leibenzone second method.

It is assumed that at each time point the temperature distribution in the solid phase is determined according to the linear law (quasistationary model):

$$t = t_{\mathrm{C}} - \frac{t_{\mathrm{C}} - t_{\mathrm{a.m.}}}{\frac{1}{Bi_{\mathrm{S}}} + 1 - \frac{\xi}{R}} \frac{x - \xi}{R}$$

where ξ is the coordinate of the boundary of phase separation.

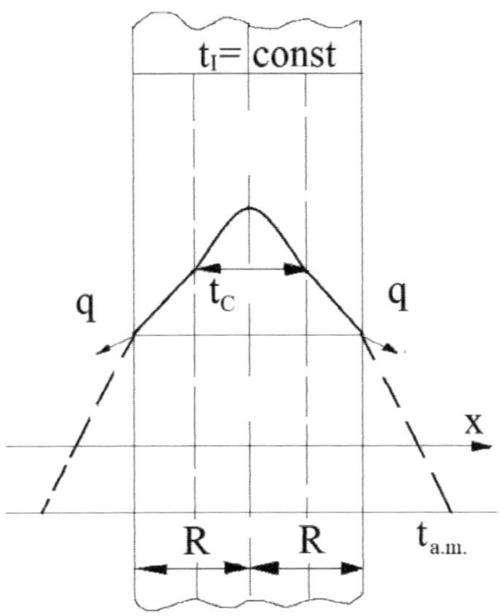

Figure 2. The temperature field of the infinite flat plate during freezing eutectic solution

The current value of the enthalpy of solid phase is defined using the next dependence:

$$W_{\mathrm{S}} = W_{\mathrm{SC}} + c_{\mathrm{S}}(t_{\mathrm{C}} - t) \quad (4)$$

where W_{SC} is the enthalpy of the eutectic solution solid phase at crystallization temperature; c_S is the heat capacity of solid phase.

Likewise, we find the current value of liquid phase enthalpy:

$$w_{\mathrm{L}} = w_{\mathrm{LC}} + c_{\mathrm{L}}(t - t_{\mathrm{C}}) \quad (5)$$

where w_L is the enthalpy of eutectic solution liquid phase at crystallization temperature; c_L is the specific heat of liquid phase.

The equation of heat balance for each time point is written as

$$\frac{d}{d\tau}\left[\int_0^{\xi} \rho_{\mathrm{L}} w_{\mathrm{L}} dx + \int_{\xi}^{R} \rho_{\mathrm{S}} w_{\mathrm{S}} dx - \lambda_{\mathrm{S}} \int_0^{\tau} \left.\frac{\partial t}{\partial x}\right|_{x=R} d\tau\right] = 0 \quad (6)$$

In equation (6) the first term specifies the change of heat content of liquid phase in time; the second one – the change of heat content of solid phase in time; the third one - the change of heat flow in time on external surfaces of the plate.

For simplicity the initial moment of time is accepted as equal to zero.

The current value of the first and second integrals is defined by equation (6):

$$\int_0^\xi \rho_L w_L dx = \rho_L \int_0^\xi (w_{LC} + \tilde{n}_{\partial L}(t - t_C))dx = \rho_L w_L \xi + \int_0^\xi \rho_L \tilde{n}_{\partial L}(t - t_C)\, dx. \quad (7)$$

The value of the second term is defined approximately as:

$$\int_0^\xi \rho_L \tilde{n}_{\partial L}(t - t_C)\, dx = \frac{\xi}{R} Q_L$$

where Q_L is a total heat quantity to be removed from the liquid phase to cool it to the crystallization temperature, i.e. the liquid phase should be in a state as at the moment of the crystallization beginning:

$$Q_L = \int_0^R \rho_L \tilde{n}_{\partial L}(t_L - t_C)\, dx$$

where t_L is the space temperature distribution in time point $\tau_1$of the crystallization process beginning that can be approximately defined by the formula:

$$t_L = t_I - (t_I - t_i)\left[1 - A_1 cos\left(\mu, \frac{x}{R}\right) exp(-\mu_1^2 Fo_{T1})\right]$$

Therefore,

$$Q_L = \tilde{n}_{\partial L}\rho_L R\left\{(t_I - t_C) - (t_I - t_i)\left[1 - \frac{A_1}{\mu_1}\sin\mu, exp(-\mu^2, Fo_{T1})\right]\right\}$$

So, the first term of equation (6) is approximately expressed by:

$$\int_0^\xi \rho_L w_L dx = \rho_L w_L \xi + \frac{\xi}{R} Q_L \quad (8)$$

The second term of equation (6) is defined by the liquid phase thickness at linear law of change of temperature distribution:

$$\int_\xi^R \rho_S w_S dx = \int_\xi^R \rho_S (w_{SC} + \tilde{n}_{\partial C}(t_C - t))dx = \rho_S w_{SC}(R - \xi) + \rho_S c_S \int_\xi^R \frac{t_C - t_{a.m.}}{\frac{1}{Bi_S} + 1 - \frac{\xi}{R}} \frac{x - \xi}{R} dx =$$

$$\rho_S w_{SC}(R - \xi) + \frac{1}{2}\rho_S c_S R \frac{t_C - t_{a.m.}}{\frac{1}{Bi_S} + 1 - \frac{\xi}{R}}\left(1 - \frac{\xi}{R}\right)^2 \quad (9)$$

After substituting expressions (8) and (9) for the first and second terms into equation (10) and making differentiation we get the next equation determining the speed of moving the boundary of phase separation:

$$\left(\rho_S H + \frac{1}{R} Q_L\right)\frac{d\xi}{d\tau} + \frac{1}{2}\rho_S c_S R(t_C - t_{a.m.})\frac{d}{d\xi}\left[\frac{\left(1 - \frac{\xi}{R}\right)^2}{\frac{1}{Bi_S} + 1 - \frac{\xi}{R}}\right]\frac{d\xi}{d\tau} + \lambda_S \frac{1}{R}\frac{t_C - t_{a.m.}}{\frac{1}{Bi_S} + 1 - \frac{\xi}{R}} = 0 \quad (10)$$

Equation (10) specifies latent heat H of phase transition (crystallization) that is defined as

$$H = \frac{1}{\rho_S}(\rho_L w_{LC} - \rho_H w_{HC})$$

Integrating equation (10) we get the formula to define the duration of the eutectic solution crystallization process:

$$\tau_2 = \frac{1}{2}\frac{R\rho_S H + Q_L}{2\frac{\lambda_I}{R}(t_C - t_{a.m.})}\left(1 + \frac{2}{Bi}\right) + \frac{1}{4}\frac{\rho_S c_S R^2}{\lambda_S}\left(\frac{2}{Bi_S^2} ln(1 + Bi_S) - 1 - \frac{2}{Bi}\right)$$

or in dimensionless form:

$$Fo_{T2} = \frac{\tau_2 a_S}{R^2} = \frac{1}{2}\frac{\rho_S H + \frac{Q_L}{R}}{2\rho_S c_S(t_C - t_{a.m.})}\left(1 + \frac{2}{Bi}\right) + \frac{1}{4}\left[\frac{2}{Bi_S^2} ln(1 + Bi_S) - 1 - \frac{2}{Bi}\right] \quad (11)$$

The first term of equation (11) defines the duration of the process that is needed for crystallization properly and liquid phase cooling to the crystallization temperature; the second one defines the duration of the process needed to the solid phase precooling.

According to allowance admitted, at the end of the crystallization process the temperature distribution along the solid phase thickness is defined by the linear dependence:

$$t = t_C - \frac{t_C - t_{\text{a.m.}}}{\frac{1}{Bi} + 1}\frac{x}{R}$$

To determine the duration of the process of liquid phase precooling τ_3 it is necessary to solve equation of heat conductivity (5) under next boundary and initial conditions:

$$\lambda_S \frac{\partial t(R,\tau)}{\partial x} = a[t(R,\tau) - t_{\text{a.m.}}]$$

$$\frac{\partial t(o,\tau)}{\partial x} = 0,$$

$$t(x,o) = t_C - \frac{t_C - t_{\text{a.m.}}}{\frac{1}{Bi}+1}\frac{x}{R} \quad (12)$$

Herewith, the coefficient of liquid phase a_L in equation of heat conductivity (1) should be substituted for coefficient of temperature conductivity of the eutectic solution solid phase a_S.

This expression is written in the next form:

$$t(x,\tau) = t_i + t_C \sum_{n-1}^{\infty} D_n \cos\left(\left(v_n \frac{x}{R}\right) exp[-v_n^2 Fo_3]\right) \quad (13)$$

Decomposition values D_n of the initial distribution of temperature are defined by the formula:

$$D_n = \frac{v_n}{v_n + sin\, v_n \cos v_n} \frac{2}{R} \int_0^R \left[1 - \frac{t_C - t_{\text{a.m.}}}{\left(\frac{1}{Bi} + 1\right) T_3} \frac{x}{R}\right] cos\left(v_n \frac{x}{R}\right) dx =$$

$$= \frac{2}{v_n + sin\, v_n \cos v_n} \left\{ \sin v_n - \frac{t_C - t_{\text{a.m.}}}{\left(\frac{1}{Bi} + 1\right) T_3} \frac{1}{v_n} [1 - \cos v_n - v_n \sin v_n] \right\}$$

where v_n is the coefficient of kinematic viscosity.

For large Fourier numbers the duration of the process of precooling of liquid phase τ_3 can be determined taking into account only the first term of the Fourier series (13):

$$\tau_3 = \frac{R^2}{a_S v_1^2} ln \frac{D_1 t_C}{t_C - t_{\text{a.m.}}} \text{ or } Fo_{T3} = \frac{1}{v_1^2} ln \frac{D_1 t_C}{t_C - t_{\text{a.m.}}}$$

An accurate value of the duration of the process of liquid phase precooling τ_3 can be defined using a numerical solution of equation (13) relative to the Fourier number Fo_{T3} at $t(0,\tau) = t_1$.

3. Conclusion

Therefore, the duration of all process of crystallization of the eutectic solution is equal to:

$$\tau_0 = \tau_1 + \tau_2 + \tau_3$$

or in dimensionless form:

$$Fo_0 = Fo_{T1} \frac{a_L}{a_O} + Fo_{T2} + Fo_{T3}$$

References

[1] Lykov A V 1978 *Heat and mass transfer* (Moscow, Energy)
[2] Farid M, Khudhair M, Razack K and Said Al-Hallaj 2004 A review on phase change energy storage: materials and applications *Energy Conversion and Management* **45** 1597–1615
[3] Dolesov A G, Danilin V N and Khrisonidi V A 2010 Cold accumulating material Water-salt system *Physical and chemical analysis of the properties of multicomponent systems* **8** 3
[4] Oro E, de Gracia A, Castell A, Farid M M and Cabeza L F 2012. Review on phase change materials (PCMs) for cold thermal energy storage applications *Applied Energy* **99** 513–533
[5] Zhai X Q, Wang X L, Wang T and Wang R Z 2013 A review on phase change cold storage in air-conditioning system: Materials and applications *Renewable and sustainable Energy Reviews* **22** 108–120

Results of temperature effect on thermo-physical properties of guava fruit of Southeast Asia

A Tun[1, 2], I V Baranov[1], V A Krylov[1], E V Tambulatova[1]

[1]ITMO University, 49 Kronverksky Pr., Saint Petersburg,197101, Russia
[2]E-mail: ayetunitmo@gmail.com

Abstract. The research presents the results of the effect of temperature on thermo-physical properties of guava fruit of Southeast Asia. Guava is an energetic fruit with high nutritional value and is considered a tropical fruit of South-east Asia. Fruit growers need to use cold storage facilities to extend the revenue period while maintaining produce quality and freshness. The storage life of fresh fruits can be extended by several days by cooling. The specific heat of foods can be used to calculate the refrigeration load calculation. Therefore, the specific heat of food plays an important role in the design of storage and transport systems. One of the most important thermo-physical properties in foods are specific heat. The authors of the article used differential scanning calorimetry (DSC 204 F1), by which the specific heat of guava fruit was determined in the temperature range from –25 to 25 °C. It was found that the specific heat of guava fruit is dependent on temperature.

1. Introduction

Tropical and Subtropical fruits are very popular among the many countries of the world. Southeast Asia, a typical tropical region, plays an important role in exporting a variety of fruits worldwide [1]. In Southeast Asia, there are tropical and subtropical fruits such as dragon fruit (pitaya or pitahaya), coconut, banana, rambutan, pomelo, mango, mangosteen, jackfruit, guava, longan, papaya, pineapple, watermelon, sapodilla, durian, avocado and carambola. Guavas (*Psidium guajava L.*) are fruits of commercial and nutritional values. Guava fruit is generally ovoid or pear shaped and depending on cultivar, their sizes vary from 2.5 to 10 cm in diameter and weight 50 to 500 g. The flesh may be pink, white or yellow, either with seed or seedless [2]. Guava (*Psidium guajava* L.) is often marketed as "super-fruits" which has a considerable nutritional importance in terms of vitamins A and C with seeds that are rich in omega-3, omega-6 poly-unsaturated fatty acids and especially dietary fiber, riboflavin, as well as in proteins, and mineral salts. The high content of vitamin C (ascorbic acid) in guava makes it a powerhouse in combating free radicals and oxidation that are key enemies that cause many degenerative diseases and that can be used to fortify children foods. The anti-oxidant virtue in guavas is believed to help reduce the risk of cancers of the stomach, esophagus, larynx, oral cavity and pancreas [3]. Guava can be consumed fresh or can be processed into juice, nectar, pulp, jam, jelly, and slices in syrup, fruit bar or dehydrated products, as well as being used as an additive to other fruit juices or pulps [4].

The export of tropical fruits has limitations such as a short shelf life and difficulty in maintaining the quality because of tropical climate conditions and undeveloped postharvest technologies in Southeast Asia [1]. An important objective for developing postharvest technologies is to extend the shelf life of fresh fruits without deterioration in fruit quality. The storage life of fresh fruits can be extended by several days by cooling and by several weeks or months by freezing. Fruits and vegetables are perishable foods with extremely rapid deterioration; this means that their stability after harvesting and during sub-sequent storage is critical [14]. Cooling at the field before the product is shipped to the market or storage warehouse is referred to as precooling [7]. Precooling is the rapid removal of field heat from freshly harvested fruits and vegetables before shipping, storage, or processing. Prompt precooling inhibits growth of microorganisms that cause decay, reduces enzymatic and respiratory activity, and reduces moisture loss. Thus, proper precooling reduces spoilage and retards loss of postharvest freshness and quality [8]. The cooling method is requirements during the transportation, storage, or processing of fruits. Cooling is one of the main tools for extending postharvest life: slow product metabolism and the activity of microorganisms responsible for quality deterioration. As a result, reserves are maintained with a lower respiration rate, ripening is retarded and vapor pressure between products and ambient is minimized, reducing water loss. During cooling and precooling operation, we need to be able to calculate processing times and heat loads of the product.

The thermo-physical properties of foods play a great role in heat transfer calculations of food processing. Thermo-physical properties of foods are included specific heat, enthalpy, thermal conductivity, thermal diffusivity and density. One of the most important thermo-physical properties in food processing is specific heat. Specific heat is an essential part of the thermal analysis of food processing or of the equipment used in heating or cooling of foods [10]. Specific heat of food product is majority of the refrigeration load calculation. Specific heat is a measure of the energy required to change the temperature of unit mass of the food by one degree [8, 12, 13]. Specific heat of foods materials needed for analysis and design processes involving heat transfer. These processes include cooling, freezing, heating and drying as found in post-harvest handlings and product processing, storage and distribution. The accurate specific heat data will allow food scientists and engineers to calculate the heat transfer precisely leading to the effective food process design including equipment design and implementation. The specific heat depends on the nature of the process of heat addition in terms of either a constant pressure process or a constant volume process. However, because specific heats of solids and liquids do not depend on pressure much, except extremely high pressures, and because pressure changes in heat transfer problems of agricultural materials are usually small, the specific heat at constant pressure is considered [10, 13].

There are several methods of determining specific heat of food and agricultural materials: Empirical equations, Method of mixtures, Method of guarded-plate, Method of comparison calorimeter, Method of calculated specific heat, adiabatic calorimeter and Method of differential scanning calorimetry. Values of specific heat determination of many food materials, using DSC have been reported for various researchers. Among the published methods for specific heat measurement, differential scanning calorimetry (DSC) has so far been the most accurate and rapid method [9]. DSC method has found use in foods and pharmaceuticals [15].

This study proposes the result of study on specific heat of guava fruit using differential scanning calorimeter (DSC). Differential scanning calorimeter (DSC), which reports heat flow as a function of temperature is an excellent tool for the measurement of temperature-dependent specific heat and phase transitions [13]. Differential scanning calorimetry means the measurement of the change of difference in the heat flow rate to the sample and to a reference sample while they are subjected to a controlled temperature program [11]. Differential scanning calorimetry determines transition temperatures and enthalpy changes in solids and liquids under controlled temperature change. DSC is a convenient, rapid technique and easy to use method of obtaining a wealth of information about a material [15, 17]. For small sample sizes (5–50 mg) the DSC gives reliable results rapidly [17]. DSC is the most frequently used method in the field of Thermal Analysis [6, 17]. Rapid analysis, high significance for research and quality control tasks, and easy handling of the measuring instrument contribute to its versatility [6, 18].

This paper presents results of the effect of temperature on specific heat of guava fruit of Southeast Asia using differential scanning calorimetry DSC 204 F1.

2. Materials and Methods

Differential scanning calorimetry DSC 204 F1 phoenix (NETZSCH Germany) was used to determine the specific heat of guava fruit of Southeast Asia. The mass of the samples was measured using an analytical balance AND HR-200, accuracy class. The mass of the crucible is 38 mg, the mass of the lid is 25 mg. The samples were kept in refrigerator at about 5°C until experiments performed. About 28-42 mg of samples is used in the experiments. The samples were cut into small pieces by a snap off blade to obtain 6 mm-diameter and thickness 1-2 mm. Samples were hermetically sealed in aluminum crucibles. The mass of the samples was determined with an accuracy of ± 0.1 mg. The measurements were carried out in the temperature range of -25 to 25 °C in an atmosphere of gaseous nitrogen. The heating rate was 2 K/min. As a thermally inert substance (reference), sapphire weighing 50.1 mg was used. The tests were carried out in triplicate, following a similar experimental procedure described by DSC method for specific heat determination.

The method for determining the specific heat of the oils under study included four steps:

1. *Setting the temperature program.* In accordance with the specified temperature range from -25 to 25 °C, a temperature program was drawn up for studying the specific heat of guava fruit. The heating/cooling rate is 2 K/min, the stationary section time is 15 min, the initial stage lasted 15 min. Nitrogen gas with a flow rate of 20 ml/min and 50 ml/min, respectively, was used as the purge and protective gases.
2. *Correction.* On both cells are placed empty crucibles. The difference of heat fluxes through two empty crucibles ($DSC = DSC_{base}$) and the temperature of the right crucible are measured. This signals ($DSC = DSC_{base}$) were dependent on time and temperature.
3. *Calibration.*Crucible cell comparison remains empty. A substance is placed in the sample crucible with a known specific heat $c_{p,\,standard}$ and mass $m_{standard}$. The difference between the heat fluxes through the empty crucible and the crucible with the reference standard $DSC_{standard}$, and the temperature of the crucible with the reference is measured. The signals ($DSC = DSC_{standard} — DSC_{base}$) were dependent on time and temperature.
4. *Measurement.* The crucible of the reference cell remains empty. In the crucible for the sample is placed the test sample mass m_{sample}. The difference between the heat fluxes through the empty crucible and the crucible with the sample DSC_{sample} is measured, and the temperature of the crucible with the sample. The signals ($DSC = DSC_{sample} — DSC_{base}$) were dependent on time and temperature.

The specific heat (c_p) was measured by DSC. The measurement was conducted using DSC method according to following equation [5, 16]:

$$\frac{DSC_{sample}-DSC_{base}}{DSC_{standard}-DSC_{base}} = \frac{c_{p\ sample} \times m_{sample}}{c_{p\ standard} \times m_{standard}} \tag{1}$$

$$c_{Psample} = \frac{m_{standard}}{m_{sample}} \frac{DSC_{sample}-DSC_{base}}{DSC_{standard}-DSC_{base}} c_{Pstandard} \tag{2}$$

DSC_{sample}, DSC_{base}, $DSC_{standard}$ are the ordinate of the DSC curves corresponding to the sample, blank specimen, and standard specimen, respectively. $c_{p,\,sample}$ and $c_{p,\,standard}$ are the specific heat of the sample and standard specimen, respectively.m_{sample} and $m_{standard}$ are the mass of the sample and sapphire, respectively [5, 16].

3. Results and discussion

The following results on specific heat of guava fruit were obtained in the experiment carried out in the thermo-physical measurements and devices laboratory (ITMO University), Saint Petersburg. The specific heat of guava fruit was deter-mined using differential scanning calorimetry (DSC 204 F1). NETZSCH Software Proteus® 6.1.0 was installed in system for evaluating the data. This study guava fruits samples (11-13 and 21-23) were used 41.1, 28.1, 30.5, 31.5, 30.7 and 36.0 mg obtained from Myanmar and Thailand. The changes in the specific heat of guava fruit at different temperature ranges are presented in Figs 1-2, respectively. From these figures, it can be seen that the specific heat for the all samples of guava fruit increased with increased temperature, then dramatically decreased with increased temperature and slightly decreased with increased temperature, respectively.

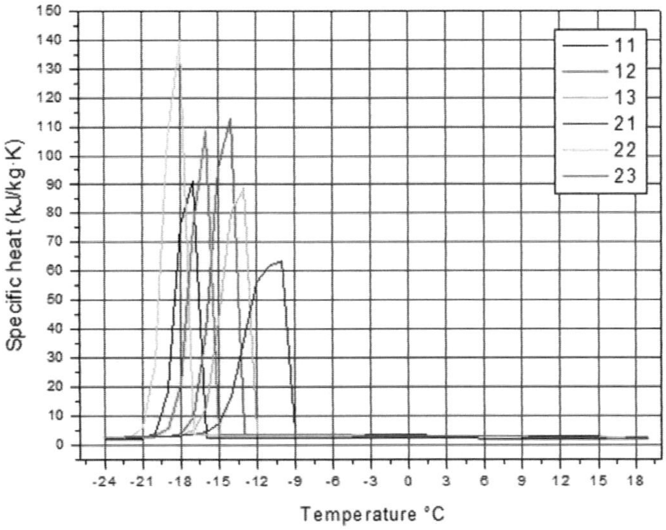

Figure 1. Effect of temperature on specific heat for the all samples of guava fruit.

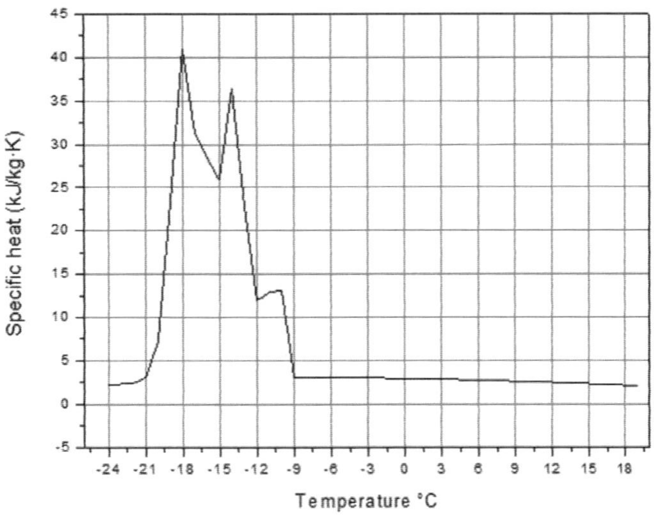

Figure 2. Approximating the arithmetic mean of specific heat of guava fruit.

The similar trend was described in the specific heat of food [8]. As in the study, the specific heat of guava is dependent on the on the temperature. Thermal properties of foods also change with time and temperature [7].

4. Conclusion

In designing food processes and processing equipment, we need numerical values for the specific heat of the food and materials to be used. Several experimental studies on determination of specific heat of foods and agricultural materials were reported in literature. However, the data depend on origin (variety, cultivar) and composition, processing conditions, and structure of the foods. The specific heat for the all samples of guava fruit increased with increased temperature, then dramatically decreased with increased temperature and slightly decreased with increased temperature, respectively. The linear decrease of specific heat for the all samples of guava fruit at temperature rises from -9°C to 19°C. It was found that the specific heat of guava fruit is dependent on temperature. Specific heat of unfrozen foods slightly lower as the temperature rises from 0°C to 20°C [8]. In literatures [10,12,13] specific heats of food depend on their composition, structure and temperature.

Acknowledgments
This research project was supported by the Research lab (Thermo-physical measurements and devices), Faculty of Physics and engineering, ITMO University.

References

[1] Kusumaningrum D, Lee S-H, Lee W-H, Mo Ch, Cho B-K 2015 A review of technologies to prolong the shelf life of fresh tropical fruits in southeast Asia *Journal of Biosystems Engineering* 40(4)345-358.
[2] Yusof S. 2003 *Encyclopedia of food sciences and nutrition*eds. B J Guavas, LTrugo,P Finglas, 2nd edition (Academic press) pp. 2985-2992.
[3] Lim Y. Y., Murtijaya J 2007 Antioxidant properties of Phyllanthus amarus extracts as affected by different drying methods *LWT -Food Science and Technology*40(9)1664-1669.
[4] Dattatreya K. M., Kaushik P, Kumar R 2012 Evaluation of guava products quality*International Journal of Food Science and Nutrition Engineering* 2(1) 7-11.
[5] Fedorov A.V., Baranov I.V., Tambulatova E.V., Volkov S.M., Prokhorova L.T. and Krylov V.A. 2019 Investigating the temperature dependence of refined sunflower oil specific heat capacity on its composition by the method of differential scanning *CalorimetryJournal of IAR*1 52-63.
[6] Netzsch-Gerätebau GmbH 2012 *Differential scanning calorimetry: Method, technique, applicationDSC 204 Phoenix F1*(Selb: Erich NETZSCH GmbH & Co.).
[7] Yunus A. C. and Afshin J. G2014 *Heat and mass transfer: fundamentals & applications* 5th edition, (New York: McGraw-Hill Education), pp. 1149-1208.
[8] ASHRAE 2018 *American society of heating, refrigerating and air-conditioning engineers. Handbook– Refrigeration* (Atlanta: SI Edition) pp. 343-445.
[9] Yang W., Sokhansanj S., Tang J., Winter P 2002 Determination of thermal conductivity. specific heat and thermal diffusivity of borage seeds *Biosystems Engineering* 82(2) 169–176.
[10] SinghR P., Heldman D. R. 2009 *Introduction to Food Engineering* 4th edition (Amsterdam: Elsevier) pp. 257-262.
[11] Höhne GWH, Hemminger WF, Flammersheim Y-J 2003 *Differential Scanning Calorimetry*(Berlin: Springer Verlag) pp. 1-27.
[12] Lewis M. J. 1987 *Physical properties of foods and food processing systems*(Cambridge) pp. 220-245.
[13] Sahin S., Sumnu S. G. 2006 *Physical Properties of Foods*(New York) pp. 118-166.
[14] Silva C. L. M., Goncalves E. M., Brandao T. R. S. 2008 Freezing of fruits and vegetables (*Frozen Food Science and Technology*) ed. Evans J. A. (Oxford: Blackwell Publishing Ltd) pp 175-208.
[15] Gabbott P. 2008 A practical introduction to differential scanning calorimetry (*Principles and Applications of Thermal Analysis*)(Oxford: Blackwell Publishing Ltd.) pp 1-50.
[16] Song N-J, Chen C-M, Lu Ch, LiuZh, Kong Q-Q, Cai R 2014 Thermally reduced graphene oxide films as flexible lateral heat spreaders *Journal of Materials Chemistry A. The Royal Society of Chemistry*.
[17] Nesvadba P. 2008. Thermal properties and ice crystal development in frozen foods (*Frozen Food Science and Technology*)ed. Evans J A(Oxford: Blackwell Publishing Ltd.) pp. 175-208.
[18] Netzsch-Gerätebau GmbH. 2012. *Advanced Materials Testing. Thermal Analysis, Methods for Determination of Thermophysical Properties, Cure Monitoring, Accelerating Rate and Isothermal Calorimetry* (Selb: Erich NETZSCH GmbH & Co.).

AUTHOR INDEX